Geometric differentiation

Geometric differentiation

for the *intelligence* of curves and surfaces

I. R. Porteous

Senior Lecturer, Department of Pure Mathematics
University of Liverpool

CAMBRIDGE
UNIVERSITY PRESS

Published by the Press Syndicate of the University of Cambridge
The Pitt Building, Trumpington Street, Cambridge CB2 1RP
40 West 20th Street, New York, NY 10011–4211, USA
10 Stamford Road, Oakleigh, Melbourne 3166, Australia

First Published 1994

Printed in Great Britain at the University Press, Cambridge

A catalogue record for this book is available from the British Library

Library of Congress cataloguing in publication data is available

ISBN 0521 39063X hardback

KT

Contents

Introduction

This book is concerned with the local differential geometry of smooth curves and surfaces in Euclidean space. This is a topic which is generally compressed into a short introductory chapter in any standard work on differential geometry, but nevertheless has a surprising richness, with new areas still being explored.

The initial impetus to look afresh at this subject was René Thom's choice ((1975), but circulating in draft for several years previous to this data) of the term 'umbilic', borrowed from differential geometry, to describe certain of the elementary catastrophes, and the realisation that Darboux's classification (1896) of the umbilics of a surface, itself not sufficiently known, was only part of the story (Porteous, 1971). In particular the *ridges* of a surface are important features that are almost entirely disregarded in the standard texts, though certain of them are familiar to structural geologists as the *hinge lines* of folds in strata (Ramsay, 1967). One of the few places where they get serious mention in the literature is in the work (1904) of A. Gullstrand, who was awarded the Nobel Prize for Physiology and Medicine in 1911 for his work on the accommodation of the eye lens, in which work he had to create the necessary fourth order differential geometry to explain the relevant optics. His Prize Lecture (1911) makes amusing reading. Recently they have been rediscovered by workers on face recognition and the interpretation of magnetic resonance scans of the surface of the brain.

The work of singularity theorists, starting with H. Whitney, R. Thom and J. Mather and extended by many others, notably the Russian School under the leadership of V. I. Arnol'd, has richly developed the geometry of Taylor series of smooth functions, making accessible and familiar the classification of the higher order critical points of families of smooth functions. What we do here is to provide a treatment of surface theory which is in the spirit of singularity theory, yet stops short of the hard and powerful theorems of that subject. For this reason some of the things that we do are heuristic only, though with reference to where a fuller treatment is to be found.

A unifying theme throughout the book is the use of the family of distance-squared functions that relate a smooth submanifold of Euclidean space to the ambient space. The description of the critical points of these functions in terms of invariant expressions involving the higher derivatives of these functions provides a high road into the interesting geometry.

The first chapter plunges straight into the interrelationship of a smooth parametric plane curve with its *evolute* or *focal curve*, a topic first studied in detail by Huygens some years before the advent of the differential calculus made this a first exercise in that subject. It is one that is still important in the practical theory of the *offsets* or *parallels* of a plane curve. As Huygens realised, these are recoverable from the evolute by unwinding a string from it or, equivalently, by rolling a ruler along it, a pregnant idea that bears fruit later in the theory of space curves and surfaces and their evolutes. Several characterisations are given of the *curvature* and the *vertices* of a curve. The latter topic is the first place where third derivatives make their entry. We learn how to locate the vertices without having to differentiate the explicit but somewhat unattractive formula for the curvature. As a digression from the main theme we apply these ideas briefly to some basic concepts of *plane kinematics* that have application in the theory of mechanisms, namely the *inflection circle*, the *Ball point*, the *cubic of stationary curvature* and the *Burmester points* of the instantaneous motion of a plane moving rigidly over the plane. A related subject, though perhaps not obviously so, is the description of the *caustics* of the light from a point source reflected from a curved mirror, a subject richly re-examined recently by Bruce, Giblin and Gibson (1981). The analogous theory of spherical kinematics is briefly sketched in a later chapter, prefaced by a short chapter in which some basic facts of elementary geometry are recalled. Spherical kinematics has application to the description of the motion of plates over the surface of the Earth in modern plate tectonics.

Next comes the theory of *space curves*, introducing the *space evolute* of a space curve and the *tangent developable* of the space evolute, the *focal surface* of the original curve. This is presented before the derivation of the classical *Serret–Frenet equations* for a space curve, usually given pride of place. The latter require for their use the theoretically convenient but practically inconvenient use of a unit-speed parametrisation for the curve. Also defined are the *parallels* to a space curve, the involutes of its evolute, and the *focal curves* of a space curve, shown to be geodesics on its focal surface.

Surface theory, depending as it does on the differential calculus of functions of several variables, requires familiarity not only with *quadratic* but also with *cubic forms* in two variables. A chapter interpolated here provides the necessary facts. Two further short chapters follow, one on probe analysis as a diagnostic technique for handling singularities of maps and the other on contact. The way is then open to describe the *focal surface* of a regular smooth parametric surface in three-dimensional Euclidean space.

Topics discussed in detail include the *first and second fundamental forms*, the *principal centres of curvature* and *principal curvatures* of the surface and the induced grid on the surface formed by the two mutually orthogonal families of *lines of curvature*. These only involve the first and second derivatives of the parametrisation. It is in the description of the *ridges* of the surface and the associated *ribs* or cuspidal edges of the focal surface and the rich geometry around the *umbilics* of the surface, those points where the principal curvatures agree and the surface is most nearly spherical, that the third and higher derivatives become important and the insights of singularity theory prove of greatest value.

Other topics discussed include various aspects of the *Gauss map* of a surface, including the *parabolic line*, the *cusps of Gauss* and Gauss' 'excellent' theorem, in Greek–Latin the *theorema egregium*.

The study of the focal surface of a surface is rounded off by a detailed examination of the inverse construction of the involutes or evolvents of a geodesic foliation on a regular surface in three-dimensional Euclidean space with emphasis on what happens when the curves of the foliation possess linear points, a subject full of surprises that has had close attention from the Russian school of singularity theorists under the leadership of V.I. Arnol'd, and where the full story has only very recently become clear.

Our treatment, based on the study of the family of distance-squared functions, has emphasised the role of spheres in surface theory. The circles associated to a surface are also of great interest. Here again the initial steps recovering the focal surface are classical, but much of the higher order material seems to be new, some of it to be found in the thesis of James Montaldi (1986a). Remarks on the bumpy circular contours round an umbilical hill-top led to the delightful work of Stelios Markatis (1980) on bumpy spheres, outlined here in a final chapter that considers some other important examples also.

The title of the book has been chosen to indicate that its purpose is to make geometrical the basic properties of the derivatives of differential maps. It is written to be accessible to final-year honours students or to first-year postgraduates, keen to add some geometry to the standard linear algebra and several variables calculus that in one form or another they will already have met. A non-standard feature which browsers must take note of is a notational one. For practical reasons that will become evident when working with third and fourth derivatives the traditional d to denote differentiation is only used occasionally. Instead differentiation is denoted by *subscripts*. Thus for a smooth map $f : X \rightarrowtail Y$ between real vector spaces X and Y (where the tail on the arrow indicates that the domain is an open subset of the source vector space) the first derivative is the map $f_1 : X \rightarrowtail L(X, Y)$, that associates to each $a \in \operatorname{dom} f$ the linear map $f_1(a) : X \to Y$ that (up to a constant) best approximates f at a, the second derivative is $f_2 : X \rightarrowtail L(X,(X, Y))$, associating to each $a \in X$ the *twice* linear (rather than *bilinear*) map $f_2(a) = (f_1)_1(a)$, and so on.

Much of the material has been class tested from time to time over the last fifteen years. It has been the basis of a number of expository lectures, for example to the Archimedeans in Cambridge on several occasions, to colleagues in singularity theory at the AMS symposium at Arcata in 1981 (Porteous 1983a,b), to specialists in computer vision at a Rank Prize Fund symposium at Liverpool in 1986 and to the Oxford group in 1988, and to specialists in computer-aided design at symposia on the Theory of Surfaces organised by the Institute of Mathematics and its Applications at Cardiff in 1986 and Oxford in 1988 (Porteous, 1987a, 1987b, 1989).

Much of the work has been done in collaboration with students at Liverpool, notably Stelios Markatis (1980), James Montaldi (1983), Alex Flegmann (1985) and Richard Morris (1990). My thanks are also due to numerous colleagues and friends for their interest and comment, especially Bill Bruce, Peter Giblin, Chris Gibson, Peter Newstead and Terry Wall and students Helen Chappell and Neil Kirk.

Especial thanks are due to Peter Ackerley, who has crafted many of the figures, and to Peter Giblin and Richard Morris for those produced on the computer. I am also most grateful to John Robinson of Yeovil for permitting me to make use of his sculpture *Eternity* to grace this book. For his remarkable story and pictures of all his abstract sculptures see Robinson (1992).

Я бпагодарен также В. И. Ариопьду, чья помошь быпа мие очень полеэна.

Ian R. Porteous
Liverpool, 1993

1

Plane curves

1.0 Introduction

Sir Christopher Wren
Went to dine with some men.
'If anyone calls,
Say I'm designing St Paul's!'

St Paul's Cathedral was designed following the Great Fire of London in 1666. Six years earlier Wren, a mathematician as well as architect, was one of the founder members of the Royal Society. At that time one of the men that he might well have been dining with was the great Dutch Scientist, Christiaan Huygens (*natus* 1629, *denatus* 1695, as a late picture of him has it! (Figure 1.1)). At the time we are speaking of Newton (*natus* 1642) and Leibniz (*natus* 1646) were still teenagers, and the Calculus had yet to be invented. Indeed the first elementary calculus textbook was published only in 1696, the year after Huygens' death. This purported to be written by an aristocratic friend of the Bernoulli family, the Marquis de l'Hôpital, and was entitled *L'analyse des infiniments petits, pour l'intelligence des courbes planes'*. Central to this first work on differential geometry are the ideas developed by Huygens and his associates thirty-five or more years previously. Curiously, de l'Hôpital did not put his name to the first edition of the work, it being added in ink in many copies (Figure 1.2). The work is in fact a fairly direct translation from the original Latin of Jean Bernoulli, which came to light many years later, neither the translator nor the writer of the unsigned preface being de l'Hôpital! For an account of this ancient scandal see Truesdell (1958).

Our aim here is to give a fresh account of these ideas which remain the basis of the whole subject.

Consider as a first example the parabola in the real plane with equation $y = x^2$. An engineer wishing to cut this curve accurately out of some sheet of material has to use a cutting tool, necessarily of finite size, whose centre has to be programmed to follow some curve *offset* the right distance from the parabola to be cut. Hasty thinking might suggest that this offset is another parabola, but this is not so – compare

Figure 1.1

Figures 1.3 and 1.4. If one examines offsets at greater and greater distances from the original curve (on the 'inner' side) one discovers that before long these are no longer regular curves but acquire sharp points or *cusps*, where the direction of the curve reverses. Moreover

ANALYSE
DES
INFINIMENT PETITS,

Pour l'intelligence des lignes courbes.

Par le Marquis de l'Hospital

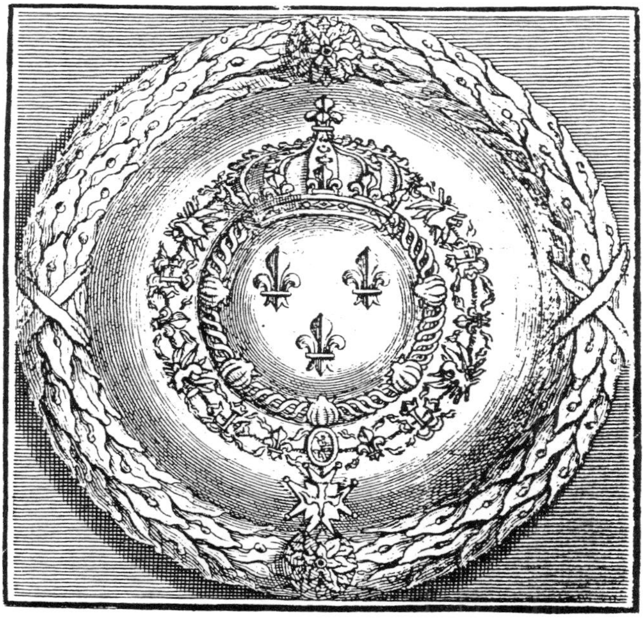

A PARIS,

DE L'IMPRIMERIE ROYALE.

M. DC. XCVI.

Figure 1.2

these cusps lie along a new curve which itself sports a cusp, pointing towards the lowest point of the original parabola – see Figure 1.5.

It is a pleasant thought to think of the parabola in another way as the shoreline of a bay in which one has gone out for a swim, swimming out normally, that is at right angles, to the shore – Figure 1.6. One's first intuition probably is that, no matter how far one swims, one's starting point * remains locally the nearest point of the shore. We say 'locally' here because if one goes far enough then clearly some point on the farther shore may well be nearer. But our local intuition is wrong, as Figures 1.7 and 1.8 illustrate. These display the same new cuspidal curve that we saw before, its tangents all being normal to the parabola.

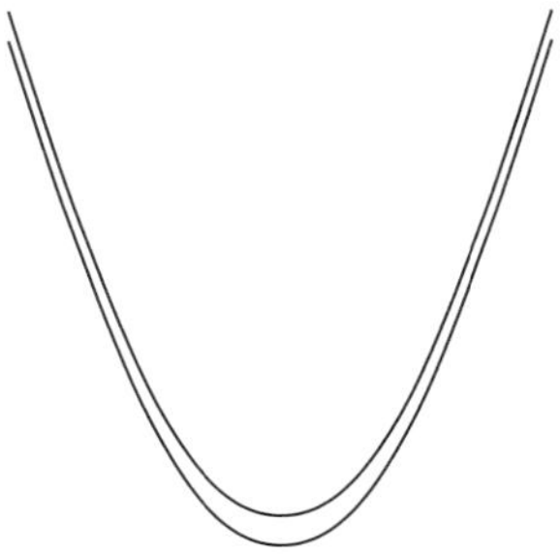

Figure 1.3

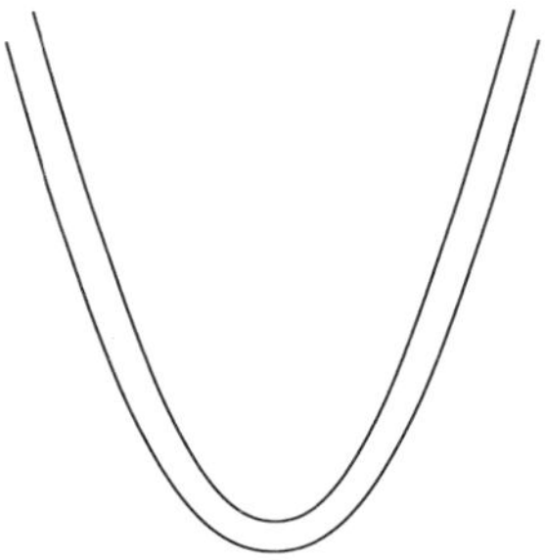

Figure 1.4

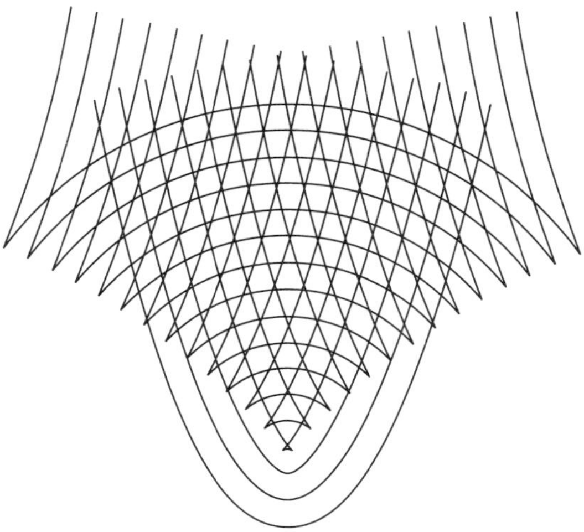

Figure 1.5

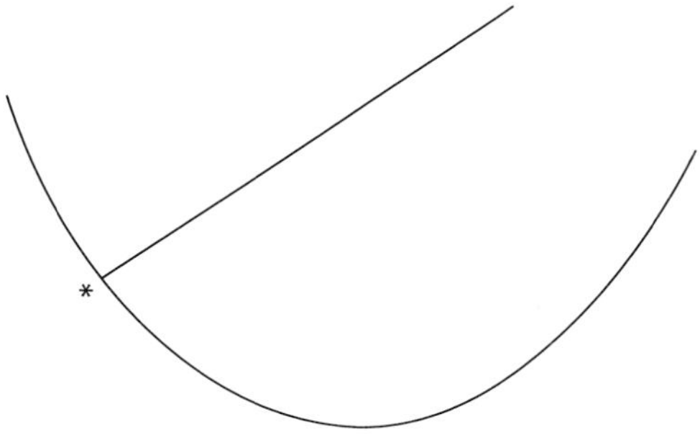

Figure 1.6

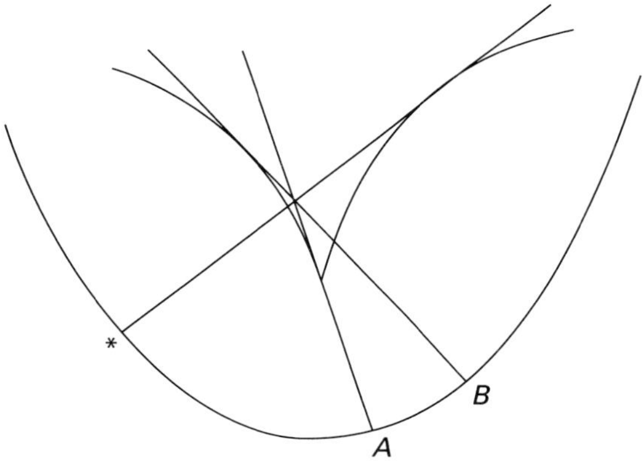

Figure 1.7

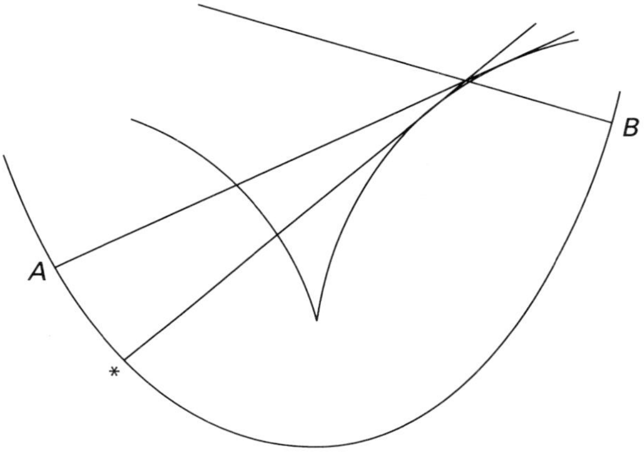

Figure 1.8

Initially one can draw only one normal to the shore from one's position $*$ in the bay, namely the path along which one has just swum, but after crossing the curve of cusps two new normals can be drawn, the three shore points $*$, A and B then being successively a local minimum at $*$, a local maximum at A and a local minimum at B, of the distance from one's position in the bay to the shoreline – Figure 1.7. As one swims on, the points A and B move round the shore in opposite directions, and as one reaches the point of tangency of the normal with the curve of cusps A comes right round to coincide with $*$. At any more distant point $*$ is a *local maximum* of distance – Figure 1.8!

The curve of cusps that falsifies both these intuitions is known as the *evolute* or *focal curve* of the original curve. In Figure 1.9 it is exhibited as the *envelope* of the family of the family of normals to the parabola. The offsets are also said to be the *parallels* or *equidistants* to the parabola.

It was Huygens who made the remarkable discovery that one can recover the original parabola from its evolute by unwinding an inextensible string laid partially along the evolute, or equivalently by rolling the tangent line to the evolute along the evolute. A bob on the string, or point of the rolling line, then describes part either of the parabola itself or, according to the position of the bob, one of the offsets to the parabola. Indeed all the offsets can be obtained in this way if one makes appropriate conventions about the unwinding pro-

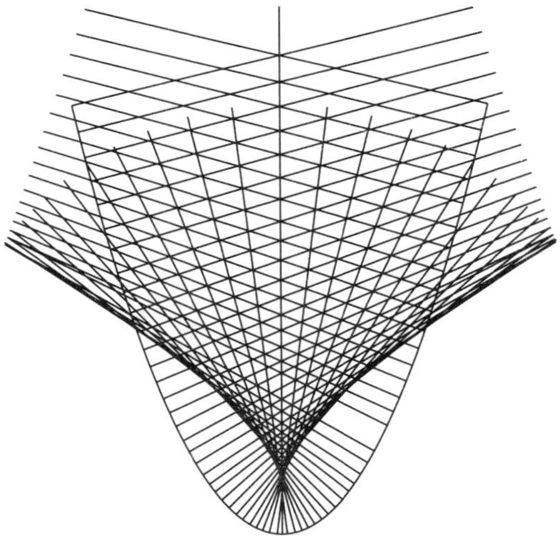

Figure 1.9

cess, especially at a cusp of the evolute. These mutually parallel curves are known as the *involutes* or *evolvents* of the evolute.

There is nothing special about the parabola in all this. Indeed a favourite curve of Huygens, and of Wren too, is the curve which features as the solution to the following take-home problem (Figure 1.10) faced by several thousand Merseyside twelve-year olds in the Spring of 1982 (Giblin and Porteous, 1990).

The curve is the *cycloid*, consisting of a series of arches supported on a series of cusps (Figure 1.11). As we shall verify later, this curve has the remarkable property that its evolute is a congruent cycloid, whose cusps this time point away from and not towards the original curve. If we turn all this upside down (Figure 1.12) and arrange for a pendulum of suitable length to be swung from one of the jaws of the evolute cycloid one obtains the Huygens cycloidal pendulum, whose period, remarkably, turns out to be independent of the amplitude.

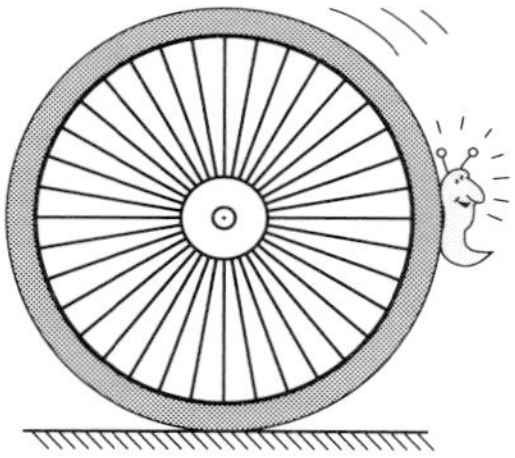

Arc Light

There was a young glow worm called Glim,
Who went for a ride on the rim
  Of a wheel that went round
  As it rolled on the ground.
Please draw the arc traced by him!

Figure 1.10

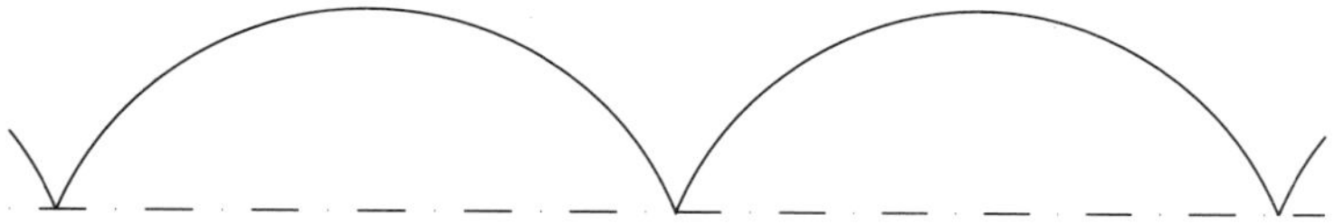

Figure 1.11

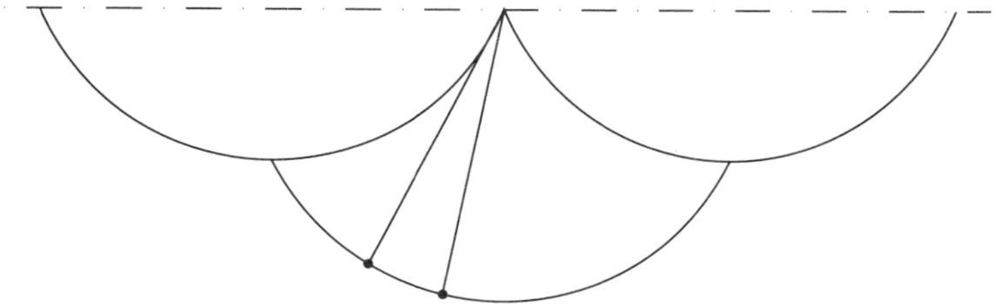

Figure 1.12

Yet a third way of regarding the evolute is as the locus of *centres of curvature* of the original curve. This is illustrated in Figure 1.13 where the circle with centre at the point of tangency of a normal to the original curve with the evolute, and passing through the base of the normal, is seen to hug the curve so closely there that it is known as the *osculating circle*, or *circle of curvature* of the curve at that point. In general, as in this example, it shares a tangent line with the original curve, but crosses the curve there. An exception to this occurs at the lowest point of the parabola, when the centre of the osculating circle lies at the cusp of the evolute and the circle lies entirely above the parabola. At this point the radius of the osculating circle, the *radius of curvature* of the curve, has a local minimum – indeed in this example an absolute minimum. In fact cusps on the evolute correspond to critical points of the radius of curvature, the cusps on the evolute pointing towards the curve at local minima and away from the curve at local maxima.

The reciprocal of the radius of curvature is known simply as the *curvature* of the curve. At a point of inflection of the curve the curvature is zero and the radius of curvature infinite, the role of osculating circle being then played by the inflectional tangent. We shall prove that the evolute of a regular plane curve does not have any points of inflection. Of course, as de l'Hôpital (or was it Jean Bernoulli?) first remarked, there is nothing to stop one swinging a pendulum from a curve with an inflection. The resulting family of non-regular involutes (see Figure 1.21) has an intimate relationship

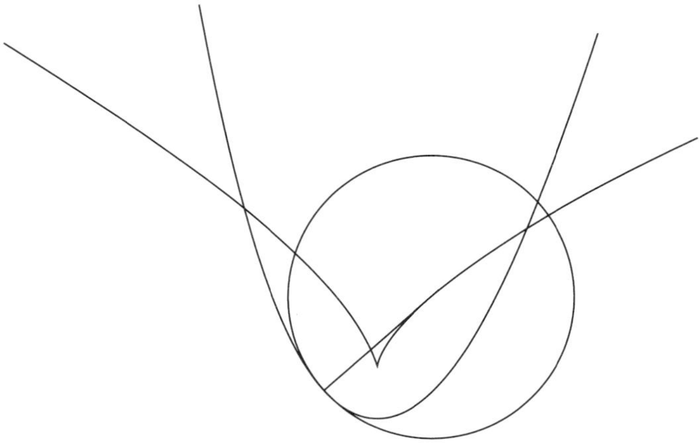

Figure 1.13

with the group of symmetries of an icosahedron – a deep and mysterious fact only recently noted by the Russian school of singularity theorists under the leadership of V.I. Arnol'd (Arnol'd, 1983, 1990b).

As we are going to be concerned in what follows with applications of the calculus to geometry we ought logically to start with reviewing the calculus. Since almost all that is required for the study of curves should already be familiar to the reader we defer this review to Chapter 4, preceded in Chapter 2 with a review of some basic frequently used facts of linear and projective geometry. For the moment it is enough to remark that the standard n-dimensional real vector space equipped with the standard Euclidean scalar product will be denoted by $\mathbb{R}^n$, the product being denoted by a dot above the line $\cdot$. The *length* of a vector $\mathbf{v} \in \mathbb{R}^n$ is $|\mathbf{v}| = \sqrt{(\mathbf{v} \cdot \mathbf{v})}$. A map $f : \mathbb{R}^n \rightarrowtail \mathbb{R}^p$ is said to be *smooth* if everywhere sufficiently many† of its derivatives exist and are continuous, the (non-standard) forked tail on the arrow indicating that the domain of definition is an open subset of $\mathbb{R}^n$ but not necessarily the whole of $\mathbb{R}^n$.

1.1 Regular plane curves and their evolutes

Curves in the plane may be presented in many different ways, for example as the zero sets of functions $\mathbb{R}^2 \rightarrowtail \mathbb{R}$, locally at least as the graphs of functions $\mathbb{R} \rightarrowtail \mathbb{R}$, or parametrically as the images of maps $\mathbb{R} \rightarrowtail \mathbb{R}^2$. For example the circle of radius 1 with centre the origin, the *unit circle*, is the zero set of the function $\mathbb{R}^2 \rightarrowtail \mathbb{R}$; $(x, y) \mapsto x^2 + y^2 - 1$, and also the image of the map $\mathbb{R} \rightarrow \mathbb{R}^2$; $\theta \mapsto (\cos\theta, \sin\theta)$. It is not globally the graph of a function from either axis to the other, but locally it is. For simplicity we begin by concentrating almost entirely on curves presented parametrically, with domains open intervals of $\mathbb{R}$. The image space will be an explicit copy of $\mathbb{R}^2$ but we occasionally will allow ourselves the luxury of choosing a fresh origin for this space, perhaps at some special point of interest of the curve, and also choosing fresh mutually orthogonal axes through this new origin. Such a change of view will, however, preserve the metric of the plane, the distance between points remaining unaltered despite the change of frame of reference.

A *smooth parametric curve* in $\mathbb{R}^2$ is a smooth map

$$\mathbf{r}: \mathbb{R} \rightarrowtail \mathbb{R}^2;\ t \mapsto \mathbf{r}(t),$$

† This usage of the word 'smooth' is slovenly but convenient. If one prefers it, take 'smooth' to mean 'infinitely differentiable', that is C^∞.

with domain an *open interval* of $\mathbb{R}$, that is an open *connected* subset of $\mathbb{R}$. It is *regular* (or *immersive*) *at t* if its first derivative $\mathbf{r}_1(t)$ is non-zero (we defy convention by using subscripts instead of ds or dots or dashes to denote differentiation with respect to the parameter). At a regular point t the vector $\mathbf{r}_1(\mathrm{t})$, which may be regarded as the *velocity* of the curve $\mathbf{r}$ at time t, generates the *tangent vector line* to $\mathbf{r}$ at t. The *tangent line* to $\mathbf{r}$ at t is then the line

$$u \mapsto \mathbf{r}(t) + u\mathbf{r}_1(t).$$

A smooth curve may be straight! But this puts strong conditions on the higher derivatives of the curve. For suppose that the image of the curve $\mathbf{r} : t \mapsto \mathbf{r}(t)$ is the line in $\mathbb{R}^2$ with equation $ax + by = k$, or part of that line. Then, for every $t \in \mathbb{R}$, $\mathbf{c} \cdot \mathbf{r}(t) = k$, where $\mathbf{c} = (a, b)$, and for every $i \geqslant 1$ we have $\mathbf{c} \cdot \mathbf{r}_i(t) = 0$, implying that each of the derived vectors is a multiple of the first non-zero one.

It is, of course, exceptional for any of the higher derivatives $\mathbf{r}_i(t)$ of a regular smooth curve $\mathbf{r}$ at a point t to be a multiple of $\mathbf{r}_1(t)$. We say that a smooth curve $\mathbf{r}$ is *linear* at t if it is regular there and its *acceleration* $\mathbf{r}_2(t)$ is a multiple of $\mathbf{r}_1(t)$. It will be said to be A_k*-linear* at t if it is regular there and $\mathbf{r}_j(t)$ is a multiple of $\mathbf{r}_1(t)$ for $1 < j \leqslant k$, but $\mathbf{r}_{k+1}(t)$ is not a multiple of $\mathbf{r}_1(t)$. According to this definition $\mathbf{r}$ is *not* linear at an A_1-linear point, but just regular there. An A_2-linear point is an *ordinary inflection* of $\mathbf{r}$ and an A_3-linear point an *ordinary undulation* of $\mathbf{r}$.

Example 1.1 The curve $t \mapsto (t, t^3)$ (Figure 1.14) has an ordinary inflection at $t = 0$, while the curve $t \mapsto (t, t^4)$ (Figure 1.15) has an ordinary undulation at $t = 0$. □

The somewhat odd term 'undulation' derives from thinking of the curve $t \mapsto (t, t^4)$ as being the curve given by the value $\varepsilon = 0$ in the family of curves $t \mapsto (t, \varepsilon t^2 + t^4)$, such a curve having no inflection for $\varepsilon > 0$, but acquiring two and a consequent wiggle when ε becomes negative.

These examples are typical:

Proposition 1.2 By suitably choosing a new origin and new mutually orthogonal axes in P^2 *the parametric equations of a smooth curve* $\mathbf{r}$ *in the neighbourhood of an ordinary inflection at* $t = 0$ *may be taken to be of the form*

$$\mathbf{r}(t) = (at + \ldots, bt^3 + \ldots), \text{ where } a \neq 0 \text{ and } b \neq 0,$$

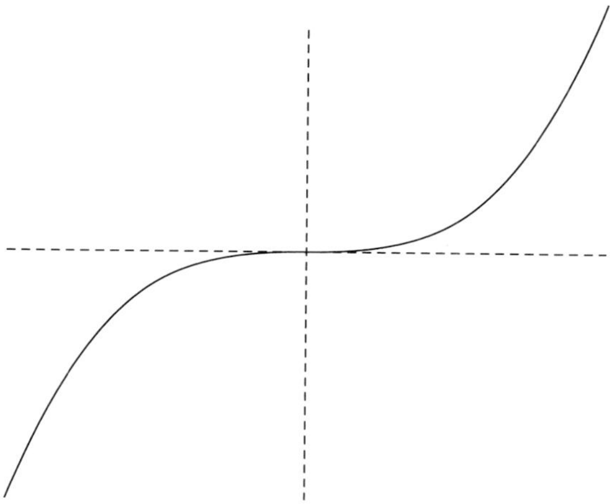

Figure 1.14

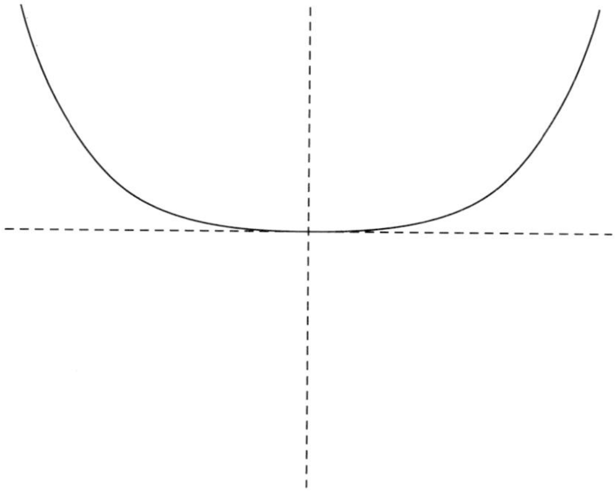

Figure 1.15

and in the neighbourhood of an ordinary undulation to be of the form

$$\mathbf{r}(t) = (at + \ldots, bt^4 + \ldots), \text{ where } a \neq 0 \text{ and } b \neq 0. \qquad \square$$

Corollary 1.3 The tangent line to a regular curve $\mathbf{r}$ *at an ordinary inflection crosses the curve, but at an ordinary undulation this is not so.*

$\square$

In the above examples the curves are actually graphs of functions.

More generally consider the curve $\mathbf{r}$ given by $\mathbf{r}(t) = (t, f(t))$, where $f : \mathbb{R} \rightarrowtail \mathbb{R}; t \mapsto f(t)$ is a smooth function. Then we have

$$\mathbf{r}_1(t) = (1, f_1(t)),$$
$$\mathbf{r}_2(t) = (0, f_2(t)),$$
$$\mathbf{r}_3(t) = (0, f_3(t)),$$
$$\mathbf{r}_4(t) = (0, f_4(t)),$$

and so on. Then $\mathbf{r}_k(t)$ is a multiple of $\mathbf{r}_1(t) \Leftrightarrow f_k(t) = 0$.

Proposition 1.4 For a regular curve $\mathbf{r}$ *that is the graph of a smooth function f,* $\mathbf{r}$ *is A_k-linear at t if and only if* $\mathbf{r}_i(t) = 0$ *for* $2 \leqslant i \leqslant k$, *and* $\mathbf{r}_{k+1}(t) \neq 0$. □

Non-regular points of smooth curves also must be considered, such a point being one where the velocity of the curve is zero. Such a point is commonly called a *cusp* of the curve. In particular a smooth curve $\mathbf{r}$ is said to have an *ordinary*, or 3/2, *cusp* at t if $\mathbf{r}_1(t) = 0$ but $\mathbf{r}_2(t) \neq 0$, with $\mathbf{r}_3(t)$ linearly independent of $\mathbf{r}_2(t)$ and to have an *ordinary kink*, or 4/3, *cusp* at t if $\mathbf{r}_1(t) = 0$ and $\mathbf{r}_2(t) = 0$, but $\mathbf{r}_3(t) \neq 0$, with $\mathbf{r}_4(t)$ linearly independent of $\mathbf{r}_3(t)$. More generally, $\mathbf{r}$ is said to have an $(n+1)/n$ *cusp* at t if $\mathbf{r}_i(t) = 0$ for $1 \leqslant i < n$ but $\mathbf{r}_n(t) \neq 0$, with $\mathbf{r}_{n+1}(t)$ linearly independent of $\mathbf{r}_n(t)$.

Example 1.5 The curve $t \mapsto (t^2, t^3)$ (Figure 1.16) has an ordinary cusp

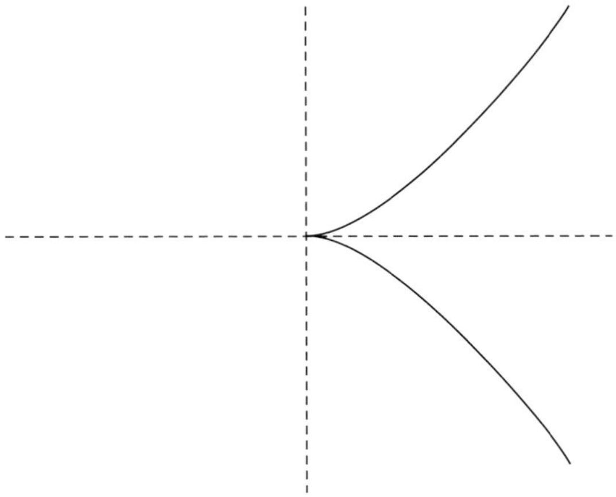

Figure 1.16

at $t = 0$, while the curve $t \mapsto (t^3, t^4)$ (Figure 1.17) has an ordinary kink at $t = 0$. □

Proposition 1.6 By suitably choosing a new origin and orthogonal axes in $\mathbb{R}^2$ the parametric equations of a smooth curve $\mathbf{r}$ *in the neighbourhood of an ordinary cusp at $t = 0$ may be taken to be of the form*

$$\mathbf{r}(t) = (at^2 + \ldots, bt^3 + \ldots), \text{ where } a \neq 0 \text{ and } b \neq 0.$$

Moreover at a cusp the vector $\mathbf{r}_2(0)$ points 'in the opposite direction' to the cusp, lying between its two 'cheeks' (Figure 1.18).

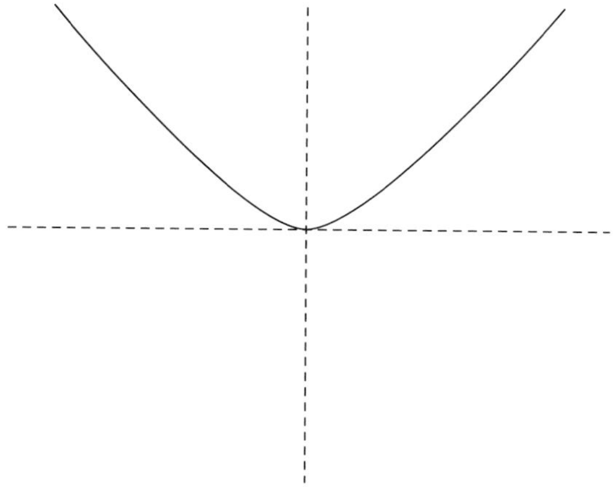

Figure 1.17

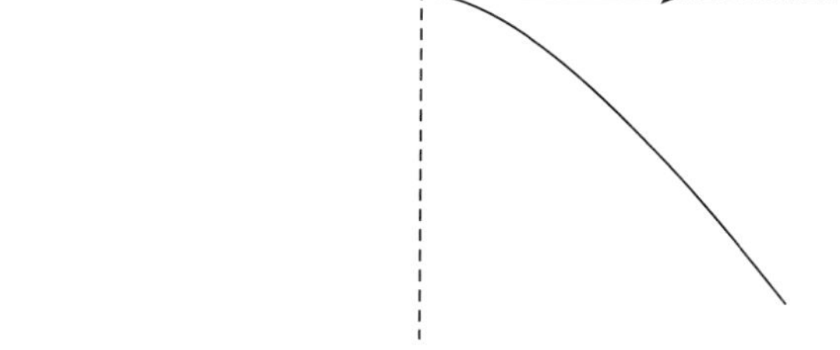

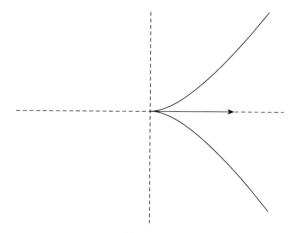

Figure 1.18

Proof Choose axes with origin at the cusp and with $\mathbf{r}_2(0) = (2a, 0)$, where $a \neq 0$. Since also $\mathbf{r}(0) = (0, 0)$ and $\mathbf{r}_1(0) = (0, 0)$ it follows from Taylor's Theorem applied to each component that $\mathbf{r}(t) = (at^2 + \ldots, bt^3 + \ldots)$. Then $\mathbf{r}_3(0) = (c, 6b)$, for some c. Since we know that $\mathbf{r}_3$ is not a multiple of $\mathbf{r}_2$, $b \neq 0$.

Suppose that $a > 0$. Then $\mathbf{r}_2(0)$ points along the x-axis in the right-hand or positive direction. Now the cusp lies entirely in the right-hand half-plane for small t. For the x-component of $\mathbf{r}(t)$ is $at^2 f(t)$, where, since $f(0) = 1$, $f(t) > 0$ for small non-zero t. Moreover, away from $t = 0$

$$\begin{aligned} \mathbf{r}_1(t) &= (2at + \ldots, 3bt^2 + \ldots) \\ &= t(2a + \ldots, 3bt + \ldots), \end{aligned}$$

which is a multiple of $(2a + \ldots, 3bt + \ldots)$, the latter tending to $(2a, 0)$ from opposite sides as t tends to 0 from either side. In particular the limit tangent direction is along the x-axis. That is $\mathbf{r}_2(0)$ points 'in the opposite direction' to the cusp, lying between its two 'cheeks'. □

We shall verify later in Example 4.20 that a smooth curve is essentially non-regular at an ordinary cusp; that is it cannot be made regular there by 'reparametrisation' of the curve.

A regular parametric curve may intersect itself; that is two or more distinct values of the parameter may have the same image point in $\mathbb{R}^2$. The common image point is said to be a *singularity* of the curve, but not a point of non-regularity of the curve.

Example 1.7 The curve $\mathbf{r} : \mathbb{R} \to \mathbb{R}^2 : t \mapsto (t^2 - 1, t(t^2 - 1))$ has a double point at the origin, for $\mathbf{r}(-1) = \mathbf{r}(1) = (0, 0)$, but $\mathbf{r}_1(t) = (2t, 3t^2 - 1)$, so that $\mathbf{r}_1(-1) = (-2, 2)$ and $\mathbf{r}_1(1) = (2, 2)$, both non-zero – Figure 1.19. □

1.2 Curvature

In studying the curvature of a regular plane curve $\mathbf{r}$ we study at each point t how closely the curve approximates there to a parametrised *circle*. Now the circle with centre $\mathbf{c}$ and radius ρ consists of all $\mathbf{r}$ of $\mathbb{R}^2$ such that $(\mathbf{r} - \mathbf{c}) \cdot (\mathbf{r} - \mathbf{c}) = \rho^2$, or equivalently such that

$$\mathbf{c} \cdot \mathbf{r} - \tfrac{1}{2}\mathbf{r} \cdot \mathbf{r} = \tfrac{1}{2}(\mathbf{c} \cdot \mathbf{c} - \rho^2),$$

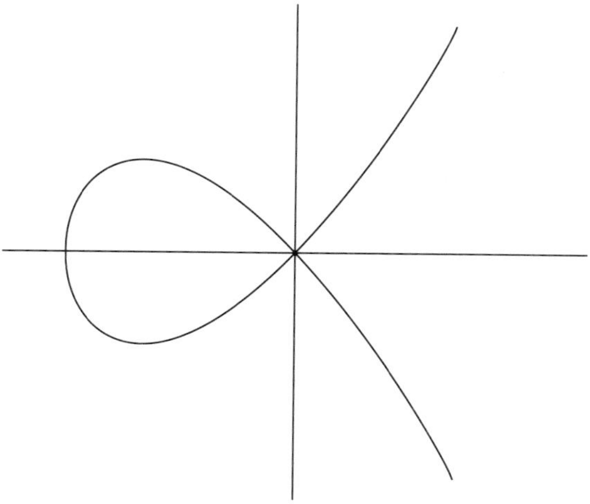

Figure 1.19

the right-hand side of this equation being constant. Accordingly for any parametrisation $t \mapsto \mathbf{r}(t)$ of this circle all the derivatives of the function

$$V(\mathbf{c})\colon t \mapsto \mathbf{c}\cdot\mathbf{r}(t) - \tfrac{1}{2}\mathbf{r}(t)\cdot\mathbf{r}(t)$$

are everywhere zero, namely

$$V(\mathbf{c})_1 = (\mathbf{c}-\mathbf{r})\cdot\mathbf{r}_1 = 0,$$

$$V(\mathbf{c})_2 = (\mathbf{c}-\mathbf{r})\cdot\mathbf{r}_2 - \mathbf{r}_1\cdot\mathbf{r}_1 = 0,$$

$$V(\mathbf{c})_3 = (\mathbf{c}-\mathbf{r})\cdot\mathbf{r}_3 - 3\mathbf{r}_1\cdot\mathbf{r}_2 = 0.$$

Now suppose that $\mathbf{r}$ is a regular parametric curve that is not everywhere circular. Clearly $V(\mathbf{c})_1(t) = 0$ whenever the vector $\mathbf{c}-\mathbf{r}$ is orthogonal to the tangent vector $\mathbf{r}_1(t)$, that is whenever the point $\mathbf{c}$ happens to lie on the *normal* to $\mathbf{r}$ at t, the line through $\mathbf{r}(t)$ orthogonal to the tangent line there. It may be that $\mathbf{r}$ is linear at t, that is that $\mathbf{r}_2(t)$ is linearly dependent on $\mathbf{r}_1(t)$. When this is not so, as will generally be the case, there will be a unique point $\mathbf{c} \neq \mathbf{r}(t)$ on the normal line such that also $V(\mathbf{c})_2(t) = 0$.

This point, which we denote by $\mathbf{e}(t)$, is called the *centre of curvature* or *focal point* of $\mathbf{r}$ at t, the curve $\mathbf{e} : t \mapsto \mathbf{e}(t)$ being called the *evolute* or *focal curve* of $\mathbf{r}$. The distance $\rho(t)$ of $\mathbf{e}(t)$ from $\mathbf{r}(t)$ is called the *radius of curvature* of $\mathbf{r}$ at t and its reciprocal $\kappa(t) = 1/\rho(t)$ the *curvature* of $\mathbf{r}$ at t.

Example 1.8 Let $\mathbf{r}$ be the parabola $t \mapsto (t, t^2)$. Then we have

$$\mathbf{r}(t) = (t, t^2),$$
$$\mathbf{r}_1(t) = (1, 2t),$$
$$\mathbf{r}_2(t) = (0, 2).$$

So the equation for $\mathbf{e}(t)$ becomes

$$\begin{bmatrix} 1 & 2t \\ 0 & 2 \end{bmatrix} \mathbf{e}(t) = \begin{bmatrix} (t, t^2)\cdot(1, 2t) \\ (t, t^2)\cdot(0, 2) + (1, 2t)\cdot(1, 2t) \end{bmatrix};$$

that is

$$\mathbf{e}(t) = \frac{1}{2}\begin{bmatrix} 2 & -2t \\ 0 & 1 \end{bmatrix}\begin{bmatrix} t + 2t^3 \\ 2t^2 + 1 + 4t^2 \end{bmatrix}$$
$$= \frac{1}{2}\begin{bmatrix} -8t^3 \\ 1 + 6t^2 \end{bmatrix} = \begin{bmatrix} 0 \\ \frac{1}{2} \end{bmatrix} + \begin{bmatrix} -4t^3 \\ 3t^2 \end{bmatrix}.$$

Now the curve $t \mapsto (-4t^3, 3t^2)$ clearly has an ordinary cusp at $t = 0$, at the origin in $\mathbb{R}^2$. So the curve $\mathbf{e}$ has an ordinary cusp at $t = 0$, at $(0, \frac{1}{2})$ in $\mathbb{R}^2$ (Figure 1.9). □

The curvature of a regular curve $\mathbf{r}$ may be defined directly by assigning to each t either of the unit normal vectors $\mathbf{n}(t)$ to $\mathbf{r}$ at t. The choice does not matter, except that it should be made continuously along the curve. For definiteness we shall generally tacitly choose $\mathbf{n}$ so that one turns through $+\frac{1}{2}\pi$ in turning from $\mathbf{r}_1(t)$ to $\mathbf{n}(t)$. (But this is not so straightforward for a curve with non-regular points. See Exercise 1.27 – the cardioid (Figure 1.25).)

In this way we have associated to the regular curve $\mathbf{r}$ a smooth circular curve $\mathbf{n}$, the image of $\mathbf{n}$ being a subset of the unit circle, the circle with centre 0 and radius 1. As one travels along the curve $\mathbf{r}$ in time the unit vector $\mathbf{n}(t)$ swings to and fro, like a pointer on a dial (Figure 1.20).

From the definition of $\mathbf{n}$ it follows that $\mathbf{n}\cdot\mathbf{r}_1 = 0$ everywhere, the vectors $\mathbf{n}(t)$ and $\mathbf{r}_1(t)$ being linearly independent for each t, any vector orthogonal to each necessarily being the zero vector, a remark that will be relevant again and again in what follows. Moreover, since $\mathbf{n}\cdot\mathbf{n} = 1$ everywhere it follows that $\mathbf{n}\cdot\mathbf{n}_1 = 0$ (cancelling a 2). So, for all t, the vectors $\mathbf{n}_1(t)$ and $\mathbf{r}_1(t)$ are linearly dependent, both being orthogonal to $\mathbf{n}(t)$. With $\mathbf{r}_1(t) \neq 0$ it follows that $\mathbf{n}_1(t)$ is a (possibly zero) multiple of $\mathbf{r}_1(t)$.

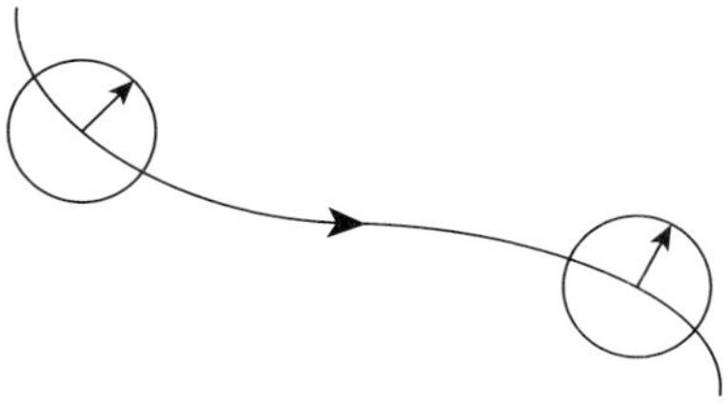

Figure 1.20

For the circle with centre **c** and radius ρ we have $\mathbf{r} + \rho\mathbf{n} = \mathbf{c}$, from which it follows that $\mathbf{r}_1 + \rho\mathbf{n}_1 = 0$, that is that $\kappa\mathbf{r}_1 + \mathbf{n}_1 = 0$, where $\kappa = \rho^{-1}$. This suggests that for a regular curve **r** the *curvature* of **r** at t, $\kappa(t)$, should be defined by the equation

$$\mathbf{n}_1(t) = -\kappa(t)\mathbf{r}_1(t),$$

the actual sign of the curvature depending on the choice of normal. That is up to sign the curvature of a curve is the ratio of the velocity of the end of the normal vector on the dial to one's velocity along the curve.

Suppose that κ is so defined. Then we have the following proposition:

Proposition 1.9 A regular curve **r** *is linear at a point t if and only if* $\kappa = 0$ *at t. Equivalently,* $\kappa \neq 0$ *at t if and only if* **r** *is not linear at t (is* A_1*-linear at t).*

Proof Since $\mathbf{r}_1(t) \neq 0$ it is clear that $\kappa = 0$ if and only if $\mathbf{n}_1 = 0$. Now $\mathbf{n} \cdot \mathbf{r}_1 = 0$. So $\mathbf{n} \cdot \mathbf{r}_2 + \mathbf{n}_1 \cdot \mathbf{r}_1 = 0$, from which it follows that if $\mathbf{n}_1 = 0$ then $\mathbf{n} \cdot \mathbf{r}_2 = 0$ and therefore that **r** is linear at t. Conversely if **r** is linear at t then $\mathbf{n}_1 \cdot \mathbf{r}_1 = 0$. But also $\mathbf{n}_1 \cdot \mathbf{n} = 0$. Thus $\mathbf{n}_1 = 0$. □

One can extend the last proposition.

Proposition 1.10 A regular curve **r** *is* A_k*-linear* $(k \geqslant 2)$ *at a point t if and only if* $\kappa_i = 0$ *at t, for* $0 \leqslant i \leqslant k-2$*, but* $\kappa_{k-1}(t) \neq 0$*. In particular* **r** *has an ordinary inflection (is* A_2*-linear) at t if and only if* $\kappa = 0$ *but* $\kappa_1 \neq 0$ *at t, and has an ordinary undulation (is* A_3*-linear) at t if and only if* $\kappa = \kappa_1 = 0$ *but* $\kappa_2 \neq 0$ *at t.* □

Note that the curvature of a regular curve changes sign as one passes through an ordinary inflection of the curve.

To relate the two definitions of curvature suppose that at t $\kappa(t) \neq 0$, with $\rho(t) = 1/\kappa(t)$. Then $\mathbf{r}_1(t) + \rho(t)\mathbf{n}_1(t) = 0$. By an earlier remark this is equivalent to the pair of equations

$$(\mathbf{r}_1(t) + \rho(t)\mathbf{n}_1(t)) \cdot \mathbf{r}_1(t) = 0 \text{ and } (\mathbf{r}_1(t) + \rho(t)\mathbf{n}_1(t)) \cdot \mathbf{n}(t) = 0,$$

the second of these being true for all t since $\mathbf{r}_1 \cdot \mathbf{n} = 0$ and $\mathbf{n}_1 \cdot \mathbf{n} = 0$. As to the first, since $\mathbf{n} \cdot \mathbf{r}_1 = 0$, $\mathbf{n}_1 \cdot \mathbf{r}_1 = -\mathbf{n} \cdot \mathbf{r}_2$ as before, so that the equation takes the form

$$\rho(t)\mathbf{n}(t) \cdot \mathbf{r}_2(t) - \mathbf{r}_1(t) \cdot \mathbf{r}_1(t) = 0,$$

in accordance with our earlier definition, with $\rho(t)\mathbf{n}(t) = \mathbf{e}(t) - \mathbf{r}(t)$. Of course ρ may now take either sign, the actual sign of ρ at any point t depending on the choice of the unit normal vector $\mathbf{n}(t)$.

We shall say that a smooth curve $\mathbf{r}$ is *circular* at a point t if it is regular and not linear at t and if also not only $V(\mathbf{c})_1(t) = V(\mathbf{c})_2(t) = 0$ but also $V_3(\mathbf{e})(t) = 0$, where $V(\mathbf{c}) = \mathbf{c} \cdot \mathbf{r} - \frac{1}{2}\mathbf{r} \cdot \mathbf{r}$ and $V_k(\mathbf{e})$, for any positive integer k, is a convenient shorthand notation for $V(\mathbf{c})_k$ with $\mathbf{c}$ *after* the differentiation put equal to $\mathbf{e}$. It has an *ordinary vertex* at t if also $V_4(\mathbf{e})(t) \neq 0$.

Several equations relate the derivatives of a regular curve $\mathbf{r}$ to the derivatives of $\mathbf{e}$, of ρ and of $\mathbf{n}$. Apart from $\mathbf{n} \cdot \mathbf{r}_1 = 0$ and $\mathbf{n} \cdot \mathbf{n}_1 = 0$ we have

$$(\mathbf{e} - \mathbf{r}) \cdot \mathbf{r}_1 = 0 \tag{1}$$

$$(\mathbf{e} - \mathbf{r}) \cdot \mathbf{r}_2 = \mathbf{r}_1 \cdot \mathbf{r}_1 \tag{2}$$

defining $\mathbf{e}$,

$$\mathbf{e}_1 \cdot \mathbf{r}_1 = 0 \tag{3}$$

obtained by differentiating (1) and using (2),

$$\mathbf{e}_1 \cdot \mathbf{r}_2 + V_3(\mathbf{e}) = 0, \tag{4}$$

obtained by differentiating (2),

$$\mathbf{r}_1 + \rho\mathbf{n}_1 = 0 \tag{5}$$

and

$$\mathbf{e}_1 = \rho_1\mathbf{n} \tag{6}$$

obtained by differentiating the equation $\mathbf{e} = \mathbf{r} + \rho\mathbf{n}$ and using (5).

We employ these in the proof of the following proposition listing some elementary properties of the evolute of a regular plane curve.

Proposition 1.11 Let $\mathbf{r}$ *be a regular curve in the plane with evolute* $\mathbf{e}$. *Then*

(*a*) *for each t at which the evolute* $\mathbf{e}$ *is regular the tangent line to* $\mathbf{e}$ *at t coincides with the normal line to* $\mathbf{r}$ *at t;*
(*b*) *the curve* $\mathbf{e}$ *has no linear points – in particular no ordinary point of inflection or undulation;*
(*c*) *if* $\mathbf{e}_1(t) = 0$ *but* $\mathbf{e}_2(t) \neq 0$ *then* $\mathbf{e}$ *has an ordinary cusp at t;*
(*d*) *the curve* $\mathbf{e}$ *is regular at t if and only if the curve* $\mathbf{r}$ *is non-circular at t;*
(*e*) *the curve* $\mathbf{e}$ *has an ordinary cusp at t if and only if the curve* $\mathbf{r}$ *has an ordinary vertex at t;*
(*f*) *the curve* $\mathbf{r}$ *has an ordinary vertex at t if and only if the radius of curvature* ρ *has an ordinary critical point at t, the cusp on* $\mathbf{e}$ *pointing towards or away from the vertex according as (the absolute value of)* ρ *has a local minimum or maximum at t.*

Proof By equation (3) in the preamble to this proposition $\mathbf{e}_1 \cdot \mathbf{r}_1 = 0$. Thus for each regular point t of $\mathbf{e}$ not only does $\mathbf{e}(t)$ lie on the normal line to $\mathbf{r}$ at t but also the tangent to $\mathbf{e}$ is normal to the tangent to $\mathbf{r}$; that is the tangent line to $\mathbf{e}$ at t coincides with the normal to $\mathbf{r}$ at t, which is assertion (a). That is the solution set of the equation for the normal to $\mathbf{r}$ at t, when put in the form

$$\text{‘}\mathbf{c} = \text{particular solution} + \text{kernel’},$$

is

$$\mathbf{c} = \mathbf{e}(t) + \lambda \mathbf{e}_1(t), \text{ for all } \lambda \in \mathbb{R}.$$

On differentiating the equation $\mathbf{e}_1 \cdot \mathbf{r}_1 = 0$ we get

$$\mathbf{e}_2 \cdot \mathbf{r}_1 + \mathbf{e}_1 \cdot \mathbf{r}_2 = 0.$$

Now, for $\mathbf{e}(t)$ to be defined, the vector $\mathbf{r}_2(t)$ is linearly independent of $\mathbf{r}_1(t)$, so that if $\mathbf{e}_1(t) \neq 0$ then $\mathbf{e}_1(t) \cdot \mathbf{r}_2(t) \neq 0$, from which it follows that $\mathbf{e}_2(t) \cdot \mathbf{r}_1(t) \neq 0$. But $\mathbf{e}_1(t) \cdot \mathbf{r}_1(t) = 0$. It follows that $\mathbf{e}_2(t)$ is linearly independent of $\mathbf{e}_1(t)$, which is assertion (b).

On differentiating the same equation a second time we get

$$\mathbf{e}_3 \cdot \mathbf{r}_1 + 2\mathbf{e}_2 \cdot \mathbf{r}_2 + \mathbf{e}_1 \cdot \mathbf{r}_3 = 0.$$

So if $\mathbf{e}_1(t) = 0$ but $\mathbf{e}_2(t) \neq 0$ then, since $\mathbf{e}_2(t) \cdot \mathbf{r}_1(t) = 0$, we must have $\mathbf{e}_2(t) \cdot \mathbf{r}_2(t) \neq 0$ and so also $\mathbf{e}_3(t) \cdot \mathbf{r}_1(t) \neq 0$. But then $\mathbf{e}_3(t)$ is linearly independent of $\mathbf{e}_2(t)$, which is assertion (c).

Next consider equation (4) of the preamble, namely

$$\mathbf{e}_1 \cdot \mathbf{r}_2 + V_3(\mathbf{e}) = 0.$$

Now at a point t where $\mathbf{r}$ is not circular $V_3(\mathbf{e})(t) \neq 0$, implying that $\mathbf{e}_1(t) \cdot \mathbf{r}_2(t) \neq 0$ and hence that $\mathbf{e}_1(t) \neq 0$. Conversely, if $\mathbf{e}_1(t) \neq 0$ then, since $\mathbf{e}_1(t) \cdot \mathbf{r}_1(t) = 0$, $\mathbf{e}_1(t) \cdot \mathbf{r}_2(t) \neq 0$, so that $V_3(\mathbf{e})(t) \neq 0$. This establishes assertion (d).

Differentiating (4) (where $V_3(\mathbf{e}) = (\mathbf{e} - \mathbf{r}) \cdot \mathbf{r}_3 - 3\mathbf{r}_1 \cdot \mathbf{r}_2$) gives

$$\mathbf{e}_2 \cdot \mathbf{r}_2 + 2\mathbf{e}_1 \cdot \mathbf{r}_3 + V_4(\mathbf{e}) = 0.$$

Now at an ordinary vertex $V_3(\mathbf{e}) = 0$ but $V_4(\mathbf{e}) \neq 0$, implying that at such a point $\mathbf{e}_1 \cdot \mathbf{r}_2 = 0$ but $\mathbf{e}_2 \cdot \mathbf{r}_2 + 2\mathbf{e}_1 \cdot \mathbf{r}_3 \neq 0$. Since also, by (3), $\mathbf{e}_1 \cdot \mathbf{r}_1 = 0$ it follows that $\mathbf{e}_1 = 0$ there. But then $\mathbf{e}_2 \cdot \mathbf{r}_2 \neq 0$, implying that $\mathbf{e}_2 \neq 0$ there and so, by assertion (c), that $\mathbf{e}$ has an ordinary cusp.

Conversely, at a point where $\mathbf{e}$ has an ordinary cusp $\mathbf{e}_1 = 0$, implying that $V_3(\mathbf{e}) = 0$ but $\mathbf{e}_2 \neq 0$, implying that $\mathbf{e}_2 \cdot \mathbf{r}_2 \neq 0$ and therefore that $V_4(\mathbf{e}) \neq 0$, so that $\mathbf{r}$ has an ordinary vertex there: Thus ordinary vertices of $\mathbf{r}$ and ordinary cusps of $\mathbf{e}$ correspond. This is assertion (e).

To prove the correspondence of each of these with ordinary critical points of ρ we note first that we may at any particular point assume that ρ is positive, by choosing the circular curve $\mathbf{n}$ appropriately near that point. Now consider equation (6) of the preamble, namely $\mathbf{e}_1 = \rho_1\mathbf{n}$. Differentiating this we obtain $\mathbf{e}_2 = \rho_2\mathbf{n} + \rho_1\mathbf{n}_1$. Clearly $\mathbf{e}_1 = 0$ if and only if $\rho_1 = 0$, with $\mathbf{e}_2 \neq 0$ also if and only if $\rho_2 \neq 0$. Finally, when $\rho_1 = 0$, it follows by Proposition 1.6 from the equation $\mathbf{e}_2 = \rho_2\mathbf{n}$ that the cusp on $\mathbf{e}$ points towards or away from the vertex of $\mathbf{r}$ according as the radius of curvature ρ has a local minimum or maximum there. This completes the proof of (f). □

Clearly the ordinary critical points of ρ are also the critical points of ρ^2. In fact earlier on when we went for a swim in the bay we were concerned with knowing at all times the points of the shore-line that were nearest to us. We develop this line of thought in Exercise 1.22.

Equally clearly the ordinary critical points of ρ are also the ordinary critical points of κ, provided that $\kappa \neq 0$.

The following proposition complements the one we have just proved.

Proposition 1.12 An ordinary undulation of $\mathbf{r}$ *is a point where the curvature has an ordinary critical point and where also the curvature is zero.* □

On occasion we shall include among the ordinary vertices of a curve its ordinary undulations.

The A-type labelling system introduced to classify the linear points of a curve extends to the circular points as well. The labels should properly be attached to the points $\mathbf{c}$ of the normals of $\mathbf{r}$. Points of each normal, as points of that normal, carry the label A_1, except for the focal points, or points of the evolute, which generally carry the label A_2, except for the ordinary cusps on the evolute, which carry the label A_3, these being the only possibilities on a generic curve. Higher orders of circularity may occur on particular curves, for example on particular curves in generic families of curves. At an A_k-linear point the label A_k should properly be attached to the point at infinity on the normal to the curve there.

Formally, a point $\mathbf{c}$ is an *A_k-centre* for $\mathbf{r}$ at t if and only if $V(\mathbf{c})_i = 0$ for $1 \leqslant i \leqslant k$ but $V(\mathbf{c})_{k+1} \neq 0$, $\mathbf{r}$ then being said to be *A_k-circular* at t, for $k \geqslant 2$.

Proposition 1.13 A regular plane curve $\mathbf{r}$ *is* A_k*-circular at* t*, where* $k \geqslant 2$*, if and only if* $\mathbf{r}$ *is not linear at* t *and* $\mathbf{r}_i + \rho\mathbf{n}_i = 0$ *for* $1 \leqslant i \leqslant k-1$ *but* $\mathbf{r}_k + \rho\mathbf{n}_k \neq 0$*. In particular, such a curve* $\mathbf{r}$ *has an ordinary vertex at t if and only if* $\mathbf{r}$ *is not linear at t and* $\mathbf{r}_i + \rho\mathbf{n}_i = 0$ *for* $i = 1$ *or* 2 *but not* 3.

Sketch of proof We give the proof in detail for an ordinary vertex, leaving the general case as an exercise.

Our starting point is the equation $\mathbf{r}_1 + \rho\mathbf{n}_1 = 0$ which holds everywhere along the curve. Differentiating it twice we find that

$$\mathbf{r}_2 + \rho_1\mathbf{n}_1 + \rho\mathbf{n}_2 = 0,$$

and

$$\mathbf{r}_3 + \rho_2\mathbf{n}_1 + 2\rho_1\mathbf{n}_2 + \rho\mathbf{n}_3 = 0.$$

Now $\mathbf{n}_1 \neq 0$, since $\mathbf{r}$ is not linear. Accordingly

$$\mathbf{r}_2 + \rho\mathbf{n}_2 = 0 \text{ if and only if } \rho_1 = 0,$$

and if so then

$$\mathbf{r}_3 + \rho\mathbf{n}_3 = 0 \text{ if and only if } \rho_2 = 0,$$

or equivalently

$$\mathbf{r}_3 + \rho\mathbf{n}_3 \neq 0 \text{ if and only if } \rho_2 \neq 0,$$

which is what had to be proved.

The formal way to set out the general case is as an induction argument. We leave the details of formulating and establishing the appropriate basis and step to the reader. □

Complementing this we have

Proposition 1.14 A smooth regular curve $\mathbf{r}$ *is* A_k*-linear at* t $(k \geqslant 2)$ *if and only if* $n_i = 0$ *for* $1 \leqslant i < k$ *but* $\mathbf{n}_k \neq 0$. □

Finally we state a characterisation of the evolute, relevant later.

Proposition 1.15 The evolute of a regular plane curve $\mathbf{r}$ *with unit normal curve* $\mathbf{n}$ *is the smooth curve* $\mathbf{e}$ *such that not only is* $\mathbf{e} - \mathbf{r}$ *everywhere a multiple of* $\mathbf{n}$, *say* $\rho\mathbf{n}$, *but also* $\mathbf{e}_1$ *is everywhere a multiple of* $\mathbf{n}$, *say* $\mu\mathbf{n}$, *the coefficients* ρ *and* μ *varying smoothly along the curve.*

Proof On differentiating the equation $(\mathbf{e} - \mathbf{r}) \cdot (\mathbf{e} - \mathbf{r}) = \rho^2$ and using $(\mathbf{e} - \mathbf{r}) \cdot \mathbf{r}_1 = 0$ we get $(\mathbf{e} - \mathbf{r}) \cdot \mathbf{e}_1 = \rho\rho_1$, from which it follows that $\rho\mathbf{n} \cdot \mu\mathbf{n} = \rho\mu = \rho\rho_1$, implying that $\mu = \rho_1$. So $\mathbf{e}_1 = \rho_1\mathbf{n}$. But also $\mathbf{e}_1 - \mathbf{r}_1 = \rho_1\mathbf{n} + \rho\mathbf{n}_1$, so also $\mathbf{r}_1 + \rho\mathbf{n}_1 = 0$. Moreover, on differentiating $(\mathbf{e} - \mathbf{r}) \cdot \mathbf{r}_1 = 0$ and using $\mathbf{e}_1 \cdot \mathbf{r}_1 = 0$ we get $(\mathbf{e} - \mathbf{r}) \cdot \mathbf{r}_2 - \mathbf{r}_1 \cdot \mathbf{r}_1 = 0$, so recovering the original definition. □

1.3 Parallels

We began this chapter by having a look at the offsets to a parabola. Given any regular curve $\mathbf{r}$ its *offsets* or *parallels* are the curves

$$t \mapsto \mathbf{r}(t) + \delta\mathbf{n}(t),$$

where $\delta \in \mathbb{R}$. The name 'parallel' is justified by the fact that for each t the velocity of each parallel to $\mathbf{r}$ is a multiple of the velocity of $\mathbf{r}$, since for each t and each δ we have

$$\mathbf{n}(t) \cdot (\mathbf{r} + \delta\mathbf{n})_1(t) = \mathbf{n}(t) \cdot (\mathbf{r}_1(t) + \delta\mathbf{n}_1(t)) = 0.$$

As we saw in the case of the parabola, the parallels to a regular curve may have points of non-regularity. In fact

$$(\mathbf{r} + \delta\mathbf{n})_1(t) = \mathbf{r}_1(t) + \delta\mathbf{n}_1(t) = 0$$

if and only if $\delta = \rho(t)$.

Thus the points of non-regularity of the parallels to $\mathbf{r}$ lie on the evolute $\mathbf{e}$ of $\mathbf{r}$. The following proposition makes this precise.

Proposition 1.16 Let $\mathbf{r}$ *be a regular smooth plane curve and* $\mathbf{r} + \delta\mathbf{n}$ *the parallel of* $\mathbf{r}$ *offset a distance* δ *along each normal, and consider those points* t *where* $\mathbf{r}(t) + \delta\mathbf{n}(t) = \mathbf{e}(t)$, $\mathbf{e}$ *being the evolute of* $\mathbf{r}$. *Then where* $\mathbf{e}$ *is regular the parallel has an ordinary* (3/2) *cusp, the limiting tangent of the cusp being orthogonal to the evolute, while where* $\mathbf{e}$ *has an ordinary cusp the parallel has an ordinary kink* (4/3 *cusp*).

Proof First we remark that, at a regular point of $\mathbf{e}$,

$$\mathbf{r}_1 + \rho\mathbf{n}_1 = 0 \text{ but } \mathbf{r}_2 + \rho\mathbf{n}_2 \neq 0,$$

while, at an ordinary cusp of $\mathbf{e}$,

$$\mathbf{r}_1 + \rho\mathbf{n}_1 = \mathbf{r}_2 + \rho\mathbf{n}_2 = 0 \text{ but } \mathbf{r}_3 + \rho\mathbf{n}_3 \neq 0.$$

Secondly, since $\mathbf{n} \cdot \mathbf{r}_1 = 0$ and $\mathbf{n} \cdot \mathbf{n}_1 = 0$,

$$\mathbf{n} \cdot (\mathbf{r}_1 + \delta\mathbf{n}_1) = 0,$$

$$\mathbf{n} \cdot (\mathbf{r}_2 + \delta\mathbf{n}_2) = -\mathbf{n}_1 \cdot (\mathbf{r}_1 + \delta\mathbf{n}_1),$$

$$\mathbf{n} \cdot (\mathbf{r}_3 + \delta\mathbf{n}_3) = -2\mathbf{n}_1 \cdot (\mathbf{r}_2 + \delta\mathbf{n}_2) - \mathbf{n}_2 \cdot (\mathbf{r}_1 + \delta\mathbf{n}_1),$$

and

$$\mathbf{n} \cdot (\mathbf{r}_4 + \delta\mathbf{n}_4) = -3\mathbf{n}_1 \cdot (\mathbf{r}_3 + \delta\mathbf{n}_3) - 3\mathbf{n}_2 \cdot (\mathbf{r}_2 + \delta\mathbf{n}_2) - \mathbf{n}_3 \cdot (\mathbf{r}_1 + \delta\mathbf{n}_1).$$

Thirdly, $\mathbf{n}$ and $\mathbf{n}_1$ are everywhere linearly independent. So if at a point t it happens that $\delta = \rho$, so that $\mathbf{r}_1 + \delta\mathbf{n}_1 = 0$, and if also $\mathbf{r}_2 + \delta\mathbf{n}_2 \neq 0$, it follows that $\mathbf{r}_3 + \delta\mathbf{n}_3$ is not a multiple of $\mathbf{r}_2 + \delta\mathbf{n}_2$ at t, while if $\mathbf{r}_2 + \delta\mathbf{n}_2 = 0$ but $\mathbf{r}_3 + \delta\mathbf{n}_3 \neq 0$ then $\mathbf{r}_4 + \delta\mathbf{n}_4$ is not a multiple of $\mathbf{r}_3 + \delta\mathbf{n}_3$ there. But this is what had to be proved.

That the ordinary cusps on a parallel are orthogonal to the evolute is intuitively obvious since the limiting tangent to the parallel must be parallel to the tangent to $\mathbf{r}$ and so normal to the evolute. Alternatively from one of the above equations we have

$$\mathbf{n} \cdot (\mathbf{r}_2 + \delta\mathbf{n}_2) = -\mathbf{n}_1 \cdot (\mathbf{r}_1 + \delta\mathbf{n}_1) = 0 \text{ when } \mathbf{r}_1 + \delta\mathbf{n}_1 = 0.$$

So the direction of the cusp is normal to $\mathbf{n}$, that is tangential to $\mathbf{e}$. □

The parallels or offsets of a smooth curve **r** all have the same unit normal curve and so the same evolute **e**. They are also known as the *involutes* or *evolvents* of **e**, being derivable from **e** by the Huygens process described in the Introduction to this chapter. As an inextensible string lying along the evolute is unwound from it, all the time being kept taut, each point of the string traces an involute of **e**, this being succinctly summarised in the equation $\mathbf{e}_1 = \rho_1\mathbf{n}$, which may be interpreted as saying that the rate of increase of the unwound length of string is equal to the speed with which the curve **e** is traversed by the point at which the unwinding takes place.

Alternatively we may imagine a straight line or ruler rolling without slipping along the curve **e**, to coincide at each moment t with the tangent line to **e** at t. Then each point of the rolling line traces out an involute of **e**. Clearly the involutes of **e** are the orthogonal trajectories of the tangent lines to **e**.

We proved earlier as part (b) of Proposition 1.11 that the evolute of a regular plane curve does not have any inflections, and it follows that each of the involutes of such a curve must have a point of non-regularity on the tangent at the inflection. Nor can this be an ordinary cusp. As we have already mentioned in the Introduction, an entertaining section in de l'Hôpital's book (1696) concerns his discussion of what actually does happen.

Before trying to solve the problem analytically it may be helpful to remark that at each point on an involute the radius of curvature of the involute is equal to the length of the unwound string, or equivalently the distance of the marked point on the ruler from the point of tangency. An ordinary cusp arises when the string lying along the curve is cut and the two pieces peeled off. This is in agreement with the readily verifiable fact that the limit of the radius of curvature of a smooth curve at an ordinary cusp is zero (Exercise 1.10). But if a point of non-regularity arises at a later stage in the unwinding then the limit of the radius of curvature of the involute there cannot be zero. If one experiments on the curve $t \mapsto (t, t^3)$ one obtains Figure 1.21. One of de l'Hôpital's diagrams is shown as Figure 1.22.

All but one of the involutes are seen to have *rhamphoid* or *hooked beak* cusps on the inflectional tangent, the limiting radius of curvature at such a non-regular point being greater than zero (the length of the unwound string). The exception is the involute that arises when the string is cut at the inflection itself. This appears to have an inflection at

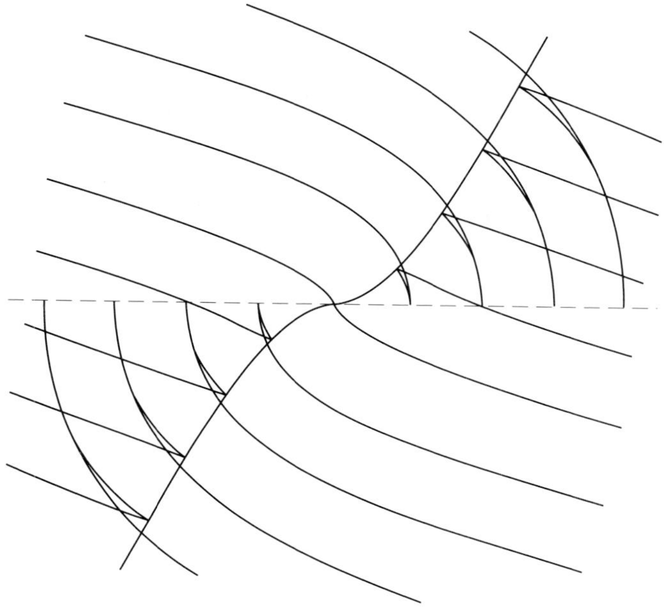

Figure 1.21

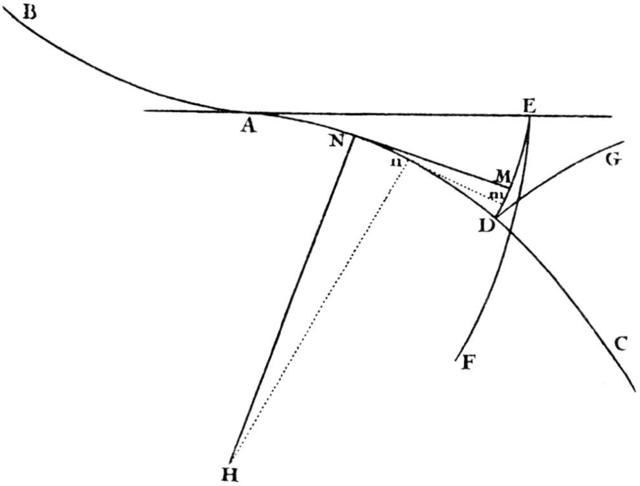

Figure 1.22

the point where it crosses the original curve, but from what we have already said the radius of curvature there must be *zero* and not infinity! Moreover, if one plays the same trick on this curve one gets a further curve with an apparent inflection, but with radius of curvature zero, and so on, for ever. So, as de l'Hôpital remarks tongue in cheek, it seems that if one encounters a curve with an inflection then it is far more likely that the radius of curvature there is not infinity but zero!

The whole situation clearly demands careful investigation. Before we do so it pays us to consider the matter of equivalent parametrisations of plane curves.

1.4 Equivalent parametric curves

Until now we have considered the given parametrisation as an essential part of the definition of a curve. But it is clear that much of the geometric interest lies in describing the image of $\mathbf{r}$ in relation to its surroundings and that different parametric curves may have the same image.

We shall say that a pair of smooth parametric curves $\mathbf{q}$ and $\mathbf{r}$ are *(right-)equivalent* if there is a smooth invertible map with smooth inverse (that is a *diffeomorphism*) $h : \mathbb{R} \rightarrowtail \mathbb{R}$ such that $\mathbf{r} = \mathbf{q}h$ $(= \mathbf{q} \circ h)$.

The *Leibniz formula* for the nth derivative of the *product* of two functions will probably be familiar to the reader, but it is likely that the nth derivative of the *composite* of two maps is not so familiar. The first few derivatives of $\mathbf{r}$ in terms of $\mathbf{q}$ and h are in fact

$$\begin{aligned}
\mathbf{r}_1 &= (\mathbf{q}_1 h) \cdot h_1, \\
\mathbf{r}_2 &= (\mathbf{q}_2 h) \cdot {h_1}^2 + (\mathbf{q}_1 h) \cdot h_2, \\
\mathbf{r}_3 &= (\mathbf{q}_3 h) \cdot {h_1}^3 + 3(\mathbf{q}_2 h) \cdot h_1 h_2 + (\mathbf{q}_1 h) \cdot h_3, \\
\mathbf{r}_4 &= (\mathbf{q}_4 h) \cdot {h_1}^4 + 6(\mathbf{q}_3 h) \cdot {h_1}^2 h_2 + 4(\mathbf{q}_2 h) \cdot h_3 + 3(\mathbf{q}_2 h) \cdot {h_2}^2 \\
&\quad + (\mathbf{q}_1 h) \cdot h_4.
\end{aligned}$$

A full discussion of the general formula, known as *Faà de Bruno's Formula*, is deferred until Chapter 4.

What should be clear from these formulas is that the *nullities* of $\mathbf{q}$ and $\mathbf{r}$ everywhere agree. That is if at a point $u = h(t)$ we have $\mathbf{q}_i = 0$ for $1 \leqslant i < k$ but $\mathbf{q}_k \neq 0$, then at the point t we have $\mathbf{r}_i = 0$ for

$1 \leqslant i < k$ but $\mathbf{r}_k \neq 0$, and conversely. In particular, $\mathbf{q}$ is regular at u if and only if $\mathbf{r}$ is regular at t.

Proposition 1.17 With $\mathbf{q}$ *and* $\mathbf{r}$ *equivalent smooth plane curves as above, and with* $u = h(t)$, *the curve* $\mathbf{r}$ *has an ordinary inflection, undulation, vertex, cusp or kink at t if and only if the curve* $\mathbf{q}$ *has an ordinary inflection, undulation, vertex, cusp or kink at u.* □

1.5 Unit-speed curves

Given a *regular* smooth curve $\mathbf{r} : \mathbb{R} \rightarrowtail \mathbb{R}^2$ with domain an interval, one can always reparametrise it, at least in theory, so that its speed is equal to 1 everywhere. To do this we simply put $\mathbf{r} = \mathbf{q}h$, where $h : \mathbb{R} \rightarrowtail \mathbb{R};\ t \mapsto s = h(t)$ is chosen such that ${h_1}^2 = \mathbf{r}_1 \cdot \mathbf{r}_1$ everywhere – in classical notation this is equivalent to requiring that

$$\frac{\mathrm{d}s}{\mathrm{d}t} = \sqrt{\left[\left(\frac{\mathrm{d}x}{\mathrm{d}t}\right)^2 + \left(\frac{\mathrm{d}y}{\mathrm{d}t}\right)^2\right]},$$

determining s as a function of t up to an additive constant.

The variable s, measured from some convenient reference point on the curve, is called the *arc-length* of $\mathbf{r}$. With ${h_1}^2 = \mathbf{r}_1 \cdot \mathbf{r}_1$ at t and with $\mathbf{r}_1(t) = \mathbf{q}_1(h(t))\mathbf{h}_1(t)$, it follows that $\mathbf{q}_1 \cdot \mathbf{q}_1 = 1$ at $s = h(t)$, for all t. The curve $\mathbf{q}$ is said to be a *unit-speed* curve.

It must be emphasised that the notion is more useful in theory than in practice since the integration of $\sqrt{(\mathbf{r}_1 \cdot \mathbf{r}_1)}$ is in general not possible in terms of elementary functions. Nor can we reparametrise a given curve as a unit-speed curve across a point of non-regularity, for example an ordinary cusp.

The letter $\mathbf{q}$ having served its purpose, is now dropped.

Proposition 1.18 Let $\mathbf{r} : s \mapsto \mathbf{r}(s)$ *be a unit-speed curve. Then* $\mathbf{r}$ *is linear at s if and only if* $\mathbf{r}_2(s) = 0$. *If* $\mathbf{r}_2(s) \neq 0$ *then also* $\mathbf{r}_3(s)$ *is not a multiple of* $\mathbf{r}_2(s)$. *At an ordinary inflection one not only has* $\mathbf{r}_2 = 0$ *but* $\mathbf{r}_3 \neq 0$, *with* $\mathbf{r}_3$ *orthogonal to* $\mathbf{r}_1$, $\mathbf{r}_4$ *a multiple of* $\mathbf{r}_3$, *but* $\mathbf{r}_5$ *not a multiple of* $\mathbf{r}_3$.

Proof All this follows from differentiating $\mathbf{r}_1 \cdot \mathbf{r}_1 = 1$. For we get $\mathbf{r}_1 \cdot \mathbf{r}_2 = 0$ (cancelling a 2 as usual), $\mathbf{r}_1 \cdot \mathbf{r}_3 + \mathbf{r}_2 \cdot \mathbf{r}_2 = 0$, $\mathbf{r}_1 \cdot \mathbf{r}_4 + 3\mathbf{r}_2 \cdot \mathbf{r}_3 = 0$, and $\mathbf{r}_1 \cdot \mathbf{r}_5 + 4\mathbf{r}_2 \cdot \mathbf{r}_4 + 3\mathbf{r}_3 \cdot \mathbf{r}_3 = 0$. So that $\mathbf{r}_2$ is a multiple of $\mathbf{r}_1$ if and only if $\mathbf{r}_2 = 0$. If $\mathbf{r}_2 \neq 0$ then $\mathbf{r}_1 \cdot \mathbf{r}_3 \neq 0$. So $\mathbf{r}_3$ is not then a multiple of $\mathbf{r}_2$. If $\mathbf{r}_2 = 0$ then

$$\mathbf{r}_1 \cdot \mathbf{r}_3 = \mathbf{r}_1 \cdot \mathbf{r}_4 = \mathbf{r}_1 \cdot \mathbf{r}_5 + 3\mathbf{r}_3 \cdot \mathbf{r}_3 = 0,$$

from which the remaining assertions are evident when $\mathbf{r}_3$ is not a multiple of $\mathbf{r}_1$. □

For a unit-speed curve $\mathbf{r}$ it is customary to denote the unit-tangent curve $\mathbf{r}_1$ by $\mathbf{t}$, the parameter, arc-length from a suitable base-point on the curve, being denoted by s.

Proposition 1.19 Let $\mathbf{r}$ *be a unit-speed curve in* $\mathbb{R}^2$ *with unit-tangent curve* $\mathbf{t}$, *unit-normal curve* $\mathbf{n}$ *and curvature* κ. *Then* $\mathbf{t}_1 = \kappa\mathbf{n}$ *and* $\mathbf{n}_1 = -\kappa\mathbf{t}$.

Proof The second of these equations is just $\mathbf{n}_1 = -\kappa\mathbf{r}_1$, an equation which holds whatever the parameter may be.

As to the first, since $\mathbf{t} \cdot \mathbf{t} = 1$, $\mathbf{t}_1 \cdot \mathbf{t} = 0$. So $\mathbf{t}_1 = \mu\mathbf{n}$, for some μ. But then, since $\mathbf{t} \cdot \mathbf{n} = 0$,

$$\mu = \mu\mathbf{n} \cdot \mathbf{n} = \mathbf{t}_1 \cdot \mathbf{n} = -\mathbf{n}_1 \cdot t = \kappa\mathbf{t} \cdot \mathbf{t} = \kappa. \qquad \square$$

1.6 Unit-angular-velocity curves

As remarked above it is not possible to take arc-length as a parameter equivalent to the given one in the neighbourhood of an ordinary cusp on a smooth curve. There are alternatives. One is to take the square root of arc-length from the cusp as parameter, though it is technically slightly easier to choose $u = \sqrt{(2s)}$. With this choice one has $\mathbf{r}_1 \cdot \mathbf{r}_1 = 2s$ in the neighbourhood of the cusp. Another possibility is to use the *angle of contingence* ψ, the angle between either the tangent or normal to the curve and some fixed direction, usually taken to be the x-axis. This is equivalent to the circular curve $\mathbf{n}$ being unit-speed, that is to having $\mathbf{n}_1 \cdot \mathbf{n}_1 = 1$. This can be arranged except in the neighbourhood of an inflection, where $\mathbf{n}_1 = 0$. With such a parameter chosen the curve $\mathbf{r}$ becomes a *unit-angular-velocity curve*, the circular curve $\mathbf{n}$ now being unit-speed.

Proposition 1.20 Up to sign the curvature $\kappa = 1/\rho$ *of a regular curve* $\mathbf{r}$ *is equal to* $\mathrm{d}\psi/\mathrm{d}s$, *where* ψ *is the angle between the unit normal* $\mathbf{n}$ *and the x-axis and s is arc-length, measured from a suitable point of reference on the curve.*

Proof Take s as parameter for the curve. Then $\mathbf{n}_1 = (\mathrm{d}\mathbf{n}/\mathrm{d}\psi)(\mathrm{d}\psi/\mathrm{d}s)$. But, with $\mathbf{n} = (\cos\psi, \sin\psi)$, the vector $\mathrm{d}\mathbf{n}/\mathrm{d}\psi = (-\sin\psi, \cos\psi)$ is of unit length, as is the velocity vector $\mathbf{r}_1$. The assertion then follows from the fact that $\mathbf{n}_1 = -\kappa\mathbf{r}_1$. □

Proposition 1.20 is just what is needed to prove that a regular plane curve is determined by its curvature, up to isometries of the ambient plane. This is the content of the next theorem.

Theorem 1.21 Let $\mathbf{r} : \mathbb{R} \rightarrowtail \mathbb{R}^2;\ \psi \mapsto \mathbf{r}(\psi)$ *be a regular plane curve with curvature* κ, *parametrised by the angle* ψ *which we take here to be the angle that the tangent to* $\mathbf{r}$ *makes with the x-axis, and let* $\mathbf{r}(\theta)$ *be specified at some particular point* $\psi = \theta$. *Then*

$$\mathbf{r}(\psi) = \mathbf{r}(\theta) + \int_\theta^\psi (\rho\cos\psi, \rho\sin\psi)\,\mathrm{d}\psi.$$

Proof This follows directly from the equations

$$\mathrm{d}x/\mathrm{d}s = \cos\psi \text{ and } \mathrm{d}y/\mathrm{d}s = \sin\psi.$$ □

1.7 Rhamphoid cusps

The Huygens unwinding procedure from a smooth curve $\mathbf{e}$ can be simply presented if, necessarily away from cusps or other points of non-regularity, we choose arc-length s from some reference point along $\mathbf{e}$ as parameter. Then $\mathbf{e}_1 \cdot \mathbf{e}_1 = 1$, with $\mathbf{e}_2 \neq 0$ except where $\mathbf{e}$ is linear, for example at an inflection. Moreover, by Proposition 1.11(c), when $\mathbf{e}_2 \neq 0$, $\mathbf{e}_3$ is not a multiple of $\mathbf{e}_2$.

This being so, each involute of $\mathbf{e}$ is of the form

$$s \mapsto \mathbf{r}(s) = \mathbf{e}(s) + (\delta - s)\mathbf{e}_1(s),$$

for some real number δ. Clearly

$$\mathbf{r}_1(s) = (\delta - s)\mathbf{e}_2(s),$$
$$\mathbf{r}_2(s) = (\delta - s)\mathbf{e}_3(s) - \mathbf{e}_2(s),$$

and, in general,

$$\mathbf{r}_k(s) = (\delta - s)\mathbf{e}_{k+1}(s) - (k - 1)\mathbf{e}_k(s).$$

We are now in a position finally to resolve de l'Hôpital's problem concerning the involutes of a regular plane curve with an inflection.

Proposition 1.22 With **e** *and* **r** *as above:*

(*a*) *at a point s where* **e** *is not linear, and where* $s \neq \delta$, **r** *is regular, intersecting the tangent to* **e** *orthogonally, but is neither linear nor circular there;*

(*b*) *at a point s where* **e** *is not linear, but where* $s = \delta$, **r** *has an ordinary cusp;*

(*c*) *at a point s where* **e** *has an ordinary inflection, and where* $s \neq \delta$, **r** *is not regular, with* $\mathbf{r}_1 = 0$ *but* $\mathbf{r}_2 \neq 0$; *but the cusp is not ordinary, for at that point* $\mathbf{r}_3$ *is a multiple of* $\mathbf{r}_2$; *however,* $\mathbf{r}_4$ *is not a multiple of* $\mathbf{r}_2$ *there;*

(*d*) *at a point s where* **e** *has an ordinary inflection, but where* $s = \delta$, *then* $\mathbf{r}_1 = \mathbf{r}_2 = 0$, $\mathbf{r}_3 \neq 0$, $\mathbf{r}_4$ *is a multiple of* $\mathbf{r}_3$ *but* $\mathbf{r}_5$ *is not a multiple of* $\mathbf{r}_3$.

Proof This follows directly from the formulas for the derivatives of the curve **r** derived above and the results of Proposition 1.18. □

The cusp in part (*c*) above is the *rhamphoid* (that is *beak-shaped*) cusp (Figure 1.23) that our earlier experiment led us to expect. An *ordinary rhamphoid* cusp on a smooth curve **r** is characterised by the conditions

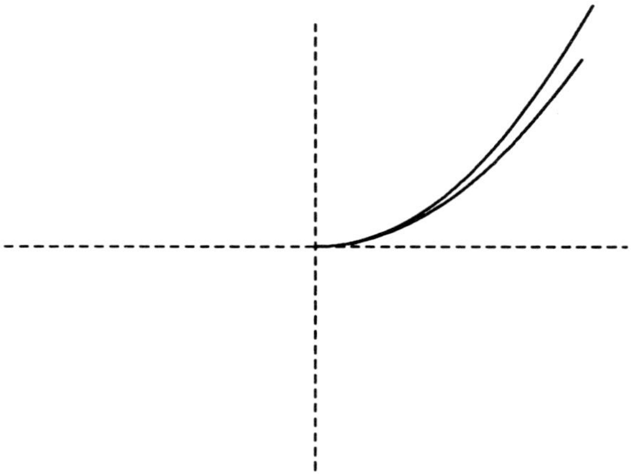

Figure 1.23

set out in (*c*) above together with the further condition, also satisfied here, that the limiting value of ρ_1 at the cusp is non-zero, that is that the radius of curvature of the curve is not critical at the cusp. Since at each regular non-linear point of the curve $\mathbf{e}_1 = \rho_1 \mathbf{n}$, an equivalent condition is that the limiting value of $\mathbf{e}_1$ at the cusp is non-zero. That this is so here follows at once from the fact that $\rho(s) = s - \delta$, so that $\rho_1 = 1$ everywhere.

A curve $\mathbf{r}$ whose derivatives are as in case (*d*) above at the origin is the curve $t \mapsto (t^3, t^5)$ (Figure 1.24, compare Figure 1.14).

Theorem 1.23 provides a characterisation of an ordinary rhamphoid cusp directly in terms of the derivatives of the parametrisation $\mathbf{r}$.

Theorem 1.23 Let $\mathbf{r} : \mathbb{R} \rightarrowtail \mathbb{R}^2$ *be a smooth plane curve. Then* $\mathbf{r}$ *has an ordinary rhamphoid cusp at t if and only if* $\mathbf{r}_1(t) = 0$, $\mathbf{r}_2(t) \neq 0$, $\mathbf{r}_3(t) = 3\lambda \mathbf{r}_2(t)$ *for some real number* λ, *and, for all* $\mu \in \mathbb{R}$, $\mathbf{r}_5(t) \neq 10\lambda \mathbf{r}_4(t) + \mu \mathbf{r}_2(t)$.

Proof Everywhere the derivatives of the curve $\mathbf{r}$ and its evolute $\mathbf{e}$ satisfy the equations

$$(\mathbf{e} - \mathbf{r}) \cdot \mathbf{r}_1 = 0$$

$$(\mathbf{e} - \mathbf{r}) \cdot \mathbf{r}_2 = \mathbf{r}_1 \cdot \mathbf{r}_1$$

$$\mathbf{e}_1 \cdot \mathbf{r}_2 + (\mathbf{e} - \mathbf{r}) \cdot \mathbf{r}_3 = 3\mathbf{r}_1 \cdot \mathbf{r}_2$$

$$\mathbf{e}_2 \cdot \mathbf{r}_2 + 2\mathbf{e}_1 \cdot \mathbf{r}_3 + (\mathbf{e} - \mathbf{r}) \cdot \mathbf{r}_4 = 4\mathbf{r}_1 \cdot \mathbf{r}_3 + 3\mathbf{r}_2 \cdot \mathbf{r}_2$$

$$\mathbf{e}_3 \cdot \mathbf{r}_2 + 3\mathbf{e}_2 \cdot \mathbf{r}_3 + 3\mathbf{e}_1 \cdot \mathbf{r}_4 + (\mathbf{e} - \mathbf{r}) \cdot \mathbf{r}_5 = 5\mathbf{r}_1 \cdot \mathbf{r}_4 + 10\mathbf{r}_2 \cdot \mathbf{r}_3,$$

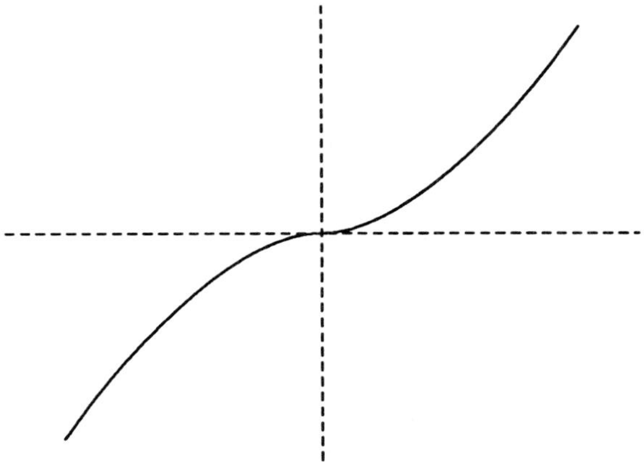

Figure 1.24

and so on, as well as the equations

$$\begin{aligned} \mathbf{e}_1 \cdot \mathbf{r}_1 &= 0 \\ \mathbf{e}_2 \cdot \mathbf{r}_1 + \mathbf{e}_1 \cdot \mathbf{r}_2 &= 0 \\ \mathbf{e}_3 \cdot \mathbf{r}_1 + 2\mathbf{e}_2 \cdot \mathbf{r}_2 + \mathbf{e}_1 \cdot \mathbf{r}_3 &= 0 \\ \mathbf{e}_4 \cdot \mathbf{r}_1 + 3\mathbf{e}_3 \cdot \mathbf{r}_2 + 3\mathbf{e}_2 \cdot \mathbf{r}_3 + \mathbf{e}_1 \cdot \mathbf{r}_4 &= 0, \end{aligned}$$

and so on.

At an ordinary rhamphoid cusp, where $\mathbf{r}_1 = 0$, $\mathbf{r}_2 \neq 0$, $\mathbf{r}_3 = 3\lambda\mathbf{r}_2$ for some $\lambda \in \mathbb{R}$, $\mathbf{r}_4$ is linearly independent of $\mathbf{r}_2$ and $\mathbf{e}_1 \neq 0$, these reduce, taking the second set first, to

$$\begin{aligned} \mathbf{e}_1 \cdot \mathbf{r}_2 &= 0 \\ 2\mathbf{e}_2 \cdot \mathbf{r}_2 &= 0 \\ 3\mathbf{e}_3 \cdot \mathbf{r}_2 + \mathbf{e}_1 \cdot \mathbf{r}_4 &= 0 \\ (\mathbf{e} - \mathbf{r}) \cdot \mathbf{r}_2 &= 0 \\ (\mathbf{e} - \mathbf{r}) \cdot \mathbf{r}_3 &= 0 \\ (\mathbf{e} - \mathbf{r}) \cdot \mathbf{r}_4 &= 3\mathbf{r}_2 \cdot \mathbf{r}_2 \\ \mathbf{e}_3 \cdot \mathbf{r}_2 + 3\mathbf{e}_1 \cdot \mathbf{r}_4 + (\mathbf{e} - \mathbf{r}) \cdot \mathbf{r}_5 &= 10\mathbf{r}_2 \cdot \mathbf{r}_3, \end{aligned}$$

from which we finally deduce that at such a cusp

$$\tfrac{8}{3}\mathbf{e}_1 \cdot \mathbf{r}_4 + (\mathbf{e} - \mathbf{r}) \cdot \mathbf{r}_5 = 10\mathbf{r}_2 \cdot \mathbf{r}_3,$$

so that

$$(\mathbf{e} - \mathbf{r}) \cdot \mathbf{r}_5 \neq 10\mathbf{r}_2 \cdot \mathbf{r}_3.$$

A necessary and sufficient condition for this is then that

$$\text{for all } \mu \in \mathbb{R},\ \mathbf{r}_5 \neq 10\lambda\mathbf{r}_4 + \mu\mathbf{r}_2 \qquad \square$$

The condition that ρ_1 be non-zero at a rhamphoid cusp of $\mathbf{r}$ is, somewhat surprisingly perhaps, invariant with respect to a *diffeomorphism* of some neighbourhood of the cusp, that is for any smooth invertible map $h : \mathbb{R}^2 \rightarrowtail \mathbb{R}^2$ with smooth inverse, with $\mathbf{r}(t) \subset \operatorname{dom} h$, the condition holds for $\mathbf{r}$ at t if and only if it holds for $h\mathbf{r}$ at t. What we shall later show, as Proposition 4.10, is that any ordinary rhamphoid cusp satisfying this condition is diffeomorphic to the cusp at 0 of the curve $t \mapsto (t^2, t^5)$, and conversely that any rhamphoid cusp diffeomorphic at 0 to the cusp at 0 of the curve $t \mapsto (t^2, t^5)$ is ordinary. The limiting centre of curvature at 0 of the latter curve is of course at infinity. So an ordinary rhamphoid cusp may be characterised as a 5/2 *cusp* with non-zero curvature. It has to be said that in the literature the condition that $\kappa \neq 0$ is not usually insisted on.

It is also the case that if at a rhamphoid cusp $\rho_1 = 0$ but $\rho_2 \neq 0$ then the cusp is diffeomorphic to the cusp at 0 of the curve $t \mapsto (t^2, t^7)$, though the converse is false. An instructive example to work through in detail (Exercise 1.33) is the curve

$$t \mapsto \left(\tfrac{1}{2}t^2, \tfrac{1}{4}t^4 + \tfrac{1}{5}at^5 + \tfrac{1}{6}bt^6 + \tfrac{1}{7}ct^7 + \tfrac{1}{8}dt^8\right).$$

1.8 The determination of circular points

In applications, for example in analysing the curvature of the paths traced by the various points of a plane lamina moving smoothly over the plane in mechanism theory, it is of importance to be able to determine efficiently which points of a curve are its circular points. The traditional route is cumbersome, involving first of all the nasty explicit formula for the curvature of the curve found in Exercise 1.12 and then the even nastier formula for its derivative. All this can be avoided!

Proposition 1.24 Let $\mathbf{r}$ *be a regular plane curve. Then* $\mathbf{r}$ *is either circular or at least* A_3*-linear at t if and only if the* 3×3 *square matrix*

$$\begin{bmatrix} \mathbf{r}_1 \cdot & \tfrac{1}{2}(\mathbf{r}\cdot\mathbf{r})_1 \\ \mathbf{r}_2 \cdot & \tfrac{1}{2}(\mathbf{r}\cdot\mathbf{r})_2 \\ \mathbf{r}_3 \cdot & \tfrac{1}{2}(\mathbf{r}\cdot\mathbf{r})_3 \end{bmatrix},$$

or equivalently the matrix

$$\begin{bmatrix} \mathbf{r}_1 \cdot & 0 \\ \mathbf{r}_2 \cdot & \mathbf{r}_1\cdot\mathbf{r}_1 \\ \mathbf{r}_3 \cdot & 3\mathbf{r}_1\cdot\mathbf{r}_2 \end{bmatrix},$$

has rank 2 at t, which will be the case if and only if its determinant is zero, since $\mathbf{r}_1 \neq 0$*, by hypothesis. Here* $\mathbf{r}_k \cdot$ *denotes the row vector that is the transpose of the column vector* $\mathbf{r}_k$.

Proof Either matrix has rank 2 if and only if either the three simultaneous linear equations

$$\mathbf{r}_1 \cdot \mathbf{c} = \tfrac{1}{2}(\mathbf{r}\cdot\mathbf{r})_1,$$

$$\mathbf{r}_2 \cdot \mathbf{c} = \tfrac{1}{2}(\mathbf{r} \cdot \mathbf{r})_2,$$
$$\mathbf{r}_3 \cdot \mathbf{c} = \tfrac{1}{2}(\mathbf{r} \cdot \mathbf{r})_3,$$

admit at t a common solution $\mathbf{c} = \mathbf{e}(t)$, this being equivalent to asserting the circularity of $\mathbf{r}$ at t, or else the matrix

$$\begin{bmatrix} \mathbf{r}_1 \cdot \\ \mathbf{r}_2 \cdot \\ \mathbf{r}_3 \cdot \end{bmatrix}$$

has rank 1 at t, which is the case if and only if the curve $\mathbf{r}$ is at least A_3-linear at $\mathbf{t}$. In either case the first of the two matrices in the statement of the proposition will have rank 2. The second matrix is obtained from the first by subtracting from the last column the scalar product of the first two columns with $\mathbf{r}$, a column operation that leaves the rank of the matrix unchanged. □

Proposition 1.25 Let Δ *denote the determinant of either of the* 3×3 *matrices in the statement of Proposition 1.24. Then, at* t, $\mathbf{r}$ *is* A_k*-circular or* A_k*-linear for* $k \geqslant 3$ *if and only if* $\Delta_i = 0$ *for* $1 \leqslant i \leqslant k-3$, *but* $\Delta_{k-2} \neq 0$.

Proof Since the one matrix is obtained from the other by an elementary column operation their determinants are equal – but the first form of the matrix is the one to use in the proof. What we do is to regard Δ as a trilinear function of the rows of the matrix. Then the first derivative is the sum of the determinants of the matrices one gets by replacing each row in turn by its derivative. Clearly the first two of the resulting matrices will have zero determinant, each having two rows equal. The third matrix will have zero determinant if and only if the new row kills the same kernel vectors as the first two rows do. Clearly this will be so if and only if either $V(\mathbf{c})_4 = 0$, where $\mathbf{c} = \mathbf{e}(t)$, or if not only $\mathbf{r}_2$ and $\mathbf{r}_3$ but also $\mathbf{r}_4$ is a multiple of $\mathbf{r}_1$ at t; that is if and only if $\mathbf{r}$ is A_k-circular or A_k-linear at t. The higher derivatives of Δ are dealt with in the same way, the details being left to the reader. □

Corollary 1.26 The A_3*-circular or* A_3*-linear points of* $\mathbf{r}$ *occur at the simple zeros of the function* $t \mapsto \Delta(t)$. □

This is clearly of great help in applications in view of the complexities of the expressions $V(\mathbf{c})_k$, for greater and greater k.

For unit-speed or unit-angular-velocity curves there are particularly simple criteria for circularity.

Proposition 1.27 Let $s \mapsto \mathbf{r}(s)$ *be a unit-speed curve. Then* $\mathbf{r}$ *is circular at* s *if and only if* $\mathbf{r}_2$ *is not a multiple of* $\mathbf{r}_1$ *but* $\mathbf{r}_3$ *is a multiple of* $\mathbf{r}_1$. *(Of course* $\mathbf{r}$ *is at least* A_3*-linear if and only if both* $\mathbf{r}_2$ *and* $\mathbf{r}_3$ *are multiples of* $\mathbf{r}_1$ *at* s.)

Proof Since $\mathbf{r}$ is unit speed, $\mathbf{r}_1 \cdot \mathbf{r}_2 = 0$. So

$$\Delta = -\det\begin{bmatrix} \mathbf{r}_1 \cdot \\ \mathbf{r}_3 \cdot \end{bmatrix},$$

or, transposing the 2×2 matrix, $\Delta = -\det[\mathbf{r}_1 \quad \mathbf{r}_3]$. From this the assertions of the proposition quickly follow. □

Proposition 1.28 Let $\psi \mapsto \mathbf{r}(\psi)$ *be a regular unit-angular-velocity curve. Then* $\mathbf{r}$ *is circular at* ψ *if and only if* $\mathbf{r}_1 \cdot \mathbf{r}_2 = 0$.

Proof A regular unit-angular-velocity curve $\mathbf{r}$ is nowhere linear, for since $\mathbf{n}_1$ is of unit length everywhere it can never be zero. Moreover, from the equations $\mathbf{n}_1 \cdot \mathbf{n}_1 = 1$ and $\mathbf{r}_1 = -\rho\mathbf{n}_1$ it follows that $\mathbf{r}_1 \cdot \mathbf{r}_1 = \rho^2$. Differentiating this we get $\mathbf{r}_1 \cdot \mathbf{r}_2 = \rho\rho_1$. By Proposition 1.11(*f*), $\mathbf{r}$ is circular at ψ *if* and only if $\rho(\psi) \neq 0$ and $\rho_1(\psi) = 0$. Now ρ is everywhere non-zero by the regularity of $\mathbf{r}$. So $\mathbf{r}$ is circular at ψ if and only if, at ψ, $\mathbf{r}_1 \cdot \mathbf{r}_2 = 0$. □

For a formula for determining the vertices of a smooth curve presented intrinsically as the zeros of a smooth map $F : \mathbb{R}^2 \rightarrowtail \mathbb{R}$ see Exercise 4.13.

1.9 The four-vertex theorem

Everything that has been said so far is local, describing curves point by point. We finish with two global theorems.

Theorem 1.29 Let $\mathbf{r}$ *be a regular plane curve with domain an interval and either linear or circular everywhere. Then* $\mathbf{r}$ *is either a straight line or a circle.*

Proof If $\mathbf{r}$ is everywhere linear then $\mathbf{n}_1 = 0$ from which it follows from the connectedness of dom $\mathbf{r}$ that $\mathbf{n}$ is constant and therefore that $\mathbf{r}$ is a straight line. If $\mathbf{r}$ is everywhere circular then ρ exists with $\rho_1 = 0$, so that ρ is constant. Moreover, not only $\rho_1 = 0$ but also $\mathbf{e}_1 = \rho_1 \mathbf{n} = 0$ so that $\mathbf{e}$ is also constant, implying that $\mathbf{r}$ is the circle with centre the point $\mathbf{e}$ and radius ρ. □

Our final theorem disproves the common intuition that an egg-shaped curve need only have two vertices, a 'sharp end' and a 'blunt end'!

Theorem 1.30 (*The Four-Vertex Theorem*) *Let* $\mathbf{r}$ *be a simple closed connected regular smooth plane curve. Then* $\mathbf{r}$ *has at least four vertices, places where* κ (*possibly* 0) *either has a local maximum or minimum.*

The theorem has a long history, outlined by Robert Osserman in 1985, to whom the following proof is due. He considers the *circumscribed circle* to the curve, defined by the following lemma:

Lemma 1.31 Let E be a compact subset of $\mathbb{R}^2$ having at least two points. Then among all the circles C with the property that the closed disk bounded by C includes E there is a unique one of minimum radius $R > 0$. Moreover, any arc of this circumscribed circle greater than a semicircle must intersect E.

Proof Either part follows from the observation that, assuming the contrary, one could find a smaller circle enclosing E. □

Lemma 1.32 Let a smooth oriented curve $\mathbf{r}$ *have the same unit tangent at a point P as a positively oriented circle C of radius R. Let κ be the curvature of* $\mathbf{r}$. *Then if $\kappa(P) > 1/R$ a neighbourhood of P on* $\mathbf{r}$ *lies inside C while if $\kappa(P) < 1/R$ a neighbourhood of P on* $\mathbf{r}$ *lies outside C.* □

The proof of the theorem then follows directly from this together with one further lemma.

Lemma 1.33 Let $\mathbf{r}$ *be a positively oriented Jordan curve, C the circumscribed circle and P_1, P_2 points of* $\mathbf{r}$ *lying on C. Then either* $\mathbf{r}$ *coincides with C between P_1 and P_2 or else there is a point Q of* $\mathbf{r}$ *between P_1 and P_2 such that $\kappa(Q) < 1/R$.*

Proof See Osserman (1985). □

Osserman concludes his account with the observation that in some sense it is more likely that such a simple closed curve has at least six vertices, rather then four, for the latter occurs only in the exceptional case that the circumscribed circle has just two points of contact with the curve **r**, these points necessarily being antipodal on the circle. We shall have occasion to recall this later when in Chapter 15 we study the vertices of the contours of a surface that is the graph of a smooth function $\mathbb{R}^2 \rightarrowtail \mathbb{R}$ near a critical point of the function.

Exercises

1.1 Compute the velocity and speed (the absolute value of the velocity), as functions of t, of each of the following parametric curves:

(i) $]-\frac{1}{2}\pi, \frac{1}{2}\pi[\rightarrow \mathbb{R}^2;\ t \mapsto (\cos^2 t, \tan t)$,

(ii) $]-\frac{1}{2}\pi, \frac{1}{2}\pi[\rightarrow \mathbb{R}^2;\ t \mapsto (\sin^2 t, \sin^2 t, \tan t)$,

(iii) $\mathbb{R} \rightarrow \mathbb{R}^2;\ t \mapsto (\cos^3 t, \sin^3 t)$.

Determine any non-regular points. Sketch the curves.

1.2 Find the coordinates of the two inflections of the curve given by $]-\frac{1}{2}\pi, \frac{1}{2}\pi[\rightarrow \mathbb{R}^2;\ t \mapsto (\cos^2 t, \tan t)$, previously discussed in Exercise 1.1(i). Then make a more accurate sketch than last time!

1.3 Show that the curve $t \mapsto (t, t^3)$ has an ordinary inflection at $t = 0$, while the curve $t \mapsto (t, t^4)$ has an ordinary undulation at $t = 0$.

1.4 Show that the curve $t \mapsto (t^2, t^3)$ has an ordinary cusp at $t = 0$, while the curve $t \mapsto (t^3, t^4)$ has an ordinary kink at $t = 0$.

1.5 Prove Proposition 1.2.

1.6 Prove Corollary 1.3.

1.7 Prove Proposition 1.4.

1.8 Find the radius of curvature $\rho(t)$ of the curve $t \mapsto (t, t^2)$ at the point t, for all $t \in \mathbb{R}$, and show that ρ has an ordinary critical point (in fact a local minimum) at 0.

1.9 Show that the curve $\mathbf{r}(x) = (x, \frac{1}{2}\kappa x^2 + \text{higher order terms})$ has curvature κ at $x = 0$.

1.10 Show that the limiting radius of curvature at an ordinary cusp is zero.

1.11 Find the limiting radius of curvature of the curve $t \mapsto (t^3, t^5)$ at 0. Sketch the latter curve near $t = 0$.

1.12 Deduce either from the equations $(\mathbf{e} - \mathbf{r}) \cdot \mathbf{r}_1 = 0$ and $(\mathbf{e} - \mathbf{r}) \cdot \mathbf{r}_2 = \mathbf{r}_1 \cdot \mathbf{r}_1$ or from the equation $\kappa \mathbf{r}_1 + \mathbf{n}_1 = 0$ that the curvature κ of a regular parametric curve $\mathbf{r}$ is, up to choice of sign, equal to

$$\frac{x_1 y_2 - x_2 y_1}{(x_1{}^2 + y_1{}^2)^{3/2}},$$

and that the evolute is given, wherever the curvature is non-zero, by

$$\mathbf{e} = \begin{bmatrix} x \\ y \end{bmatrix} + \frac{x_1{}^2 + y_1{}^2}{x_1 y_2 - x_2 y_1} \begin{bmatrix} -y_1 \\ x_1 \end{bmatrix}.$$

Deduce a formula for the derivative of the curvature at t.

1.13 Find the vertices, if any, of the parametric curves

(i) $t \mapsto (\frac{1}{2}t^2, \frac{1}{5}t^5)$

(ii) $t \mapsto (\frac{1}{3}t^3, \frac{1}{4}t^4)$

(iii) $t \mapsto (t - \sin t, 1 - \cos t)$

and sketch the curves and their evolutes on the same diagram, for suitable intervals of t, a different diagram for each of the three curves, of course!

(Since you want $\mathbf{e}$ as well as the vertices of $\mathbf{r}$ it is sensible to use the formula for $\mathbf{e}$ from Exercise 1.12. At a vertex $\mathbf{e}_1 = 0$.)

1.14 Find the evolute of the curve $t \mapsto (t^2, t^3)$.

1.15 A curve is given in terms of polar coordinates by the formula $r = f(\theta)$, where f is smooth. Derive a formula for the curvature in terms of the derivatives of f.

(The curve is the curve $\theta \mapsto (r \cos \theta, r \sin \theta)$.)

1.16 Verify that the plane curve with polar equation $r = a + b \cos \theta$, where $a \geqslant 0$ and $b \geqslant 0$, that is the curve

$$\theta \mapsto ((a + b \cos \theta) \cos \theta, (a + b \cos \theta) \sin \theta),$$

has an inflection if and only if $b < a < 2b$. (You are on the right road if you have shown that inflections occur only when $a^2 + 2b^2 + 3ab \cos \theta = 0$. Don't forget that $\cos^2(t) = \frac{1}{2}(1 + \cos 2t)$ and that $\cos t \sin t = \frac{1}{2} \sin 2t$.)

(These curves are known as the *limaçons of Pascal*. When

$a = b$ the curve is heart-shaped and is known as the *cardioid* (Figure 1.25). These curves will reappear in Chapter 3.)

1.17 Verify that:

(i) the cardioid with polar equation $r = 1 + \cos\theta$ has no inflections, but has one ordinary cusp and one ordinary vertex $(\mathrm{mod}\, 2\pi)$;

(ii) the curve with polar equation $r = 2 + \cos\theta$ has no inflections, but has an ordinary undulation and three further vertices $(\mathrm{mod}\, 2\pi)$, the four points in question being where either $\sin\theta = 0$ or $\cos\theta = -\frac{1}{2}$.

1.18 Find the evolute of the cardioid with polar equation

$$r = 1 + \cos\theta.$$

Sketch both curves on the same diagram.

1.19 Determine the vertices of the curve $\mathbf{r}: \mathbb{R} \to \mathbb{R}^2$; $t \mapsto (t^2 - 1, \alpha t(t^2 - 1))$ for different values of α. For which value of α does the curve have an A_5 vertex at $t = 0$?

1.20 Prove that the evolute of the cycloid $\theta \mapsto (\theta - \sin\theta, 1 - \cos\theta)$ is a congruent cycloid, that passes through the cusps of the original cycloid. Sketch both the curve and its evolute for $-2\pi \leqslant t \leqslant 4\pi$.

(Cycloids will be studied in Chapter 3.)

1.21 For which values of α does the smooth curve $\mathbf{r}$ defined by

$$\mathbf{r}(t) = (\alpha t + \sin t, \cos t)$$

have ordinary inflections and for which values does it have

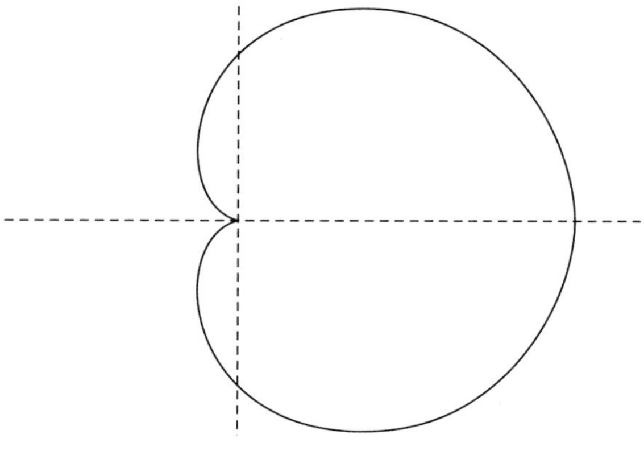

Figure 1.25

ordinary cusps? Determine for which values of t these inflections or cusps occur. Sketch the curve for relevant values of α.

1.22 On page 4 we imagined ourselves swimming out normally from a smooth shore, measuring at all times the distance ρ back from our position $\mathbf{c}$ to points $\mathbf{r}$ of the shore. By considering the sign of the second derivative of $\rho^2 = (\mathbf{c} - \mathbf{r}) \cdot (\mathbf{c} - \mathbf{r})$, or otherwise, determine at what point of our path the point of departure changes from being a local minimum of ρ to being a local maximum of ρ.

1.23 Let $\mathbf{r}$ be a regular plane curve and $\mathbf{n}$ a smooth circular curve such that, at each point t, $\mathbf{n}(t)$ is a unit normal to $\mathbf{r}$. Prove that $\mathbf{r}$ has an ordinary undulation at t if and only if $\mathbf{n}_1(t) = \mathbf{n}_2(t) = 0$, but $\mathbf{n}_3(t) \neq 0$.

1.24 Prove Proposition 1.12.

1.25 Complete the proof of Proposition 1.13.

1.26 Let $\mathbf{r} : t \mapsto \mathbf{r}(t)$ and $\mathbf{r}' : u \mapsto \mathbf{r}'(u)$ be equivalent smooth, but necessarily regular, plane curves, with $\mathbf{r}' = \mathbf{r}h$ say, where $h : \mathbb{R} \rightarrowtail \mathbb{R}$ is an invertible smooth map with smooth inverse (or *diffeomorphism*). Prove that $\mathbf{r}$ has an ordinary cusp at $t = a = h(b)$ if, and only if, $\mathbf{r}'$ has an ordinary cusp at $u = b$.

1.27 Verify that the curve $\theta \mapsto ((1 - \cos\theta)\cos\theta, (1 - \cos\theta)\sin\theta)$ is a cardioid with cusp at the origin. Verify also that for this curve the unit circular curve $\mathbf{n}$ is described at constant speed. What is the angle turned through by $\mathbf{n}$ in a single circuit of the cardioid, from the cusp round to the cusp again?

1.28 Find the maximal and minimal values of the curvature of the ellipse with equation

$$\frac{x^2}{a^2} + \frac{y^2}{b^2} = 1.$$

(You could use the parametric equations $x = a\cos\varphi$, $y = b\sin\varphi$.)

1.29 Locate any inflections and find the limiting centre of curvature at the cusp for each of the following curves:

(i) $$t \mapsto \left(\tfrac{1}{2}t^2, \tfrac{1}{4}t^4 + \tfrac{1}{5}t^5\right)$$

(ii) $$t \mapsto \left(\tfrac{1}{2}t^2 + \tfrac{1}{3}t^3, \tfrac{1}{4}t^4\right).$$

(The vertices are a lot harder to find in these cases.)

Sketch the curves.

1.30 Verify explicitly all the results of Proposition 1.16 in the particular case of the parabola $t \mapsto \mathbf{r}(t) = (t, \frac{1}{2}t^2)$.

$$\left[\frac{1}{\sqrt{(1+t^2)}} = 1 - \frac{1}{2}t^2 + \frac{3}{8}t^4 + O(t^6)\right].$$

1.31 Reparametrise the cycloidal arch

$$]0, 2\pi[\to \mathbb{R}^2 : t \mapsto (t - \sin t, 1 - \cos t)$$

so that:

(i) it is described at unit speed,
(ii) the unit normal traverses the unit circle at unit speed.

1.32 Let $\mathbf{r} : \mathbb{R} \rightarrowtail \mathbb{R}^2$ be a smooth plane curve, with $\mathbf{r}(0) = 0$, having an ordinary rhamphoid cusp at 0, and let the axes in $\mathbb{R}^2$ be chosen such that $\mathbf{r}_2(0) = (a, 0)$, with $a \neq 0$. Prove that, for some $b \neq 0$,

$$\mathbf{r}(t) = \left(\frac{1}{2}at^2 f(t), \frac{1}{24}bt^4 g(t)\right),$$

where f and g are smooth functions, with $f(0) = 1,\ g(0) = 1$, and moreover, in the case that

$$\mathbf{r}(t) = \left(\frac{1}{2}t^2 + \frac{u}{6}t^3, \frac{1}{24}t^4 + \frac{v}{120}t^5\right),$$

that $10u \neq 3v$. □

1.33 Classify the cusp at the origin of the curve

$$t \mapsto (\tfrac{1}{2}t^2, \tfrac{1}{4}t^4 + \tfrac{1}{5}at^5 + \tfrac{1}{6}bt^6 + \tfrac{1}{7}ct^7 + \tfrac{1}{8}dt^8)$$

for different values of a, b, c, d.

1.34 Let $s \mapsto \mathbf{r}(s)$ be a unit speed curve. Proposition 1.27 gives conditions for the curve $\mathbf{r}$ to be circular at s. Prove that the extra condition for $\mathbf{r}$ to have an ordinary vertex is

$$\det[\mathbf{r}_2 \quad \mathbf{r}_3] + \det[\mathbf{r}_1 \quad \mathbf{r}_4] \neq 0.$$

1.35 Let $\psi \mapsto \mathbf{r}(\psi)$ be a regular unit angular velocity curve. Proposition 1.28 gives a condition for the curve $\mathbf{r}$ to be circular at ψ. Prove that the extra condition for $\mathbf{r}$ to have an ordinary vertex is

$$\mathbf{r}_1 \cdot \mathbf{r}_3 + \mathbf{r}_2 \cdot \mathbf{r}_2 \neq 0.$$

2

Some elementary geometry

2.0 Introduction

We assume that the reader is familiar with the standard elementary concepts and theorems of linear algebra, but it would be rash for us to assume equal familiarity with the elementary concepts and theorems of projective geometry. In this chapter we look briefly at some of the things in both these areas that we shall later take for granted. In Chapter 7 we look in greater detail at quadratic and cubic algebra.

2.1 Some linear facts

The terms *vector space* and *linear space* are synonymous. A *vector space X over a field* $\mathbb{K}$ consists of a set X, with operations of *addition* $X^2 \to X$; $(a, b) \mapsto a + b$ and *scalar multiplication* $\mathbb{K} \times X \to X$; $(\lambda, a) \mapsto \lambda a$, satisfying the most obvious axioms, namely addition is an additive group structure, while for any λ, $\mu \in \mathbb{K}$ and any a, $b \in X$, $1a = a$, $\lambda(\mu a) = (\lambda\mu)a$, $(\lambda + \mu)a = \lambda a + \mu a$, and $\lambda(a + b) = \lambda a + \lambda b$, it then being a consequence that $0a = 0$, where 0 here denotes both the zero of the field $\mathbb{K}$ and the zero of the group structure, and $(-1)a = -a$. A *linear subspace* of a vector space X is a subset W of X that inherits a vector space structure by restriction from that of X. An *affine subspace* of X is a translate of a linear subspace.

Let X and Y be vector spaces over $\mathbb{K}$. Then a map $t : X \to Y$ is said to be ($\mathbb{K}$-)*linear* if it respects the linear structures on X and Y in the sense that, for any a, $b \in X$, $t(a + b) = t(a) + t(b)$, and for any $a \in X$ and any $\lambda \in \mathbb{K}$, $t(\lambda a) = \lambda t(a)$. The *image* of t, $\operatorname{im} t$, is a linear subspace of Y and the *kernel* of t, $\ker t$, the set of elements of X mapped to 0 by t, is a linear subspace of X. An *affine map* is the sum of a linear map and a constant map, the linear map being known as the *linear part* of the affine map. The levels or fibres of an affine map are affine subspaces of its domain, the parallels to the kernel of its linear part.

A *basis* for a vector space X consists of elements of X that not only span X but are also linearly independent, X being said to be *finite-dimensional* if it has a finite basis. The *Basis Theorem* states that any two finite bases for a vector space X have the same number of elements. This number is the *dimension* of the space.

Almost all vector spaces considered here will be finite-dimensional *real* vector spaces, though *complex* vector spaces will make occasional appearances. That is normally $\mathbb{K}$ will be $\mathbb{R}$ but occasionally will be $\mathbb{C}$. Examples are $\mathbb{R}^n$, which is of dimension n, and the space of homogeneous polynomials in two variables x and y of degree n, with real coefficients, together with the zero polynomial, which is of dimension $n + 1$.

A linear subspace of $\mathbb{K}^n$ of dimension k is said to have *codimension* $n - k$. An *affine* subspace of $\mathbb{K}^n$ is of *dimension* k if the linear subspace to which it is parallel has dimension k. Linear or affine subspaces of codimension 1 are commonly called *hyperplanes*.

Any linear map $t : \mathbb{K}^n \to \mathbb{K}^p$ is representable by its $p \times n$ *matrix*. The convention we adopt is the analyst's convention that the elements of $\mathbb{K}^n$ may be represented as *column* $n \times 1$ matrices on which t operates on the left. The matrix of the composite ut of two linear maps $t : \mathbb{K}^m \to \mathbb{K}^n$ and $u : \mathbb{K}^n \to \mathbb{K}^p$ is the product of the matrices of the constituent maps, this being the *raison d'être* of the matrix multiplication rule.

The *equation* of a linear map $t : X \to Y$ is the equation $t(x) = y$. The map is *surjective* if it has a solution $x \in X$ for each $y \in Y$ and is *injective* if whenever $t(x) = t(x')$ then $x = x'$. A linear map t is injective if and only if $\ker t = \{0\}$. The full solution set of the equation in the case that $y \in \operatorname{im} t$ is $x =$ particular solution + kernel. We have already had an example of this in Chapter 1, on page 19.

The dimension of the image of a linear map $t : X \to Y$ between finite-dimensional vector spaces X and Y is called the *rank* $\operatorname{rk} t$ of t. The dimension of the kernel is called the *kernel rank* $\operatorname{kr} t$, or *nullity* of t. Then $\dim X = \operatorname{rk} t + \operatorname{kr} t$. It follows at once from this that a linear map between finite-dimensional vector spaces *of the same dimension* is injective if and only if it is surjective.

A *bijective* map $t : X \to Y$ is one that is both injective and surjective, and is so if and only if it is invertible, that is if and only if there is a map $t^{-1}: Y \to X$, necessarily unique, such that both $t^{-1}t = 1_X$ and $tt^{-1} = 1_Y$. The inverse of a bijective linear map is itself linear, such a map being called a *linear isomorphism*. Any choice of basis for a finite-dimensional vector space X over $\mathbb{K}$ induces an isomorphism $\mathbb{K}^n \to X$ in which the elements of the basis for X are the images of the standard basis vectors for $\mathbb{K}^n$. Choices of basis for finite-dimensional vector spaces X and Y over $\mathbb{K}$ induce matrix representations of linear maps $t : X \to Y$.

The *determinant* $\det t$ of a linear map $t : X \to X$, X being a finite-dimensional vector space, is the determinant of any of its matrix representations, provided that the same choice of basis is made in both the source and target copies of the vector space X. The determinant determines many things. In particular it is non-zero if and only if the map t is invertible if and only if t is injective if and only if t is surjective if and only if the columns (or rows) of any matrix representation of t are linearly independent if and only if the columns (or rows) of any matrix representation of t span $\mathbb{K}^n$. Put another way, the determinant of t is zero if and only if, for some non-zero $x \in X$, $t(x) = 0$.

2.2 Some bilinear facts

Let X, Y and Z be vector spaces over a field $\mathbb{K}$. A map $\beta: X \times Y \to Z;\ (x, y) \mapsto \beta(x, y)$ is said to be *bilinear* if for each $a \in X$, $b \in Y$ the partial maps $\beta(-, b) : X \to Z;\ x \mapsto \beta(x, b)$ and $\beta(a, -) : Y \to Z;\ y \mapsto \beta(a, y)$ are linear. The standard scalar product on the real vector space $\mathbb{R}^n$, namely the map $\mathbb{R}^n \times \mathbb{R}^n \to \mathbb{R};\ (a, b) \mapsto a \cdot b$, is bilinear.

The set $L(X, Y)$ of linear maps from X to Y is generally assigned the linear structure defined, for all t, $u \in L(X, Y)$, all $x \in X$ and all $\lambda \in \mathbb{K}$, by

$$(t + u)(x) = t(x) + u(x) \text{ and } (\lambda t)(x) = \lambda t(x).$$

When $\dim X = n$ and $\dim Y = m$ then $\dim L(X, Y) = mn$. The maps

$$L(X, Y) \times L(Y, Z) \to L(X, Z);\ (t, u) \mapsto ut \text{ and}$$

$$X \times L(X, Y) \to (x, t) \mapsto t(x)$$

are bilinear maps.

A bilinear map $\beta : X \times X \to \mathbb{R}$ on a real vector space X induces a map $\gamma : X \to \mathbb{R};\ x \mapsto \gamma(x) = \beta(x, x)$ called the *quadratic form* of β, and every such quadratic form is induced by a unique symmetric bilinear map β. Clearly if $\gamma(x) = 0$ for some non-zero $x \in X$ then $\gamma(\lambda x) = 0$ for every $\lambda \in \mathbb{R}$.

Quadratic forms on $\mathbb{R}^2$ play an important part in the theory of smooth surfaces studied in the later half of this book and we have more to say about them in Chapter 7.

A quadratic form $\gamma : \mathbb{R}^n \to \mathbb{R}$ has a unique representation by a *symmetric* real $n \times n$ matrix B with $\gamma(x) = x^{\tau} Bx = x \cdot Bx$, for all $x \in \mathbb{R}^n$.

A map $t : \mathbb{R}^n \to \mathbb{R}^n$ is said to be an *orthogonal transformation* of $\mathbb{R}^n$ if it respects the standard scalar product on $\mathbb{R}^n$, that is if, for any points a and b of $\mathbb{R}^n$, $t(a) \cdot t(b) = a \cdot b$. Such a map is necessarily linear, being orthogonal if and only if the columns of its matrix form an *orthonormal basis* for $\mathbb{R}^n$; that is the scalar product of any column vector with itself is 1 and with any other column vector is zero. Equivalently a linear map $t : \mathbb{R}^n \to \mathbb{R}^n$ is an orthogonal transformation if and only if the inverse of its matrix coincides with the transpose of the matrix. From the product theorem for determinants it follows that the determinant of any orthogonal transformation is either 1 or -1.

An orthogonal transformation of $\mathbb{R}^n$ of determinant 1 is said to be a *rotation*. One that is of determinant -1 will be said to be an *antirotation*. The orthogonal transformations of $\mathbb{R}^n$ form a group known as the *orthogonal group* $O(n)$. The rotations of $\mathbb{R}^n$ form a subgroup of $O(n)$ known as the *special orthogonal group* $SO(n)$. The eigenvalues of an orthogonal matrix are possibly non-real roots of unity. An orthogonal matrix each of whose eigenvalues is equal to 1 or to -1, with at least one equal to -1, is called a *reflection*. It is a common error to confuse antirotations with reflections. They coincide only for $\mathbb{R}^2$.

Any quadratic form on $\mathbb{R}^n$ is *reducible to a sum of squares* by an orthogonal transformation of $\mathbb{R}^n$ or equivalently by a fresh choice of orthonormal basis for $\mathbb{R}^n$, the matrix of the reduced form being a diagonal matrix with entries the necessarily real eigenvalues of the matrix of the quadratic form. The number of non-zero eigenvalues, being equal to the rank of the matrix of the form, is called the *rank* of the quadratic form, while the division of the non-zero eigenvalues into positive and negative numbers is called the *signature* or *index* of the form. There is no agreement in the literature as to which function of the numbers of positive and negative eigenvalues this is taken to be!

2.3 Some projective facts

In Chapter 1 curves were presented parametrically as smooth maps $r : \mathbb{R} \rightarrowtail \mathbb{R}^2$, even though in practice it was the image of the map rather than the map itself that was of real interest. In practice it is just as likely that a curve will be presented as the zeros of a smooth map $F : \mathbb{R}^2 \rightarrowtail \mathbb{R}$, and then it is frequently the case that the map F is a non-homogeneous polynomial of some degree. As an example, the circle with centre (a, b) and radius ρ is the set of zeros of the polynomial $(x - a)^2 + (y - b)^2 - \rho^2$. The projective geometer then has two tricks to play. First he identifies $\mathbb{R}^2$ with the plane with equation $z = 1$ in $\mathbb{R}^3$ and then, when he looks across this plane, he apparently sees new points at infinity at the horizon. In fact to every point $(x, y, 1)$ of the plane $z = 1$ there is a line through the origin consisting of all the scalar multiples of $(x, y, 1)$. Lines that apparently intersect the plane at infinity are then the lines through the origin in the plane $z = 0$. The *real projective plane* $\mathbb{R}P^2$ is defined as the set of lines through the origin in $\mathbb{R}^3$, but thought of as an ordinary plane with an additional (projective) 'line at infinity', where of course a point at infinity in front of one has to be identified with a point at infinity behind one. The *n-dimensional real projective space* $\mathbb{R}P^n$ is likewise defined to be the set of lines through the origin in $\mathbb{R}^{n+1}$, thought of as an ordinary n-dimensional vector space with an additional (projective) 'hyperplane at infinity'.

Conversely, given the real projective space $\mathbb{R}P^n$, any of its projective hyperplanes may be regarded as the 'hyperplane at infinity'.

The *projective point* $[x, y, z]$ of $\mathbb{R}P^2$ is the line through the origin consisting of all scalar multiples of the vector (x, y, z). Clearly then $[\lambda x, \lambda y, \lambda z] = [x, y, z]$ for any non-zero $\lambda \in \mathbb{R}$. The numbers x, y, z are then called the *homogeneous coordinates* of the projective point.

2.4 Projective curves

A *curve* in $\mathbb{R}P^2$ is given as the set of zeros of a homogeneous polynomial of some degree. For example, instead of the circle with equation $(x - a)^2 + (y - b)^2 - \rho^2 = 0$ we have the curve defined by the equation $(x - az)^2 + (y - bz)^2 - \rho^2 z^2 = 0$.

The curve given parametrically as $\mathbf{r} : \mathbb{R} \rightarrow \mathbb{R}^2 : t \mapsto (t^2 - 1,\ t(t^2 - 1))$ in the previous chapter is also presentable as the zeros of the polynomial $f(x, y) = y^2 - x^2(1 + x)$, or projectively as the curve defined by the equation $F(x, y, z) = y^2 z - x^2(z + x) = 0$. The singularity at the origin is the only point where the partial derivatives of f, namely $\partial f / \partial x$ and $\partial f / \partial y$ are both zero, or equivalently, by Euler's theorem on homogeneous polynomials, where $\partial F / \partial x = \partial F / \partial y = \partial F / \partial z = 0$. A simple crossing point, where the tangent lines at the crossing are distinct, such as this curve has at the origin, is known as a *crunode*, or *crossing node*, of the curve. By contrast the curve defined by the equation $y^2 - x^2(1 - x) = 0$, or projectively by the equation $y^2 z - x^2(z - x) = 0$, has a singularity at the origin which is isolated from the rest of the curve, the main part admitting the parametrisation $\mathbb{R} \to \mathbb{R}^2 : t \mapsto (t^2 + 1, t(t^2 + 1))$. Such an isolated singularity is known as an *acnode*, or *spot node*, of the curve.

It is not long before one gets round to computing the number of points of intersection of a pair of polynomial curves, and here the other trick is brought to play, namely to *complexify* everything, replacing the reals $\mathbb{R}$ throughout by the complex numbers $\mathbb{C}$. It is then an old theorem of Bézout that the number of intersections properly counted of a polynomial curve of degree m in $\mathbb{C}P^2$ with one of degree m' is exactly mm', provided that the two curves do not have a common component in common. For a precise statement and proof see for example Brieskorn and Knörrer (1986). In fact we shall not need the full force of this since in certain special cases the result is easily proved. In particular this is so if one of the curves is either of degree one or of degree two.

Consider as a special case the circle with centre (a, b) and radius ρ with homogeneous polynomial equation $(x - az)^2 + (y - bz)^2 - \rho^2 z^2 = 0$. This intersects the line at infinity $z = 0$ in the same two points $[1, i, 0]$ and $[1, -i, 0]$, regardless of which circle we start from. These points are known as the *circular points at infinity*, first studied by the young French mathematician Jean-Victor Poncelet when incarcerated in a Russian prison camp in the period 1812–15.

Of course some of the mn points of intersection of curves of degrees m and n may coincide. For example for a line tangent to a polynomial curve of degree m at least two of the n points of intersection are coincident at the point of tangency.

In the case of an acnode of a curve as introduced above the tangents at the acnode are determinable as complex conjugate lines that intersect in the real point that is the acnode itself.

2.5 Spaces of polynomials

The space of polynomials in one variable x of degree at most some positive integer n and with coefficients in $\mathbb{R}$ is clearly a vector space over $\mathbb{R}$ of dimension $n + 1$. Likewise the space of real homogeneous polynomials in two variables x and y of degree n, together with the zero form, is a vector space over $\mathbb{R}$ of dimension $n + 1$.

In much of what follows non-zero homogeneous polynomials in two variables that differ by a non-zero real factor only will be regarded as equivalent. The equivalence classes of such polynomials of degree n then form a real projective space of dimension n. Accordingly much of the classification theory of such polynomials can be and will be expressed in projective terms. In Chapter 7 we study in detail the spaces of quadratic and cubic forms on $\mathbb{R}^2$.

2.6 Inversion and stereographic projection

Throughout this book there is interest in the contact of curves and surfaces with circles or spheres or with lines or planes. Accordingly there is interest in any transformation of $\mathbb{R}^n$ furnished with its standard scalar product, or of part of it, that maps hyperplanes or hyperspheres to hyperplanes or hyperspheres, and therefore planes or spheres of any dimension to planes or spheres of the same dimension.

The simplest is *inversion* in the unit (hyper)sphere, S^{n-1}, with centre the origin, the map $\mathbb{R}^n\backslash\{0\} \to \mathbb{R}^n\backslash\{0\} : \mathbf{r} \mapsto \mathbf{r}/|\mathbf{r}|^2$, where $|\mathbf{r}|^2 = \mathbf{r} \cdot \mathbf{r}$. This map, which maps the inside of the sphere to the outside and the outside to the inside, leaving the sphere itself fixed, is its own inverse.

Let $\mathbf{r}' = \mathbf{r}/|\mathbf{r}|^2$. The equation $\alpha\mathbf{r}' \cdot \mathbf{r}' + \mathbf{b} \cdot \mathbf{r}' + \gamma = 0$, with $\mathbf{b} \cdot \mathbf{b} - \alpha\gamma > 0$, is the equation of a hypersphere in $\mathbb{R}^n$ for $\alpha \neq 0$, and of a hyperplane for $\alpha = 0$. It passes through the origin if and only if $\gamma = 0$. It is the image by inversion of the hyperspace with equation $\alpha + \mathrm{b} \cdot \mathrm{r} + \gamma\mathbf{r} \cdot \mathbf{r} = 0$, a hypersphere if $\gamma \neq 0$ and a hyperplane if $\gamma = 0$, passing through the origin if and only if $\alpha = 0$.

The hypersphere with respect to which we invert need not be the unit hypersphere. Any hypersphere, with any point as centre, will do, and in each case it is clear that spheres and planes of any dimension will invert to spheres and planes of the same dimension.

Analogous properties are possessed by the *stereographic projection* of the unit hypersphere S^n in $\mathbb{R}^{n+1} = \mathbb{R}^n \times \mathbb{R}$ from the *South pole* $(0, -1)$ of S^n to the *equatorial plane* $\mathbb{R}^n \times \{0\}$, mapping any point of the sphere other than the South pole to the point of intersection of the line that passes through that point and the pole with the equatorial plane. The image by this projection of the intersection of S^n with any affine hyperplane in $\mathbb{R}^{n+1}$ is either a hypersphere in $\mathbb{R}^n$ or, if the hyperplane passes through the pole, a hyperplane in $\mathbb{R}^n$. For the details see Exercises 2.8 and 2.9.

Exercises

2.1 (The Cauchy–Schwarz *equality*) Let $\mathbf{u}$ and $\mathbf{v}$ be vectors in $\mathbb{R}^n$. Then $\mathbf{u}$ and $\mathbf{v}$ are linearly dependent if and only if

$$(\mathbf{u}\cdot\mathbf{u})(\mathbf{v}\cdot\mathbf{v}) = (\mathbf{u}\cdot\mathbf{v})^2.$$

(Consider the expression $(\lambda\mathbf{u} + \mu\mathbf{v})\cdot(\lambda\mathbf{u} + \mu\mathbf{v})$ as a nowhere negative quadratic form in λ and μ.)

This exercise has application in Chapter 14.

2.2 Two linear subspaces U and V of a finite-dimensional vector space X are said to be *transversal* in X if $X = U + V$, that is if each element x of X is equal to $u + v$ for some $u \in U$ and $v \in V$. Prove that the image of a linear injection $i : W \to X$ and the kernel of a linear surjection $p : X \to Y$ are transversal in X if and only if $pi : W \to Y$ is surjective.

2.3 The *dual* X^* of a finite-dimensional real vector space is defined to be the real vector space $L(X, \mathbb{R})$ of real linear maps from X to $\mathbb{R}$. Prove that $\dim X^* = \dim X$. Prove also that the map ev: $X \to X^{**} = L(L(X, \mathbb{R}), \mathbb{R})$, where $ev(x)\omega = \omega(x)$, for all $x \in X$, $\omega \in L(X, \mathbb{R})$, is linear and injective and therefore bijective. ('ev' stands for 'evaluation'.)

2.4 Let $t : X \to Y$ be a real linear map, X and Y being finite-dimensional real vector spaces. Let the dual map $t^* : Y^* \to X^*$ be defined by $t^*(\omega) = \omega t$, for all $\omega \in Y^*$. Prove that t^* is linear. Prove also that if t is injective then t^* is surjective while if t is surjective then t^* is injective.

2.5 Prove that if the quadratic form $\gamma : \mathbb{R}^3 \to \mathbb{R}$ of a symmetric bilinear map $\beta : \mathbb{R}^3 \times \mathbb{R}^3 \to \mathbb{R}$ is zero on a linear subspace of $\mathbb{R}^3$ of dimension 2 then, for some non-zero $u \in \mathbb{R}^3$, $\beta(u) = 0$ or, equivalently, the matrix of γ has determinant zero. Show also that the converse is not true.

(There is no loss of generality in assuming that the quadratic form is a sum of squares.)

2.6 Some feeling for the projective plane may be acquired as follows:

Imagine the coordinate plane pictured as the plane $z = 1$ with the x-axis running across the bottom of the picture from left to right and the y-axis running away to the horizon, shown as a line running across the centre of the picture. Then all the lines $x = \text{const}$, $z = 1$ will appear to meet with the y-axis at a point of the horizon. Now draw on the plane $z = 1$ the parabola with equation $y = x^2$. What does it look like?!

2.7 The parabola drawn in Exercise 2.6 is the intersection of the cone $yz = x^2$ with the plane $z = 1$. Show how by tilting this cone up or down one can produce the other *conic sections*, the ellipse and the hyperbola.

2.8 Let S^n denote the unit hypersphere, with centre the origin, in $\mathbb{R}^{n+1}$, identified with the Cartesian product $\mathbb{R}^n \times \mathbb{R}$. Prove that the map $S^n \rightarrowtail \mathbb{R}^n$, $(\mathbf{r}, \lambda) \mapsto (1 + \lambda)^{-1}\mathbf{r}$, undefined only at the 'South pole' $(0, -1)$, is bijective, the three points $(0, -1)$, $(\mathbf{r}, \lambda)$ and $(1 + \lambda)^{-1}\mathbf{r}$ being collinear.

(This is *stereographic projection from the South pole.*)

2.9 Prove that the image by stereographic projection from the South pole of the intersection of S^n with any affine hyperplane in $\mathbb{R}^{n+1}$ is either a hypersphere in $\mathbb{R}^n$ or, if the hyperplane passes through the South pole, a hyperplane in $\mathbb{R}^n$.

3
Plane kinematics

3.0 Introduction

In mechanism theory one is concerned with the motion of objects in $\mathbb{R}^2$ or $\mathbb{R}^3$, moving smoothly in such a way that the distance between any two points of a moving object remains constant in time. When this occurs each point of the moving object describes a smooth curve in the fixed space. This application motivates our concern in what follows in studying *rigid* (that is *smooth distance preserving*) motions of the plane over the plane. It is a big subject whose edges we only scratch! Nevertheless our approach to the linearity or circularity of curves at special points leads directly to several of the basic theorems of the theory. Standard references to the extensive engineering literature are Hunt (1978) and Bottema and Roth (1979).

3.1 Instantaneous rotations and translations

Consider a rigid motion of a plane over the plane. We shall normally use vectorial notations, but there are times when it is also useful to employ complex numbers, $\mathbb{R}^2$ being identified with $\mathbb{C}$ in the usual way. Then, for any $\mathbf{v} \in \mathbb{R}^2$ and any $\theta \in \mathbb{R}$, $e^{i\theta}\mathbf{v}$ will denote the vector obtained by rotating $\mathbf{v}$ through the angle θ positively, that is anticlockwise. The vector $i\mathbf{v}$ will also be denoted by $\mathbf{v}^{\perp}$, with $\mathbf{v}^{\perp\perp} = -\mathbf{v}$. Points of the moving plane will be denoted in what follows by bold capital letters and their smooth paths in time t in the fixed plane by the corresponding bold lower-case letters. In particular the origin in the moving plane is denoted by $\mathbf{O}$ and its path in the fixed plane by $\mathbf{o}$. For any point $\mathbf{R}$ we then have

$$\mathbf{r}(t) = \mathbf{o}(t) + \mathbf{R}e^{i\theta(t)}, \text{ for all } t.$$

The velocity of $\mathbf{R}$ at t is $\mathbf{r}_1(t) = \mathbf{o}_1(t) + \mathbf{R}^\perp \theta_1(t) e^{i\theta(t)}$.

There are two cases, according as $\theta_1(t) = 0$ or $\neq 0$. In the first case $\mathbf{r}_1(t) = \mathbf{o}_1(t)$ for all $\mathbf{r}$. The plane is said to be *instantaneously translating* at t, all points at that moment having the same velocity. In the second, more general case, $\mathbf{r}_1(t) = 0$ if and only if

$$\mathbf{o}_1(t) + \mathbf{R}^\perp \theta_1(t) e^{i\theta(t)} = 0,$$

that is if and only if $\mathbf{R} = (\theta_1(t))^{-1} e^{-i\theta(t)} (\mathbf{o}_1(t))^\perp$, a unique point at rest, called the *instantaneous centre of rotation or pole* of the motion at t.

3.2 The motion of a plane at $t = 0$

In studying the rigid motion of a plane at a particular moment of time it simplifies the discussion to take $t = 0$ as the moment of interest and to select the pole as the base point $\mathbf{O}$, supposing, of course, that the plane is not instantaneously translating at that time. It further simplifies the theory if we then reparametrise the motion so that the angle θ becomes the parameter. In what follows we suppose this done. We also suppose that we are in the most general case, where the pole is at an ordinary cusp on its path. That is $\mathbf{o}_1 = 0$ but $\mathbf{o}_2$ and $\mathbf{o}_3$ are linearly independent. From the equation

$$\mathbf{r}(\theta) = \mathbf{o}(\theta) + \mathbf{R} e^{i\theta},$$

differentiating the equation several times and then setting $\theta = 0$, we find that, at $\theta = 0$, not only $\mathbf{r} = \mathbf{R}$ but also

$$\begin{aligned}\mathbf{r}_1 &= \mathbf{r}^\perp,\\ \mathbf{r}_2 &= \mathbf{o}_2 - \mathbf{r},\\ \mathbf{r}_3 &= \mathbf{o}_3 - \mathbf{r}^\perp,\\ \mathbf{r}_4 &= \mathbf{o}_4 + \mathbf{r},\end{aligned}$$

and so on. The centre of curvature $\mathbf{c}$ of $\mathbf{r}$ at 0 is then determined by the equations

$$\mathbf{r}^\perp \cdot \mathbf{c} = \mathbf{r}^\perp \cdot \mathbf{r} = 0$$

and

$$\begin{aligned}(\mathbf{o}_2 - \mathbf{r}) \cdot \mathbf{c} &= (\mathbf{o}_2 - \mathbf{r}) \cdot \mathbf{r} + \mathbf{r}^\perp \cdot \mathbf{r}^\perp\\ &= \mathbf{o}_2 \cdot \mathbf{r},\end{aligned}$$

since

$$\mathbf{r}^{\perp} \cdot \mathbf{r}^{\perp} = \mathbf{r} \cdot \mathbf{r}.$$

From the first of these it follows that $\mathbf{c}$ is linearly dependent on $\mathbf{r}$. Put in another way, the points $\mathbf{r}$ and $\mathbf{c}$ are collinear with the pole $\mathbf{o}$. From the second, known as the *Euler–Savary equation*, it follows that $\mathbf{c}$ coincides with the pole, that is $\mathbf{c} = 0$, if and only if $\mathbf{r}$ lies on a particular line of the moving plane, namely the line with equation $\mathbf{o}_2 \cdot \mathbf{r} = 0$, generated by the vector $\mathbf{o}_2{}^{\perp}$. This assumes, of course, that $\mathbf{o}_2 \neq 0$. When $\mathbf{o}_2 = 0$ the origin is indeed the centre of curvature of every moving point at $\theta = 0$.

The theorems of the following sections tell us which points of the moving plane at $\theta = 0$ are at inflections or at vertices of their paths.

3.3 The inflection circle and Ball point

Theorem 3.1 Let a plane be moving rigidly over the plane in such a way that at a given moment it is not instantaneously translating, the pole being at an ordinary cusp on its path. Then the points of the moving plane that at that moment are passing through linear points of their paths lie on a circle through the pole, the inflection circle, *these points all being ordinary inflections save one which is, in general, an ordinary undulation.*

Proof We adopt the conventions discussed above. Then, for each θ,

$$\mathbf{r}(\theta) = \mathbf{o}(\theta) + \mathbf{R}e^{i\theta},$$

where $\mathbf{o}(0) = 0$, $\mathbf{o}_1(0) = 0$ and $\mathbf{o}_2(0)$ and $\mathbf{o}_3(0)$ are linearly independent. Moreover at $\theta = 0$, not only $\mathbf{r} = \mathbf{R}$ but also

$$\mathbf{r}_1 = \mathbf{R}^{\perp},$$

$$\mathbf{r}_2 = \mathbf{o}_2 - \mathbf{R},$$

$$\mathbf{r}_3 = \mathbf{o}_3 - \mathbf{R}^{\perp},$$

$$\mathbf{r}_4 = \mathbf{o}_4 - \mathbf{R},$$

and so on. In particular $\mathbf{r}_2$ is a multiple of $\mathbf{r}_1$ at $t = 0$ if and only if

$$(\mathbf{o}_2 - \mathbf{R}) \cdot \mathbf{R} = 0.$$

In more familiar notations, with $\mathbf{R} = (x, y)$ and $\mathbf{o}_2(0) = (a, b)$ this is just

$$x^2 + y^2 = ax + by,$$

the equation of a circle through the origin, with tangent line there the line with equation

$$ax + by = \mathbf{o}_2 \cdot \mathbf{R} = 0,$$

the vector $\mathbf{o}_2$ (more strictly $\mathbf{o}_2(0)$) being non-zero by hypothesis. Provided that

$$(\mathbf{o}_3 - \mathbf{R}^\perp) \cdot \mathbf{R} = \mathbf{o}_3 \cdot \mathbf{R} \neq 0$$

the point $\mathbf{R}$ of the inflection circle is at an ordinary inflection, a point of A_2 linearity.

The equation $\mathbf{o}_3 \cdot \mathbf{R} = 0$ is the equation of a line, since $\mathbf{o}_3$ is non-zero by hypothesis. This line passes through the origin and intersects the circle in one further point distinct from the pole, since $\mathbf{o}_3$ is not a multiple of $\mathbf{o}_2$. This point will in general will be an ordinary undulation, it only being a point of linearity of higher order if the circle $(\mathbf{o}_4 + \mathbf{R}) \cdot \mathbf{R} = 0$ passes through it. However, for points common to this circle and the inflection circle $(\mathbf{o}_4 + \mathbf{o}_2) \cdot \mathbf{R} = 0$, and so the exceptional case arises only when the vector $\mathbf{o}_4 + \mathbf{o}_2$ is a multiple of the vector $\mathbf{o}_3$. In general this is not so. □

The unique point on the inflection circle marking an undulation of the motion is generally denoted by U and called the *Ball point* of the motion, named after one of the founders of mechanism theory Sir Robert S. Ball (1871). (So Ball points are older than you might have thought!)

3.4 The cubic of stationary curvature

Theorem 3.2 Let a plane be moving rigidly over the plane in such a way that at a given moment it is not instantaneously translating, the pole being at an ordinary cusp on its path. Then at that moment the points of the moving plane passing through circular points of their paths lie on a cubic curve with a node at the pole, and passing through the Ball point, the tangents at the node being mutually orthogonal, one of them being the tangent to the inflection circle (Figure 3.1).

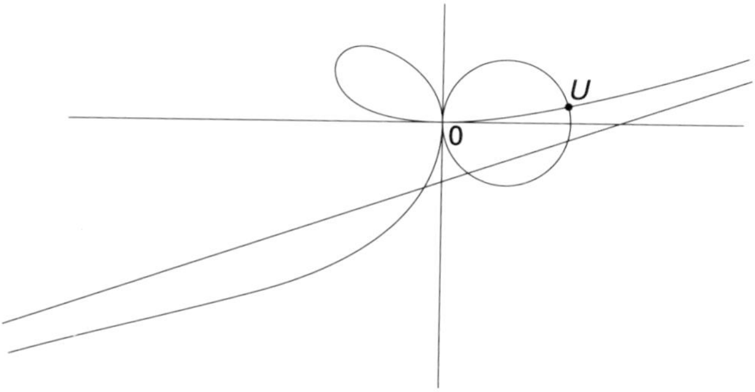

Figure 3.1

Proof We have as before, at 0, the equations

$$\mathbf{r}_1 = \mathbf{R}^\perp,$$

$$\mathbf{r}_2 = \mathbf{o}_2 - \mathbf{R},$$

$$\mathbf{r}_3 = \mathbf{o}_3 - \mathbf{R}^\perp,$$

$$\mathbf{r}_4 = \mathbf{o}_4 + \mathbf{R},$$

and so on. For the curve $\mathbf{r}$ to be circular at 0 we therefore have that the matrix

$$\begin{bmatrix} \mathbf{r}_1 \cdot & 0 \\ \mathbf{r}_2 \cdot & \mathbf{r}_1 \cdot \mathbf{r}_1 \\ \mathbf{r}_3 \cdot & 3\mathbf{r}_1 \cdot \mathbf{r}_2 \end{bmatrix} = \begin{bmatrix} \mathbf{R}^\perp \cdot & 0 \\ (\mathbf{o}_2 - \mathbf{R}) \cdot & \mathbf{R}^\perp \cdot \mathbf{R}^\perp \\ (\mathbf{o}_3 - \mathbf{R}^\perp) \cdot & 3\mathbf{R}^\perp \cdot (\mathbf{o}_2 - \mathbf{R}) \end{bmatrix}$$

has rank 2, or equivalently has a non-zero kernel vector, necessarily of the form (R, λ), where $\lambda \in \mathbb{R}$. But then we must have

$$(\mathbf{o}_2 - \mathbf{R}) \cdot \mathbf{R} + \lambda \mathbf{R}^\perp \cdot \mathbf{R}^\perp = 0,$$

and

$$\mathbf{o}_3 \cdot \mathbf{R} + 3\lambda \mathbf{R}^\perp \cdot \mathbf{o}_2 = 0,$$

or equivalently

$$(\mathbf{o}_2 - \mathbf{R}) \cdot \mathbf{R} + \lambda \mathbf{R} \cdot \mathbf{R} = 0,$$

and

$$\mathbf{o}_3 \cdot \mathbf{R} - 3\lambda {\mathbf{o}_2}^\perp \cdot \mathbf{R} = 0.$$

Eliminating λ we then have

$$\mathbf{R}\cdot\mathbf{R}\,\mathbf{o}_3\cdot\mathbf{R} + 3(\mathbf{o}_2 - \mathbf{R})\cdot\mathbf{R}\,\mathbf{o}_2{}^{\perp}\cdot\mathbf{R} = 0,$$

that is

$$\mathbf{R}\cdot\mathbf{R}\,(3\mathbf{o}_2{}^{\perp} - \mathbf{o}_3)\cdot\mathbf{R} = 3\mathbf{o}_2\cdot\mathbf{R}\,\mathbf{o}_2{}^{\perp}\cdot\mathbf{R}.$$

If we choose axes in $\mathbb{R}^2$ so that $\mathbf{o}_2 = (a, 0)$ then this may be written as

$$(x^2 + y^2)(lx + my) = 3a^2xy,$$

where $\mathbf{R} = (x, y)$ and $3\mathbf{o}_2{}^{\perp} - \mathbf{o}_3 = (l, m)$.

The equation has all the properties asserted of it. In particular it passes through the Ball point since the equation is satisfied when both $(\mathbf{o}_2 - \mathbf{R})\cdot\mathbf{R} = 0$ and $\mathbf{o}_3\cdot\mathbf{R} = 0$.

It exceptionally degenerates to its pair of tangent lines at the origin when $\mathbf{o}_3 = 3\mathbf{o}_2{}^{\perp}$. Less drastically it degenerates to a circle and diameter whenever $\mathbf{o}_3$ is a multiple of $\mathbf{o}_2{}^{\perp}$, other than $3\mathbf{o}_2{}^{\perp}$. As we shall see, this is in practice quite an important case. It likewise degenerates to a circle and diameter whenever $\mathbf{o}_3 = 3\mathbf{o}_2{}^{\perp} + \alpha\mathbf{o}_2$ for any $\alpha \in \mathbb{R}$. □

Since circular points of a curve are points of stationary curvature this cubic curve is known as the *cubic of stationary curvature*. It is also known as the *circling point curve*. One of the first to study its properties was Schönflies (1886). Our method of deriving it is close to that of Krause (1920), Bottema (1961) and Veldkamp (1964). For further historical references see Hunt (1978), pp. 143 and 148. Note that we have been able to derive the cubic without having first to obtain a messy formula for the curvature from the Euler–Savary equation and then differentiate the formula so obtained. There is a differentiation involved, of course, but awkward denominators are avoided along our route.

Before proceeding to discuss special cases of the cubic it is worthwhile to look in greater detail at cubic curves of the form

$$(lx + my)(x^2 + y^2) = xy.$$

Such a cubic is said to be *circular*, since it intersects the line at infinity in the same two imaginary points in which any circle intersects it, the *circular points at infinity*. It also has a double point at the origin, the tangents to the two branches there being the x- and y-axes. In terms of polar coordinates it takes the form

$$r = \frac{\cos\theta\sin\theta}{l\cos\theta + m\sin\theta},$$

where the right-hand side is of period 2π, but simply changes sign when θ is replaced by $\theta + \pi$. This form of the cubic shows clearly that on each line through the origin with three exceptions, there is just one point of the cubic other than the origin. The exceptions are the two tangent lines at 0, corresponding to $\cos\theta = 0$ and $\sin\theta = 0$ and the line corresponding to $l\cos\theta + m\sin\theta = 0$, parallel to the asymptote, with equation

$$(lx + my)(l^2 + m^2) + lm = 0,$$

this being the only line to have a single simple intersection with the cubic.

With a little knowledge of projective geometry we can usefully do some bookkeeping here. One of the oldest theorems of that subject, Bézout's Theorem (1779) (see for example Brieskorn and Knörrer (1986)) states that two curves of the *complex* projective plane given by the zeros of mutually prime polynomials, of degrees m and n say, have exactly mn points of intersection, provided that multiple intersections are properly counted. So the number of real intersections not at infinity is at most mn. In particular a conic and a cubic in the complex projective plane should intersect each other in six points, provided that multiple intersections are properly accounted for, and provided that the two curves do not share a common component. We have already remarked that the cubic of stationary curvature and the inflection circle pass through the circular points at infinity. This accounts for two of the six points. A third is the Ball point. The remainder are all at the origin. In fact the circle intersects one branch of the cubic orthogonally there, with tangent line $\mathbf{o}_2{}^\perp \cdot \mathbf{r} = 0$, and shares the tangent line $\mathbf{o}_2 \cdot \mathbf{r} = 0$ with the other branch.

Two circles have four-point intersection with the cubic

$$(lx + my)(l^2 + m^2) = xy$$

at the origin, namely those with equations

$$m(x^2 + y^2) = x \text{ and } l(x^2 + y^2) = y.$$

These are therefore the circles of curvature at the origin of the two branches of the cubic curve.

As already remarked, there are various special cases in which the cubic of stationary curvature takes a degenerate form. For example it reduces to a circle and diameter line if either $\mathbf{o}_2 \cdot \mathbf{o}_3 = 0$ or $\mathbf{o}_2{}^{\perp} \cdot \mathbf{o}_3 = 3\mathbf{o}_2 \cdot \mathbf{o}_2$. In particular, if $\mathbf{o}_3 = 0$ but $\mathbf{o}_2 \neq 0$ (when the pole is not at an *ordinary* cusp on its path) it reduces to the line $\mathbf{o}_2{}^{\perp} \cdot \mathbf{r} = 0$ and the circle $\mathbf{r} \cdot \mathbf{r} = \mathbf{o}_2 \cdot \mathbf{r}$, which in this case is more appropriately called the *undulation circle* of the motion.

3.5 Burmester points

The points of the cubic of stationary curvature will in general be ordinary vertices of their respective curves, that is A_3 circular points. For the circularity to be of higher order one requires that the rank of the matrix

$$\begin{bmatrix} \mathbf{r}_1 \cdot & 0 \\ \mathbf{r}_2 \cdot & \mathbf{r}_1 \cdot \mathbf{r}_1 \\ \mathbf{r}_3 \cdot & 3\mathbf{r}_1 \cdot \mathbf{r}_2 \\ \mathbf{r}_4 \cdot & 4\mathbf{r}_1 \cdot \mathbf{r}_3 + 3\mathbf{r}_2 \cdot \mathbf{r}_2 \end{bmatrix}$$

$$= \begin{bmatrix} \mathbf{R}^{\perp} \cdot & 0 \\ (\mathbf{o}_2 - \mathbf{R}) \cdot & \mathbf{R}^{\perp} \cdot \mathbf{R}^{\perp} \\ (\mathbf{o}_3 - \mathbf{R}^{\perp}) \cdot & 3\mathbf{R}^{\perp} \cdot (\mathbf{o}_2 - \mathbf{R}) \\ (\mathbf{o}_4 + \mathbf{R}) \cdot & 4\mathbf{R}^{\perp} \cdot (\mathbf{o}_3 - \mathbf{R}^{\perp}) + 3(\mathbf{o}_2 - \mathbf{R}) \cdot (\mathbf{o}_2 - \mathbf{R}) \end{bmatrix}$$

is two, leading to the supplementary equation

$$(\mathbf{o}_2 + \mathbf{o}_4) \cdot \mathbf{R} - \lambda(4\mathbf{R} \cdot \mathbf{o}_3{}^{\perp} + 6\mathbf{R} \cdot \mathbf{o}_2 - 3\mathbf{o}_2 \cdot \mathbf{o}_2) = 0.$$

Eliminating λ between this and $\mathbf{o}_3 \cdot \mathbf{R} + 3\lambda \mathbf{o}_2{}^{\perp} \cdot \mathbf{R} = 0$ we find that

$$3\mathbf{o}_2{}^{\perp} \cdot \mathbf{R}(\mathbf{o}_2 + \mathbf{o}_4) \cdot \mathbf{R} + \mathbf{o}_3 \cdot \mathbf{R}(4\mathbf{R} \cdot \mathbf{o}_3{}^{\perp} + 6\mathbf{R} \cdot \mathbf{o}_2 - 3\mathbf{o}_2 \cdot \mathbf{o}_2) = 0,$$

a conic through the origin with tangent there $\mathbf{o}_3 \cdot \mathbf{R} = 0$. Now, by Bézout's Theorem, stating that the number of intersections of curves of degrees m and n without common components is at most mn, it follows that this conic intersects the cubic of stationary curvature in at most six points, two of which occur at the pole of the motion, the node of the cubic curve. The remaining four points, two or all four of which may be non-real, are known as the *Burmester points* of the motion. At a Burmester point the point of the moving plane is at least A_4-circular. For references see Hunt (1978) pp. 156–8.

3.6 Rolling wheels

The inflection circle and the cubic of stationary curvature can be written down directly once we know the first few derivatives of the reference point **o** with respect to the turning angle θ at $\theta = 0$.

Consider a circle of radius 1 rolling without slipping on a fixed circle of radius 1, the rolling circle carrying the rest of the plane with it. Then, as should be clear from Figure 3.2,

$$\mathbf{o}(\theta) = -1 + 2e^{\frac{1}{2}i\theta} - e^{i\theta}, = 0 \text{ at } \theta = 0,$$

$$\mathbf{o}_1(\theta) = ie^{\frac{1}{2}i\theta} - ie^{i\theta}, = 0 \text{ at } \theta = 0,$$

$$\mathbf{o}_2(\theta) = -\tfrac{1}{2}e^{\frac{1}{2}i\theta} + e^{i\theta}, = \tfrac{1}{2} \text{ at } \theta = 0.$$

So at $\theta = 0$ the circle of inflection has equation $x^2 + y^2 = \frac{1}{2}x$, centre at $(\frac{1}{4}, \theta)$ and radius $\frac{1}{4}$, while, since $3\mathbf{o}_2{}^{\perp} - \mathbf{o}_3 = \frac{3}{2}i - \frac{3}{4}i = \frac{3}{4}i$, the equation of the cubic of stationary curvature there is $(x^2 + y^2 - x)y = 0$, consisting of the circle with equation $x^2 + y^2 - x = 0$, centre $(\frac{1}{2}, 0)$ and radius $\frac{1}{2}$, together with the x-axis. The Ball point is at $(\frac{1}{2}, 0)$ (Figure 3.3).

From these results one can give very good descriptions of the various curves described by different points of the rolling wheel or attached to it. Clearly the centre of the wheel simply describes a circle of radius 2. If one moves a short distance towards the rim the circle deforms to an oval (no points of inflection) with four vertices. This flattens until at a distance $\frac{1}{2}$ from the centre it acquires a point of undulation, the moving point coinciding with the Ball point of the motion once on each revolution. For points at a distance from the centre between $\frac{1}{2}$ and 1 the curve has, in addition to the four vertices, two inflections. As one nears the rim these inflections merge with three of the vertices to form in the limit an ordinary cusp. The path of the point of the rim of the

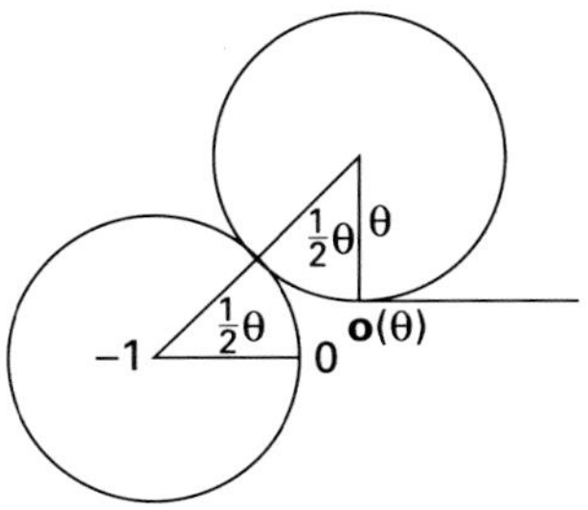

Figure 3.2

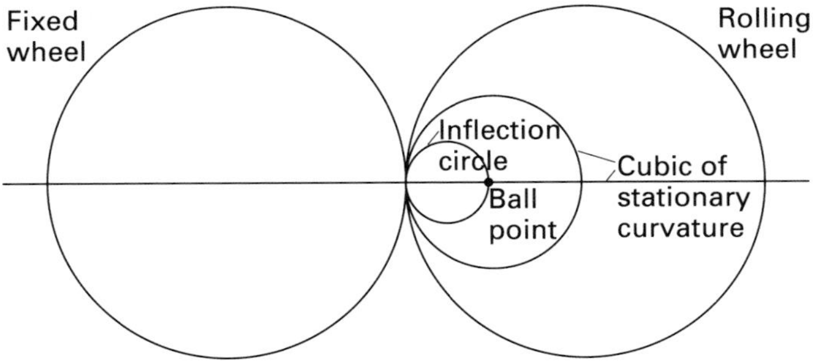

Figure 3.3

rolling wheel is the *cardioid* – a curve that each year at the University of Liverpool we contrive to get our students to trace on polar graph paper on the 14th of February. It has one ordinary cusp and one ordinary vertex. Proceeding further beyond the rim a point attached to the rolling wheel describes a curve that loops round twice with only two vertices – this is not contrary to the four-vertex theorem as it is not a *simple* closed curve. The entire sequence of curves thus described are known as the *limaçons of Pascal*, *'limaçon'* being a French word for a (non-edible) type of snail.

By similar methods to those employed in the previous example it can be shown that if a wheel, of unit radius, rolls to the left along a horizontal rail, namely the x-axis, then

$$\mathbf{o}(\theta) = -\theta + \mathrm{i} - \mathrm{i}e^{\mathrm{i}\theta}.$$

So at $\theta = 0$ one has $\mathbf{o}_1 = 0$, $\mathbf{o}_2 = \mathrm{i}$, $\mathbf{o}_3 = -1$ and $\mathbf{o}_4 = -\mathrm{i}$.

The inflection circle has equation $x^2 + y^2 = y$.

The cubic of stationary curvature is the circle $x^2 + y^2 = \frac{3}{2}y$, together with the line $x = 0$.

The Ball point, naturally enough, is at i, the centre of the wheel at $\theta = 0$, and this clearly is a Burmester point. The only other Burmester point is at $\frac{3}{2}\mathrm{i}$. It is in fact an A_5 point, a local maximum of curvature.

There are traditional names for the various curves traced by the points of a plane carried by a rolling wheel. Points inside the wheel describe *curtate cycloids* (or *shortened* cycloids). These have two inflections per revolution, but no crossing points. When the moving point is half-way from the rim to the centre there are two vertices per revolution, the higher one being an A_5-vertex, as we have seen. When the moving point is closer to the centre than this there are two ordinary vertices per revolution while when it is farther away there are four, the highest one being a local minimum of curvature. Points outside the wheel describe *prolate cycloids* (or *extended* cycloids). These have crossing points, but no inflections. There are two vertices per revolution, the higher one being a minimum of curvature and the lower one, below the rail, a maximum of curvature. When the point is on the rim of the wheel the curve described is called, simply, a cycloid. It has one ordinary cusp per revolution and one ordinary vertex per revolution, the latter being a point of minimum curvature. Curtate and prolate cycloids are also known as *trochoids*.

Cycloids have an extensive literature, going right back, as we remarked in the Introduction, to the work of Huygens (and Christopher Wren also). Consult, for example, the bibliography in Lawrence (1972). A quaint reference is Procter (1878).

3.7 Polodes

Chasles (1830) remarked that any smooth motion of the plane on the plane, free from instantaneous translations, may be induced by the rolling of a curve of the moving plane along a curve in the fixed plane, the point of contact of the two curves at any time t being the pole of the motion at t. The two curves, being the loci of the pole of the motion in the moving and the fixed plane, are known as the *moving* and *fixed polodes* or *polhodes* (‘*odos*’ or ‘*hodos*’ being the Greek word for ‘path’) of the motion.

The equations of these polodes are readily found from the equation

$$\mathbf{r}(\theta) = \mathbf{o}(\theta) + \mathbf{R}e^{i\theta}.$$

Indeed at θ the pole satisfies the equation

$$0 = \mathbf{o}_1(\theta) + e^{i\theta}\mathbf{R}^{\perp},$$

θ being taken as the parameter, as before. So

$$\mathbf{R} = e^{-i\theta}(\mathbf{o}_1(\theta))^{\perp}$$

is the equation of the moving polode, while

$$\mathbf{r} = \mathbf{o}(\theta) + (\mathbf{o}_1(\theta))^{\perp}$$

is the equation of the fixed polode.

Note that the derivative of $\mathbf{R}$ at θ is $e^{-i\theta}(\mathbf{o}_1(\theta) + (\mathbf{o}_2(\theta))^{\perp})$, equal to

$\mathbf{o}_2{}^{\perp}$ at 0, while the derivative of $\mathbf{r}$ at θ is $\mathbf{o}_1(\theta) + (\mathbf{o}_2(\theta))^{\perp}$, also equal to $\mathbf{o}_2{}^{\perp}$ at 0.

So the polodes are singular at 0 if and only if $\mathbf{o}_2 = 0$ at 0.

3.8 Caustics

Suppose light from a point source in space is reflected from a smooth mirror. For simplicity we confine our attention to what happens in a plane containing the light source and intersecting the mirror in a regular smooth curve. At a point of the mirror the ray of light impinging there reflects as if from the tangent line to the mirror at that point. The reflected ray is then easily seen (Exercise 3.9) to be normal to the curve consisting of the reflections of the light source in successive tangent lines to the mirror – a curve known as the *orthotomic* of the mirror with respect to the chosen light source, since it *cuts* each reflected ray at *right angles*. In the case that the mirror is a circle the orthotomic turns out to be a limaçon, being a cardioid in the particular case that the light source is on the mirror itself. The theory of such orthotomics and their evolutes, the *caustics* of the reflected light, has long been studied, most recently in a series of papers by Bruce, Giblin and Gibson, for example (1981). The link with kinematics resides in the remark that as one looks from the light source at each successive tangent line to the mirror so the reflected 'world' appears to roll, the reflected image of the mirror rolling on the mirror itself. The inflection circle and cubic of stationary curvature of this rolling motion then play a part in the study of the caustics of the reflected light.

The identification of evolutes with caustics may be made the basis of a whole fresh approach to the differential geometry of curves on the plane in which a curve and its parallels are thought of as the *fronts* of a *wave* of light travelling at a constant speed along *rays* that are the normals to the curves and which focus on their evolute, the *caustic* of the wave. This intuition, whose historical origins go back to the earliest work on differential geometry, lies at the basis of much contemporary work on singularity theory, especially that of the Russian school under the direction of V.I. Arnol'd, but is one that we do not propose to employ except incidentally. For a recent account of Russian work during the last twenty years see Arnol'd (1990b).

In what we have just been saying we have taken the simplistic view of geometric optics that light travels along rays. For a study of actual light caustics see Berry and Upstill (1980). Berry's remarkable experiments at Bristol have been greatly influenced by the work of René Thom (1975), the original inspiration of the present book.

Exercises

3.1 Suppose that a plane lamina moving over $\mathbb{R}^2$ is instantaneously translating. Prove that the inflections of the motion lie along a line, all in general ordinary except one, the Ball point of the motion, where the path of the point is linear to a higher order.

3.2 A wheel of unit radius rolls from right to left along the x-axis with parameter θ the angle through which the circle has rotated. The point of the wheel **O** in contact with the x-axis at the origin at $\theta = 0$ has path $\mathbf{o}$. Verify that, for all θ, $\mathbf{o}(\theta) = -\theta + \mathrm{i} - \mathrm{i}e^{\mathrm{i}\theta}$, so that at $\theta = 0$ one has $\mathbf{o}_1 = 0$, $\mathbf{o}_2 = \mathrm{i}$, $\mathbf{o}_3 = -1$ and $\mathbf{o}_4 = -\mathrm{i}$.

Find the equation of the inflection circle and the position of the Ball point at $\theta = 0$ and verify that the cubic of stationary curvature consists of the circle $x^2 + y^2 = \frac{3}{2}y$, together with the line $x = 0$. Verify also that $\frac{3}{2}\mathrm{i}$ is the only Burmester point (apart from the Ball point) and that it is in fact an A_5 point, a local maximum of curvature.

Draw the inflection circle, the cubic of stationary curvature and the Ball point, as well as the wheel, all on the same diagram. What can you deduce about the occurrence of inflections or of vertices on the paths described by different points attached to the wheel?

Sketch the paths of the points which when $\theta = 0$ lie at the positions i, $\frac{5}{4}\mathrm{i}$, $\frac{3}{2}\mathrm{i}$, $\frac{7}{4}\mathrm{i}$, $2\mathrm{i}$ and also $\frac{5}{2}\mathrm{i}$, beyond the wheel, but fixed to it.

3.3 Let the plane move rigidly over the plane in such a way that the origin **O** describes the cardioid with polar equation $r = 1 - \cos\theta$, where θ is the angle through which the plane has turned. Determine the inflection circle, the Ball point and the locus of points of stationary curvature of the motion at $\theta = 0$.

(The '3-jet' at 0 (see Theorem 4.31 for the definition) of the parametric equation of the cardioid is $(\frac{1}{2}\theta^2, \frac{1}{2}\theta^3)$, from which it follows that $\mathbf{o}_2(0) = (1, 0)$, $\mathbf{o}_2(0)^\perp = (0, 1)$ and $\mathbf{o}_3(0) = (0, 3)$. So the inflection circle has equation $x^2 + y^2 = x$, the Ball point is $(0, 1)$ and the locus of points of stationary curvature reduces to the coordinate axes.)

3.4 Let a plane move rigidly over the plane in such a way that at a given moment it is not instantaneously translating, the pole being at an ordinary rhamphoid cusp on its path. Show that the Ball point coincides with the pole, that is there is no point passing through an undulation at that moment. (Such occurrences occur generically from time to time in a generic isometric motion of the plane over the plane.)

3.5 A plane is moving rigidly over a fixed plane with one of its points **A** in circular motion about a fixed point **P** and another point **B** in circular motion about a fixed point **Q**. Show how at almost every moment the instantaneous centre or pole of the motion can be determined by a very simple geometrical construction. Discuss in detail any exceptional cases.

3.6 Suppose we have three movable bars in the plane AB, BC and CD, with A and D fixed points and with hinges at B and C (such an arrangement is called a *four*-bar mechanism!), and suppose that the bars are in motion in plane. At any given moment of time where will the instantaneous centre of rotation be for the bar BC?

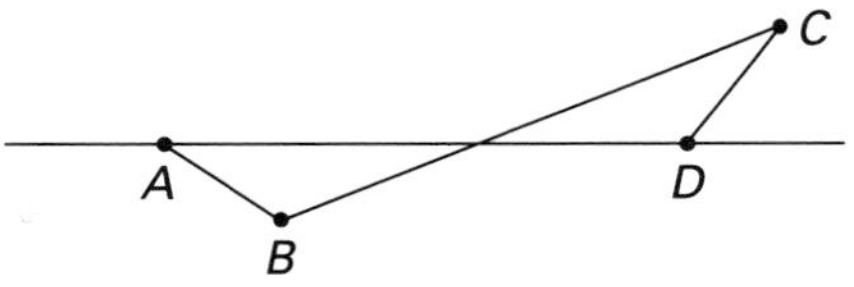

Figure 3.4

3.7 Several years ago 'le petit train Poulhain' was used by the go-ahead French cocoa firm Poulhain to market its product! Study the motion of the coupler rods, assuming that the length of each is equal to the distance between the centres of the wheels.

Figure 3.5

At significant stages of the motion of either rod determine as many as possible of the following: the instantaneous centre, the inflection circle, the Ball point, the cubic of stationary curvature and any Burmester points.

What sort of track performance would you expect from this engine?

The easiest position to analyse is the one in which a coupling rod and the two cranks are all horizontal:

Figure 3.6

where A and D are the centres, at $(-1, 0)$ and $(1, 0)$ say, BC is the coupler, of length 2, and the cranks each have length ρ. This position slightly disturbed is

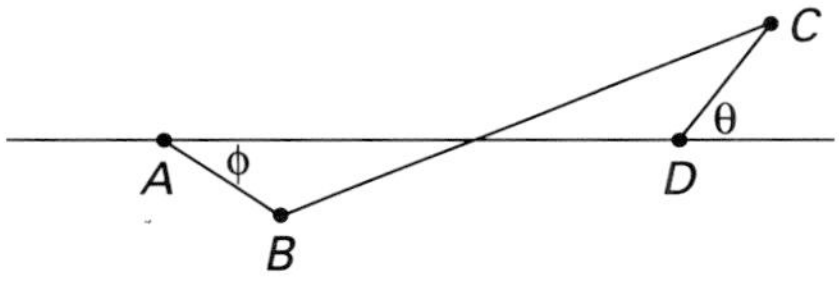

Figure 3.7

Using complex numbers one proves that, for any point $\mathbf{r}$ of the coupler initially at $\mathbf{r}_0$,

$$\mathbf{r} = \mathbf{r}_0 + \tfrac{1}{2}\rho\{(1 + \mathbf{r}_0)e^{i\theta} + (1 - \mathbf{r}_0)e^{-i\phi}\},$$

where $\sin\tfrac{1}{2}(\theta + \phi)\{2\sin\tfrac{1}{2}(\theta - \phi) - \rho\sin\tfrac{1}{2}(\theta + \phi)\} = 0$. The first factor put equal to zero gives the 'natural' motion and the second the 'Poulhain' motion of the coupler (the one illustrated).

One may then verify that for $\theta = \phi = 0$ the instantaneous centre of the 'Poulhain' motion is $(-\tfrac{1}{2}\rho, 0)$, the inflection circle has its centre at $(-(4 + 3\rho^2)/8\rho, 0)$, with radius $(4 - \rho^2)/8\rho$, the Ball point is $(-(4 + \rho^2)/4\rho, 0)$, the cubic of stationary curvature decomposes into the x-axis and a circle with centre on that axis, whose radius is one and a half times the radius of the inflection circle, while the Burmester points turn out to be three in number, namely $A = (-1, 0)$ and $B = (1, 0)$ and the extra node on the cubic of stationary curvature.

Another case that one can discuss in detail is that in which $\theta + \phi = \pi$. In this case the motion is instantaneously a translation and the locus of inflections is a line, which you should determine.)

3.8 A particularly non-generic motion is that of a wheel rolling round the inside of a circle of double the radius of the wheel. Verify that in this case each point of the rim of the rolling wheel moves to and fro along a straight line segment, all other points describing ellipses, the inflection circle at any moment coinciding with the rim of the rolling wheel and the cubic of stationary curvature consisting of this same circle and the line through the centres of the rolling and fixed circles.

(With respect to appropriate axes, $\mathbf{o}(\theta) = (2(\cos\theta - 1), 0)$.)

3.9 The *orthotomic* of a parametrised smooth mirror $\mathbf{m}$ in the plane with respect to a *pole* $\mathbf{p}$ is the curve $\mathbf{r}$ determined by the equations

$$(\mathbf{r} - \mathbf{m}) \cdot (\mathbf{r} - \mathbf{m}) = (\mathbf{m} - \mathbf{p}) \cdot (\mathbf{m} - \mathbf{p})$$

and

$$(\mathbf{r} - \mathbf{p}) \cdot \mathbf{m}_1 = 0.$$

Deduce from these the further equations

$$(\mathbf{r} - \mathbf{m}) \cdot \mathbf{r}_1 = 0,$$

$$(\tfrac{1}{2}(\mathbf{r} + \mathbf{p}) - \mathbf{m}) \cdot (\mathbf{r} - \mathbf{p}) = 0$$

and

$$(\mathbf{r} + \mathbf{p} - 2\mathbf{m}) \cdot \mathbf{m}_1{}^{\perp} = 0,$$

(the first of these justifying the term ‘orthotomic’).

(Light emitted from the pole and travelling in straight lines reflects off the mirror along normals to the orthotomic. The evolute of the orthotomic is then the *caustic* of the reflected light.)

3.10 Derive parametric equations for the orthotomic of the parabolic mirror $\mathbf{m}(t) = (t^2, 2t)$ with respect to each of the points $(0, 0)$, $(1, 0)$, $(-1, 0)$.

3.11 Derive parametric equations for the orthotomic of the unit circle with centre the origin, with respect to the pole $(a, 0)$, where $a = 0, \frac{1}{4}, \frac{1}{2}, \frac{3}{4}, \frac{7}{8}$, 1 and 2. Sketch the caustics in each case.

4
The derivatives of a map

4.0 Introduction

In Chapter 1 we took for granted knowledge by the reader of the calculus, but to continue to do so is unwise as we proceed to the detailed study of curves and surfaces in $\mathbb{R}^3$. The matter will become acute in the study of surfaces, for in the parametric presentation $\mathbf{s} : \mathbb{R}^2 \rightarrowtail \mathbb{R}^3$ of a surface we shall be involved essentially with functions of more than one variable. While the first derivative of such a map is simple enough, being represented by the Jacobian matrix of the partial derivatives of the several components of the map, there are subtleties involving the higher derivatives which are less familiar but will nevertheless play an essential part in our story. The formal definition of the higher derivatives leads directly to the concept of a *k-times linear map* from one vector space to another. Among the simplest examples are twice linear and thrice linear maps from $\mathbb{R}^2$ to $\mathbb{R}$, intimately related to quadratic and cubic forms on $\mathbb{R}^2$.

There is no need to master all of this material at a first reading. Indeed the impatient reader may well wish to proceed directly to Chapters 5 and 6, returning to the relevant parts of this chapter as the need arises. Most proofs are omitted, being readily available in standard texts. All vector spaces are finite-dimensional real vector spaces.

4.1 The first derivative and C^1 submanifolds

The *first derivative* of a map $f : \mathbb{R}^n \rightarrowtail \mathbb{R}^p$ *at* a point a where $f(a)$ is defined is the linear map $f_1(a)$(or dfa) : $\mathbb{R}^n \to \mathbb{R}^p$ that best approximates f near a in the sense that

$$\lim_{x \to a} \frac{|f(x) - f(a) - f_1(a)(x - a)|}{|x - a|} = 0,$$

or equivalently, in the case that $n = p = 1$, is the number

$$f_1(a) = \lim_{x \to a} \frac{f(x) - f(a)}{x - a}.$$

Such a map $f_1(a)$, if it exists, is unique, its matrix, the *Jacobian matrix* of f, being the matrix of partial derivatives at a of the p components of f, each of the latter being a real-valued function of n real variables. The *first derivative* or *differential* of f, $f_1 : \mathbb{R}^n \rightarrowtail L(\mathbb{R}^n, \mathbb{R}^p)$, more commonly denoted by $\mathrm{d}f$, is then the map that assigns to each a the linear map $f_1(a)$, continuous if each of its pn components is continuous as a function of n real variables.

There is one linear map $f_1(a)$ for each $a \in \operatorname{dom} f$. It is important not to confuse the linear maps $f_1(a)$, each a linear map $\mathbb{R}^n \to \mathbb{R}^p$, with the map $f_1 : \mathbb{R}^n \rightarrowtail L(\mathbb{R}^n, \mathbb{R}^p)$, the latter being a map whose values are themselves maps.

In the definition $\mathbb{R}^n$ and $\mathbb{R}^p$ may be replaced everywhere by any finite-dimensional real vector spaces X and Y. When $X = \mathbb{R}^n$ and $Y = \mathbb{R}^p$ the modulus signs may be taken to denote Euclidean (or Pythagorean) length. In the general case they could be any norms, there being a theorem that any two norms on a finite-dimensional vector space are equivalent in defining the continuity or differentiability of maps from or to the space.

Our first proposition is obvious from the definition of the derivative.

Proposition 4.1 Let X and Y be finite-dimensional vector spaces and $t : X \to Y$ a linear map. Then, for all $a \in X$, $t_1(a) = t$. □

The next is essentially Leibniz's rule.

Proposition 4.2 Let X, Y and Z be finite-dimensional vector spaces and $\beta : X \times Y \to Z$ a bilinear map. Then for all (a, b), $(x, y) \in X \times Y$,

$$\beta_1(a, b)(x, y) = \beta(x, b) + \beta(a, y).$$ □

The existence of the various partial derivatives of a map at a point is not enough to establish its differentiability there, as examples show. For our present purpose it is enough to remark that if the partial derivatives, each regarded as a map $\mathbb{R}^n \rightarrowtail \mathbb{R}$, are all continuous at a, then f is differentiable at a, and, moreover, f_1 is then continuous at a. A map $f : X \rightarrowtail Y$ differentiable in a neighbourhood of a with f_1 continuous at a is said to be *continuously differentiable*, or C^1, *at* the point a, and to be C^1 if it is C^1 at every point of its domain.

Theorem 4.3 (The Chain Rule) Let X, Y and Z be finite-dimensional vector spaces and let $f : X \rightarrowtail Y$ and $g : Y \rightarrowtail Z$ be differentiable maps. Then $gf : X \rightarrowtail Z$ is differentiable, with $(gf)_1 = (g_1 f) \cdot f_1$, where the $\cdot$ indicates composition of values. Explicitly, at each point $a \in$ dom gf, *$(gf)_1(a) = g_1(b) f_1(a)$, where $b = f(a)$. Moreover, if f and g are C^1 then so also is gf.* □

Special case 4.4 Let X, Y and Z be finite-dimensional vector spaces and $f : X \rightarrowtail Y$ be a differentiable map and $g : Y \to Z$ a *linear* map. Then $gf : X \rightarrowtail Z$ is differentiable, with $(gf)_1 = g f_1$. □

The familiar formula of elementary calculus $(\mathrm{d}/\mathrm{d}x)(1/x) = -1/x^2$ generalises as follows:

Proposition 4.5 Let X be a finite-dimensional vector space. Then the derivative at an invertible linear map $u \in L(X, X)$ of the map $L(X, X) \rightarrowtail L(X, X)$; $u \rightarrowtail u^{-1}$ is the linear map $L(X, X) \to$ $L(X, X)$: $t \mapsto -u^{-1} t u^{-1}$.

Indication of proof For t with $|t|$ small

$$u + t \mapsto (u + t)^{-1} = [u(1_X + u^{-1}t)]^{-1} = [1_X - u^{-1}t + O(|t|^2)]u^{-1}.$$

□

Strictly speaking a map $h : A \to B$ between two sets A and B is invertible if and only if it is bijective, that is if and only if it is not only injective but also surjective. In the calculus it is convenient to extend the concept of invertibility somewhat. Let $h : X \rightarrowtail Y$ be a map between finite-dimensional vector spaces X and Y, the domain of h being an open subset of the source vector space. We say that such a map is *locally surjective* if its image is an open subset of the vector space Y. What this means is that every $y \in Y$ sufficiently close to an image point (or value) of h also is an image point (or value) of h. The map $Y \rightarrowtail X$; $h(x) \mapsto x$, with domain the image of h, that associates to any $y \in \operatorname{im} h$ the unique solution $x \in \operatorname{dom} h$ of the equation

$h(x) = y$, will then be denoted by h^{-1} and regarded as the *inverse* of the map h.

Proposition 4.6 Let $h : X \rightarrowtail Y$ *be an injective, locally surjective, differentiable map between finite-dimensional vector spaces* X *and* Y, *with the inverse map* h^{-1} *also differentiable. Then* $\dim X = \dim Y$, *and for any* $a \in \operatorname{dom} f$, $(h^{-1})_1(b) = (h_1(a))^{-1}$, *where* $b = f(a)$. □

The differentiability of an injective map h does not guarantee the differentiability of its inverse, as the example of the map $\mathbb{R} \to \mathbb{R}$; $x \mapsto x^3$ shows. The inverse is continuous at the origin but is not differentiable there. Nor is it enough at a point a for the derivative $h_1(a)$ to be invertible at $b = f(a)$; one must also have h^{-1} continuous at a.

Theorem 4.7 Let X *and* Y *be finite-dimensional vector spaces and* $h : X \rightarrowtail Y$ *an injective, locally surjective, map that is differentiable at a point* a *of* $\operatorname{dom} h$, *the linear map* $f_1(a)$ *being invertible and the inverse map* $h^{-1} : Y \rightarrowtail X$ *being defined on a neighbourhood of* $b = h(a)$ *and continuous at* b. *Then* h^{-1} *is differentiable at* b *and* $(h^{-1})_1(b) = (h_1(a))^{-1}$. □

This is a detail in the proof of the next theorem which is much more substantial and requires that the map h involved is not just differentiable but is *continuously* differentiable at the point of interest.

Theorem 4.8 (The Inverse Function Theorem) Let X *and* Y *be finite-dimensional vector spaces, let* $h : X \rightarrowtail Y$ *be a continuously differentiable map, and let* $a \in \operatorname{dom} h$ *be such that the linear map* $h_1(a) : X \to Y$ *is invertible, a necessary condition for this being that* $\dim Y = \dim X$. *Then there exists an open neighbourhood* A *of* a *such that* $B = h(A)$ *is an open neighbourhood of* $b = h(a)$ *and* $(h|A)^{-1}$ *with domain* B *is continuously differentiable. Moreover if* C *is any connected subset of* B *containing* b *then there is a unique continuous map* $g : C \to X$ *with* $g(b) = a$ *such that the map* $C \to C$; $y \mapsto hg(y)$ *is the identity map on* C, *namely* $g = (h|C)^{-1}$.

The proof depends on two lemmas, the *Contraction Lemma* and a generalisation of the familiar mean-value theorem of one variable calculus known as the *Increment Inequality*. See Porteous (1981) for details. □

In traditional notations the theorem asserts that the equation $h(x) = y$ is locally continuously differentiably solvable for x in terms of y near $x = a$ provided that the approximating linear equation $(\mathrm{d}y/\mathrm{d}x)_{x=a}\,\mathrm{d}x = \mathrm{d}y$ is solvable for $\mathrm{d}x$ in terms of $\mathrm{d}y$.

An invertible continuous map whose inverse is continuous is said to be a *homeomorphism*, the prefix '*homeo*' being derived from a Greek word meaning 'similar' and the root '*morphé*' being a Greek word meaning 'form'. An invertible differentiable map whose inverse is differentiable is (rather horribly from the etymological point of view!) called a *diffeomorphism*. In particular an invertible C^1 map whose inverse also is C^1 is called a C^1 *diffeomorphism*. The inverse function theorem asserts that the map f of the theorem is 'locally a diffeomorphism' at the point a.

The behaviour of a map $f : X \rightarrowtail Y$ near some special point a, say, of its domain is the same as the behaviour of any other map $f' : X \rightarrowtail Y$ that coincides with f on some neighbourhood of a. The maps f and f' are said to define the same *map-germ* $X,a \rightarrowtail Y,b$, where $b = f(a)$. A map-germ is therefore an equivalence class of maps. It is usual to denote a map-germ in a somewhat lax manner by one of its representatives, together with specification of the point at which the germ is defined, as we have just done. The terminology extends in obvious ways. For example, a C^1 map-germ is an equivalence class of C^1 maps, a *curve-germ* in $\mathbb{R}^2$ is a map-germ $\mathbf{r} : \mathbb{R}, a \rightarrowtail \mathbb{R}^2, f(a)$, and so on.

The next proposition provides a canonical form for an ordinary cusp.

Proposition 4.9 Let $\mathbf{r} : \mathbb{R},0 \rightarrowtail \mathbb{R}^2,0$ *be a smooth curve-germ in* $\mathbb{R}^2$ *with an ordinary cusp at the origin. Then there are diffeomorphism-germs* $h : \mathbb{R},0 \rightarrowtail \mathbb{R},0$ *and* $k : \mathbb{R}^2,0 \rightarrowtail \mathbb{R}^2,0$ *such that* $k^{-1}\mathbf{r}h(u) = (u^2, u^3)$.

Proof We may by a linear transformation of $\mathbb{R}^2$ arrange to start with that $\mathbf{r}_2(0) = (1, 0)$ and $\mathbf{r}_3(0) = (0, 1)$, and there is then a diffeomorphism-germ $h:\mathbb{R},0 \rightarrowtail \mathbb{R},0$ such that $\mathbf{r}h(u) = (u^2, u^3\varphi(u))$, where $\varphi(0) \neq 0$. In fact we can write $u^3\varphi(u)$ as $\psi(u^2) + u^3\chi(u^2)$ where $\chi(0) \neq 0$. Then define $k : \mathbb{R}^2,0 \rightarrowtail \mathbb{R}^2,0$ by $k(x, y) = (x, y\chi(x) + \psi(x))$. This has invertible differential at the origin, and so by the inverse function theorem is a diffeomorphism-germ. The diffeomorphism-germs h and k do the trick. □

A similar trick leads to the characterisation of ordinary rhamphoid cusps referred to following Theorem 1.23.

Proposition 4.10 *Let* $\mathbf{r} : \mathbb{R},0 \rightarrowtail \mathbb{R}^2,0$ *be a smooth curve in* $\mathbb{R}^2$ *with a rhamphoid cusp at the origin. Then the cusp is an ordinary rhamphoid cusp if and only if there are diffeomorphisms* $h : \mathbb{R},0 \rightarrowtail \mathbb{R},0$ *and* $k : \mathbb{R}^2,0 \rightarrowtail \mathbb{R}^2,0$ *such that* $k^{-1}\mathbf{r}h(u) = (u^2, u^5)$.

Proof Exercise 4.12. □

Theorem 4.11 (The Implicit Function Theorem) Let X, Y *and* Z *be finite-dimensional vector spaces and* $F : X \times Y \rightarrowtail Z$ *a continuously differentiable map. Moreover let* $(a, b) \in \operatorname{dom} F$ *be such that the second partial of the linear map*

$$F_1(a, b) : X \times Y \to Z; (x, y) \mapsto F_{1_{(1)}}(a, b)x + F_{1_{(2)}}(a, b)y,$$

namely $F_{1_{(2)}}(a, b) : Y \to Z$, *is invertible, so that necessarily* $\dim Z = \dim Y$. *Then there exists an open neighbourhood* A *of* (a, b) *in* $X \times Y$ *and a continuously differentiable map* $f : X \rightarrowtail Y$ *such that*

$$\operatorname{graph} f = A \cap \{(x, y) \in X \times Y;\ F(x, y) = F(a, b)\}.$$

Moreover, if C *is any connected subset of* $\operatorname{dom} f$ *containing* a, *there is a unique continuous map* $g : C \mapsto Y$, *namely* $g = f|C$, *such that, for all* $x \in C$, $F(x, g(x)) = F(a, b)$, with $g(a) = b$.

Sketch of proof Apply the Inverse Function Theorem to the map

$$X \times Y \rightarrowtail X \times Z; (x, y) \mapsto (x, F(x, y)),$$

at the point (a, b). Its inverse locally is then of the form

$$X \times Z \rightarrowtail X \times Y; (x, z) \mapsto (x, G(x, z)),$$

and the map $x \mapsto g(x) = G(x, F(a, b))$ has the required properties. □

In traditional notations the theorem asserts that the equation $F(x, y) = F(a, b)$ is locally continuously differentiably solvable for y in terms of x near (a, b) provided that the approximating linear equation

$$(\partial F/\partial x)_{(x,y)=(a,b)}\, \mathrm{d}x + (\partial F/\partial y)_{(x,y)=(a,b)}\, \mathrm{d}y = 0$$

is solvable for $\mathrm{d}y$ in terms of $\mathrm{d}x$.

Example 4.12 Consider a regular plane curve $\mathbf{r} : \mathbb{R} \rightarrowtail \mathbb{R}^2$. Then, provided that $(\mathbf{c} - \mathbf{r}(t)) \cdot \mathbf{r}_2(t) - \mathbf{r}_1(t) \cdot \mathbf{r}_1(t) \neq 0$, the equation $(\mathbf{c} - \mathbf{r}(t)) \cdot \mathbf{r}_1(t) = 0$ is solvable locally for the parameter t in terms of $\mathbf{c}$, so that locally, provided that the inequality holds, $\rho = u(\mathbf{c})$, where $\rho^2 = (\mathbf{c} - \mathbf{r}) \cdot (\mathbf{c} - \mathbf{r})$. Then $\rho u_1 = (\mathbf{c} - \mathbf{r}) \cdot$, so that $\rho \nabla u = \mathbf{c} - \mathbf{r}$, where $\nabla u \cdot\; = u_1$. Accordingly, away from the evolute of $\mathbf{r}$, $(\nabla u)^2 = 1$. □

The partial differential equation $(\nabla u)^2 = 1$ is known as the (here the *two-dimensional*) *Hamilton–Jacobi equation*. It plays a central role in the classical theory of geometrical optics.

The Inverse Function Theorem has direct application to the description of *submanifolds* of a finite-dimensional vector space X, the jargon term for non-singular curves in the plane, non-singular curves or surfaces in three-dimensional space and their analogues in spaces of higher dimensions.

An *affine subspace* of a finite-dimensional vector space X is a parallel in X to a linear subspace, its dimension being the dimension of the linear subspace to which it is parallel. A subset M of X is said to be C^1 at a point $a \in M$ if there is an affine subspace T of M passing through a, an open neighbourhood A of a and a C^1 diffeomorphism $h : X, a \rightarrowtail X, a$ with domain A and $h_1(a) = 1_X$, the identity map on X, such that $h(A \cap T) = h(A) \cap M$. The affine space T, uniquely determined if it exists, is called the *tangent space* to M at a. It is generally given the structure of a vector space by taking a as origin. The subset M is said to be a C^1 *submanifold* of X if it is C^1 at each of its points. For a connected submanifold the dimension of the tangent space is constant. This dimension is said to be the *dimension* of the submanifold.

In practice a subset of X is often presented either *explicitly* parametrically as the image of a map or *implicitly* as a *fibre* or level set (most frequently the set of zeros) of a map.

Example 4.13 Consider a map $f : \mathbb{R} \rightarrowtail \mathbb{R};\ x \mapsto f(x)$. Then graph f is both the image of the map

$$\mathbb{R} \rightarrowtail \mathbb{R}^2;\ x \mapsto (x, f(x))$$

and the fibre over 0 of the map

$$\mathbb{R}^2 \rightarrowtail \mathbb{R};\ (x, y) \mapsto y - f(x).$$

□

Example 4.14 The image of the map

$$\mathbb{R} \to \mathbb{R}^2;\ t \mapsto (t^2 - 1,\ t(t^2 - 1))$$

is also the fibre over 0 of the map

$$\mathbb{R}^2 \to \mathbb{R};\ (x, y) \mapsto y^2 - (1 + x)x^2.$$

(See Figure 1.19.) □

Two corollaries of the Inverse Function Theorem relevant to the determination of submanifolds are as follows.

Theorem 4.15 (The injective criterion*) Let $f : W \rightarrowtail X$ be a C^1 map, with $f_1(c)$ injective for some $c \in \operatorname{dom} f$, W and X being finite-dimensional vector spaces. Then there exists an open neighbourhood C of c in W such that the image of $f|C$ is C^1 at $a = f(c)$, with tangent space the parallel through a of the image of $f_1(c)$.*

Indication of proof Consider the case that c is the origin in W and a the origin in X and $f_1(0) : W \to Y$ is the inclusion in X of a linear subspace W. Let Y be a complementary linear subspace, and identify X with the product space $W \times Y$. Then consider the map $h : W \times Y \rightarrowtail W \times Y;\ (w, y) \mapsto f(w) + (0, y)$, with derivative at $(0, 0)$ the identity on $X = W \times Y$. Accordingly, by the Inverse Function Theorem, h is a local C^1 diffeomorphism at $(0, 0)$.

It follows at once that the image of the restriction of f to some open neighbourhood of the origin in W is C^1 at the origin in X with $W \times \{0\}$ as tangent space. This completes the proof in this special case. The general case reduces at once to this one if the origin in X is set at a and new bases for the vector spaces W and X are chosen appropriately. □

By the above 'injective' corollary of the Inverse Function Theorem the image of a regular parametric curve $\mathbf{r} : \mathbb{R} \rightarrowtail \mathbb{R}^2$ is locally everywhere a one-dimensional submanifold of $\mathbb{R}^2$ and likewise the image of a regular parametric curve $\mathbf{r} : \mathbb{R} \rightarrowtail \mathbb{R}^3$ is locally everywhere a one-dimensional submanifold of $\mathbb{R}^3$. It is necessary to insist on the word 'locally' here, for a regular curve need not be injective.

Example 4.16 The C^1 map $\mathbf{r} : \mathbb{R} \to \mathbb{R}^2;\ t \mapsto (t^2 - 1,\ t(t^2 - 1))$ has derivative at t the linear map with matrix

$$\begin{pmatrix} 2t \\ 3t^2 - 1 \end{pmatrix},$$

which is injective for all t, since there is no t for which $2t$ and $3t^2 - 1$ are both equal to zero. So the criterion is everywhere applicable. The map f itself is not injective, for $\mathbf{r}(1) = \mathbf{r}(-1) = (0, 0)$. (See Figure 1.19.) □

More subtly, even injectivity is not enough, as the restriction of the above map to the interval $]-1, \infty[$ shows. (Again see Figure 1.19.)

The following jargon is in common use. A C^1 map $f : W \rightarrowtail X$ is said to be *immersive* at a point a of its domain if $f_1(a)$ is injective, to be an *immersion* if it is everywhere immersive, and to be an *embedding* if also the map $\operatorname{dom} f \to \operatorname{im} f;\ w \mapsto f(w)$ is a homeomorphism.

Theorem 4.17 (The surjective criterion*) Let $f : X \rightarrowtail Y$ be a C^1 map, with $f_1(a)$ surjective for some $a \in \operatorname{dom} f$, X and Y being finite-dimensional vector spaces. Then there exists an open neighbourhood A of a in X such that the $A \cap f^{-1}\{f(a)\}$ is C^1 at a, with tangent space the parallel through a of the kernel of $f_1(a)$.*

Indication of proof Consider the case that Y is a subspace of X with a and $f(a)$ both the origin in X, $f_1(0)$ being the projection of X on to Y with kernel a linear subspace W. Then let us identify X with the product space $W \times Y$.

Consider the map $h : W \times Y \rightarrowtail W \times Y;\ (w, y) \mapsto (w, f(w, y))$. This has as derivative at $(0, 0)$ the identity on $X = W \times Y$. Accordingly, by the Inverse Function Theorem, h is a local C^1 diffeomorphism at $(0, 0)$. It follows at once that $f^{-1}\{0\}$ is C^1 at the origin in X with $W \times \{0\}$ as tangent space.

This completes the proof in this special case. The general case reduces at once to this one if the origin in X is set at a and new bases for X and Y are chosen appropriately. □

Example 4.18 The C^1 map $F;\ \mathbb{R}^2 \to \mathbb{R};\ (x, y) \mapsto y^2 - (1 + x)x^2$ has derivative at (x, y) the linear map with matrix $[-2x - 3x^2 \quad 2y]$, which is surjective for all (x, y) except the origin, which is a point where F is zero. Everywhere else the criterion is applicable. So $F^{-1}\{0\}$ is C^1 except at (0.0). □

Notice the contrast between the two ways of dealing with the same example. The parametric form of the curve is regular at both points mapping to the origin while the representation of the curve as a fibre is singular at the origin.

A C^1 map $f : X \rightarrowtail Y$ is said to be *submersive* at a point a of its domain if $f_1(a)$ is surjective and to be a *submersion* if it is everywhere submersive.

We shall often have occasion to employ the surjective criterion in the sequel. It is worth remarking that to prove that a linear map $t : \mathbb{R}^n \to \mathbb{R}^p$ is surjective it is enough to verify that the map $\mathbb{R}^p \to \mathbb{R}^n$ whose matrix is the transpose of the matrix of t is injective.

Example 4.19 Let $\mathbf{r} : \mathbb{R} \rightarrowtail \mathbb{R}^2$ be a regular plane curve and for each $\mathbf{c} \in \mathbb{R}^2$, $t \in \operatorname{dom} \mathbf{r}$ let $V(\mathbf{c})(t) = -\frac{1}{2}(\mathbf{c} - \mathbf{r}(t)) \cdot (\mathbf{c} - \mathbf{r}(t))$. Then the subsets of $\mathbb{R}^2 \times \mathbb{R}$

$$\{(\mathbf{c}, t) : V(\mathbf{c})_1 = 0\},$$

$$\{(\mathbf{c}, t) : V(\mathbf{c})_1 = 0,\ V(\mathbf{c})_2 = 0\},$$

$$\{(\mathbf{c}, t) : V(\mathbf{c})_1 = 0,\ V(\mathbf{c})_2 = 0,\ V(\mathbf{c})_3 = 0,\ V(\mathbf{c})_4 \neq 0\},$$

are smooth submanifolds of $\mathbb{R}^2 \times \mathbb{R}$ of dimensions 2, 1, 0 respectively.

Proof We use the surjective criterion each time.

In the first case the Jacobian matrix at $(\mathbf{c}, t)$ of the map $(\mathbf{c}, t) \mapsto V(\mathbf{c})_1(t)$ is the 1×3 matrix $[\mathbf{r}_1(t) \cdot \quad V(\mathbf{c})_2(t)]$, of rank 1 and so surjective (regardless of whether or not $V(\mathbf{c})_2(t) = 0$) since the row vector $\mathbf{r}_1(t) \cdot$ is non-zero.

In the second case the Jacobian matrix at $(\mathbf{c}, t)$ of the map $(\mathbf{c}, t) \mapsto (V(\mathbf{c})_1(t), V(\mathbf{c})_2(t))$ is the 2×3 matrix

$$\begin{bmatrix} \mathbf{r}_1(t) \cdot & V(\mathbf{c})_2(t) \\ \mathbf{r}_2(t) \cdot & V(\mathbf{c})_3(t) \end{bmatrix} = \begin{bmatrix} \mathbf{r}_1(t) \cdot & 0 \\ \mathbf{r}_2(t) \cdot & V(\mathbf{c})_3(t) \end{bmatrix},$$

of rank 2 and so surjective (regardless of whether or not $V(\mathbf{c})_3(t) = 0$) since wherever $\mathbf{c}$ exists the vectors $\mathbf{r}_1$ and $\mathbf{r}_2$ are linearly independent.

In the third case the Jacobian matrix at $(\mathbf{c}, t)$ of the map $(\mathbf{c}, t) \mapsto (V(\mathbf{c})_1(t), V(\mathbf{c})_2(t), V(\mathbf{c})_3(t))$ is the 3×3 matrix

$$\begin{bmatrix} \mathbf{r}_1(t) \cdot & V(\mathbf{c})_2(t) \\ \mathbf{r}_2(t) \cdot & V(\mathbf{c})_3(t) \\ \mathbf{r}_3(t) \cdot & V(\mathbf{c})_4(t) \end{bmatrix} = \begin{bmatrix} \mathbf{r}_1(t) \cdot & 0 \\ \mathbf{r}_2(t) \cdot & 0 \\ \mathbf{r}_3(t) \cdot & V(\mathbf{c})_4(t) \end{bmatrix},$$

of rank 3 and so surjective, since $V(\mathbf{c})_4(t) \neq 0$ and since wherever $\mathbf{c}$ exists the vectors $\mathbf{r}_1$ and $\mathbf{r}_2$ are linearly independent. □

The following example is typical of many.

Example 4.20 Let $\mathbb{R}(n)$ denote the vector space of $n \times n$ matrices with real entries, and let $O(n)$ denote the subset of $\mathbb{R}(n)$ consisting of matrices a such that $a^\tau a = 1$, where a^τ is the transpose of the matrix a and 1 is the unit $n \times n$ matrix. Then $O(n)$ is a submanifold of $\mathbb{R}(n)$ of dimension $\frac{1}{2}n(n-1)$.

Proof Let $\mathbb{R}^+(n)$ denote the linear subspace of $\mathbb{R}(n)$ consisting of all the symmetric $n \times n$ matrices, that is matrices c such that $c = c^\tau$. Clearly this is a linear subspace of $\mathbb{R}(n)$ of dimension $\frac{1}{2}n(n+1)$. Consider the map $\mathbb{R}(n) \to \mathbb{R}^+(n) : a \mapsto a^\tau a$. By Proposition 4.2 the derivative of this map at a is the linear map $\mathbb{R}(n) \to \mathbb{R}^+(n) : b \mapsto b^\tau a + a^\tau b$. This is surjective when a belongs to $O(n)$, for let c be any element of $\mathbb{R}^+(n)$ and consider the equation $a^\tau b = \frac{1}{2}c$; since $a^\tau a = 1$ it follows that $a^{-1} = a^\tau$ implying that $b = \frac{1}{2}ac$. But then also $b^\tau a = \frac{1}{2}c^\tau = \frac{1}{2}c$ so that $b^\tau a + a^\tau b = c$. Now apply the surjective criterion, implying that $O(n)$ is C^1 at a, the tangent space being the kernel of the linear map, of dimension $n - \frac{1}{2}n(n+1) = \frac{1}{2}n(n-1)$. □

Note in particular that the tangent space to $O(n)$ at the identity matrix 1 consists of all matrices b such that $b^\tau 1 + 1^\tau b = b^\tau + b = 0$, the vector space of *skew-symmetric* $n \times n$ matrices. The submanifold $O(n)$ is an example of a *Lie* (pronounced Lee) *group*, the orthogonal group of order n, the tangent space at the origin being its *Lie algebra*. For more on Lie groups see Chapter 20 of Porteous (1981).

Proposition 4.21 A skew-symmetric matrix of odd order is not invertible.

Hint at proof The determinants of a square matrix and its transpose are equal. □

One word of warning. Just because the surjective criterion fails in a particular case does not of itself imply that the set under consideration fails to be a submanifold. For example one might conclude from the fact that the matrix $[3x^2 \quad -2y]$ is not surjective at the origin that the subset consisting of all points (x, y) such that $y^2 = x^3$ is not a submanifold of $\mathbb{R}^2$. Of course it is not, as we shall prove shortly in Example 4.21, but the failure of the surjectivity of the derivative is not enough to prove this. Counterexamples are provided by the equations $(x - y)^2 = 0$ and $(x - y)(x^2 + y^2) = 0$.

It follows from the continuity of the derivative that a C^1 map $f : X \rightarrowtail Y$ that is submersive at a point a of its domain is submersive everywhere near a and accordingly *foliates* X near a. To quote from a book on foliations (Camacho and Neto, 1985) a *foliation* of a manifold is intuitively a decomposition of the manifold as a union of connected disjoint submanifolds of the same dimension, called *leaves*, which pile up locally like pages of a book. The simplest example of a foliation of dimension k is the foliation of $\mathbb{R}^n = \mathbb{R}^k \times \mathbb{R}^{n-k}$ where the leaves are k-planes of the form $\mathbb{R}^k \times \{c\}$ with $c \in \mathbb{R}^{n-k}$.

It follows directly from the definition of a submanifold that for any point a of a C^1 submanifold M of a finite-dimensional vector space X the intersection of M with some neighbourhood of a in X is expressible either as the image of an immersion or as the zero set of a submersion. Not quite so obvious, but a corollary of the Inverse Function Theorem, is that M is also expressible locally as the graph of a C^1 map from an appropriate linear subspace of X to any complementary subspace.

Theorem 4.22 Let M be a C^1 submanifold of a finite-dimensional vector space X, let T be the tangent space at some point $a \in M$, and let U be any *complementary affine subspace of X through a. Then, locally near a, M is expressible as the graph of a C^1 map f from T to U, with $f_1(a) = 0$.*

Indication of proof Consider the case that a is the origin and identify X with $T \times U$, and let π be the projection of X to U with kernel T. Then there is a local diffeomorphism h of X at 0, with derivative there the identity on X, such that, near 0, M is the zero set of the map $F = \pi h : X \rightarrowtail U$. The derivative at 0 of the C^1 map $T \times U \rightarrowtail T \times U$; $(x, y) \mapsto (x, z) = (x, F(x, y))$ is the identity, and so by the Inverse Function Theorem this map is a local diffeomorphism of X at 0. Let the local C^1 inverse be $(x, z) \mapsto (x, G(x, z))$ and define $f : T \rightarrowtail U$ by $f(x) = G(x, 0)$. Then, locally near 0, M is the graph of f. The general case is easily reduced to this. □

This last theorem can be used to show that a set is *not* C^1 at some particular point. The following example complements Proposition 4.9, and finally settles a question raised following Proposition 1.6.

Example 4.23 Consider the set $M = \{(x, y) \in \mathbb{R}^2 : x^3 = y^2\}$. This is not C^1 at the origin, since there is no line T through the origin such that for *any* complementary line U the set locally is the graph of a continuous map from T to U, let alone a differentiable one. This is clearly so if T is taken to be the x-axis, while if T is taken to be any other line all we need do to make the point is to choose U to be any line other than T or the x-axis, since for any line other than the x-axis the curve in the neighbourhood of the origin lies entirely to one side of the line, the only point of intersection of line and curve being the origin itself. □

When X is the Euclidean space $\mathbb{R}^n$ and U is taken to be the normal subspace N to M at a the presentation of M as graph f is called the *Monge form* of M near a.

The injective and surjective criteria for a subset M of a vector space X to be C^1 at a point a of M are both special cases of the following more general 'double-ended' criterion which, however, requires rank information on the derivative of a map in a neighbourhood of a and not just at a itself.

Theorem 4.24 (The general rank criterion*) Let X and Y be finite-dimensional vector spaces and let $f : X \rightarrowtail Y$ be a C^1 map, with f_1 of constant rank in some neighbourhood of $a \in X$. Then the fibre or level set of f through a is C^1 at a in X and the image of f is C^1 at $f(a)$ in Y.* □

Corollary 4.25 Let X and Y be finite-dimensional vector spaces and let $f : X \rightarrowtail Y$ be a C^1 map, with f_1 of constant rank in some neighbourhood of $a \in X$. Then there are C^1 diffeomorphisms $h : X, a \rightarrowtail X, a$ and $k : Y, f(a) \rightarrowtail Y, f(a)$ such that, for x near a,

$$kfh(x) = f(a) + f_1(a)(x - a),$$

or, if one prefers it, C^1 diffeomorphisms $h : X, 0 \rightarrowtail X, a$ and $k : Y, f(a) \rightarrowtail Y, 0$ such that, for x near 0, $kfh(x) = f_1(a)(x)$. □

We shall make further comment about this corollary presently.

Example 4.26 The unit sphere in $\mathbb{R}^3$, with centre the origin, is most commonly proved to be a two-dimensional submanifold of $\mathbb{R}^3$ by the remark that the map $\mathbb{R}^3 \to \mathbb{R} : \mathbf{r} \mapsto \mathbf{r} \cdot \mathbf{r}$ is submersive everywhere except at the origin, for $2\mathbf{r} \cdot \in L(\mathbb{R}^3, \mathbb{R})$ is zero if and only if $\mathbf{r} = 0$. Quite a different proof is to remark that the differential of the map $\pi : \mathbb{R}^3 \rightarrowtail \mathbb{R}^3 : \mathbf{r} \mapsto \mathbf{r}/|\mathbf{r}|$, defined except at the origin, with image the sphere, has rank 2 everywhere. For

$$\pi_1(\mathbf{r})\mathbf{u} = \frac{\mathbf{u}}{|\mathbf{r}|} - \frac{\mathbf{u} \cdot \mathbf{r}\mathbf{r}}{|\mathbf{r}|^3} = 0 \Leftrightarrow \mathbf{u} = \lambda \mathbf{r} \text{ for some } \lambda \in \mathbb{R}. \qquad \square$$

Two linear subspaces U and V of a finite-dimensional vector space X are said to be *transversal* in X if $X = U + V$. Two C^1 submanifolds M and M' of X are said to be *transversal* at a point a of $M \cap M'$ if their tangent vector spaces at a are transversal in X. They are said to be *transversal* if they are transversal at every point of their intersection. As a special case, two submanifolds with empty intersection are transversal.

Proposition 4.27 Let M and M' be transversal C^1 submanifolds of a finite-dimensional vector space X. Then their intersection $L = M \cap M'$ is a C^1 submanifold of X, the tangent space to L at a point of L being the intersection of the tangent spaces to M and M' at a. $\square$

The concept of transversality is central to an understanding of what we loosely term *generic* phenomena, linear subspaces of a vector space being *generically related* if and only if they are mutually transversal. We duck the task of developing this theme here, but return briefly to it at the very end of the book.

4.2 Higher derivatives and C^k submanifolds

The first derivative of a differentiable map $f : X \rightarrowtail Y$, $f_1 : X \rightarrowtail L(X, Y)$, is the map that assigns to each a the linear map $f_1(a)$, that up to a constant best approximates the function f at a. It frequently is the case that f_1 is itself differentiable. The *second derivative* of f at a, $f_2(a)$, is then defined to be the first derivative of f_1

at a. It is an element of the vector space $L(X, L(X, Y))$, which we term the space of *twice* linear maps from X to Y. We prefer to maintain the formal distinction between this space and the space of bilinear maps $X \times X \to Y$. Note that $f_2(a)$ is a map whose values also are maps. It has two *slots* in either of which any element of X may be inserted. Once both are filled the outcome is an element of Y.

Theorem 4.28 Let $f : X \rightarrowtail Y$ be twice differentiable at a. Then the twice linear map $f_2(a)$ is symmetric *in the sense that, for all $\mathbf{u}, \mathbf{v} \in X$,*

$$f_2(a)\mathbf{vu} = f_2(a)\mathbf{uv}.$$

It does not matter which of the vectors $\mathbf{u}$ and $\mathbf{v}$ goes in which slot. □

In the case that $X = \mathbb{R}^n$ and $Y = \mathbb{R}$ this implies and is implied by the fact that the second partial derivatives of f are symmetric. Consider for example a twice differentiable map $f : \mathbb{R}^2 \to \mathbb{R};\ (x, y) \mapsto z = f(x, y)$. Then

$$\frac{\partial^2 z}{\partial y \partial x} = \frac{\partial^2 z}{\partial x \partial y},$$

with

$$f_2(x, y)\begin{bmatrix} \mathrm{d}x \\ \mathrm{d}y \end{bmatrix} = \begin{bmatrix} \dfrac{\partial^2 z}{\partial x^2}\mathrm{d}x + \dfrac{\partial^2 z}{\partial x \partial y}\mathrm{d}y & \dfrac{\partial^2 z}{\partial x \partial y}\mathrm{d}x + \dfrac{\partial^2 z}{\partial y^2}\mathrm{d}y \end{bmatrix}.$$

The reader bemused by all this might usefully at this point skip to Chapter 7 and then return having become familiar there with twice linear and thrice linear forms in a purely algebraic setting.

The space of all symmetric twice linear maps from X to Y is clearly a linear subspace of the space of all twice linear maps from X to Y. We denote this subspace by $L_2(X, Y)$. In the case that $\dim X = n$ and $\dim Y = p$ its dimension is $\frac{1}{2}n(n-1)p$.

Associated to the twice linear map $f_2(a)$ is the associated *quadratic* map $X \to Y : \mathbf{u} \mapsto f_2(a)\mathbf{u}^2 = f_2(a)\mathbf{uu}$. For the twice differentiable map $f : \mathbb{R}^2 \to \mathbb{R};\ (x, y) \mapsto z = f(x, y)$ we have, in traditional notations,

$$f_2(x, y)\begin{bmatrix} \mathrm{d}x \\ \mathrm{d}y \end{bmatrix}^2 = \frac{\partial^2 z}{\partial x^2}\mathrm{d}x^2 + 2\frac{\partial^2 z}{\partial x \partial y}\mathrm{d}x\,\mathrm{d}y + \frac{\partial^2 z}{\partial y^2}\mathrm{d}y^2.$$

The *second derivative* or *differential* of a twice differentiable map $f : X \rightarrowtail Y$ is the map $f_2 : X \rightarrowtail L_2(X, Y)$ associating to any point a in the domain of f the symmetric twice linear map $f_2(a)$. The map f is said to be C^2 if f_2 exists and is continuous.

The *third derivative* of f at a, $f_2(a)$, is the first derivative of f_2 at a and is an element of the vector space $L(X, L(X, L(X, Y)))$, the space of *thrice linear* maps from X to Y, and indeed of the linear subspace $L_3(X, Y)$ of fully symmetric thrice linear maps, that is symmetric in all three slots. Thrice linear maps that are symmetric in only two of the three slots do occur in the further development of the subject. A map f is said to be C^3 if f_3 exists and is continuous.

Proceeding in like fashion, the $(k + 1)$*th derivative* of a map f at $a \in \operatorname{dom} f$ is the first derivative of the kth derivative at a. The map f is said to be C^k if f_k exists and is continuous. It is said to be C^∞ if it is C^k for every integer $k \geqslant 1$.

A map $f : X \rightarrowtail Y$ between finite-dimensional vector spaces will be said to be *smooth* if it is C^∞ or, more loosely, if it is C^k for k sufficiently large for the matter in hand.

Theorem 4.29 Let W, X and Y be finite-dimensional vector spaces and $f : W \rightarrowtail X$ and $g : X \rightarrowtail Y$ be C^k maps, for any k, $1 \leqslant k \leqslant \infty$. Then gf is C^k. □

Theorem 4.30 Let $h : X \rightarrowtail Y$ be an injective C^k map, for any k, $1 \leqslant k \leqslant \infty$, where $\dim Y = \dim X$, and suppose that h^{-1} is C_1. Then h^{-1} is C^k. □

Such a map h is said to be a C^k *diffeomorphism*.

A subset M of a finite-dimensional vector space X is said to be C^k for some k, $1 \leqslant k \leqslant \infty$, at a point a of M, if it is locally the image of an affine subspace T by some C^k diffeomorphism $h : X \rightarrowtail X$, with $h_1(a) = 1_X$. The subset M is said to be a C^k *submanifold* of X if it is C^k at each of its points. The various corollaries of the Inverse Function Theorem still apply in determining whether or not a subset of X is a C^k submanifold of X provided only that the defining maps are C^k.

Taylor's Theorem is a theorem about smooth map-germs. The familiar theorem of one variable calculus generalises to smooth map-germs $f : X, a \rightarrowtail Y, b = f(a)$ as follows:

Theorem 4.31 (W.H. Young's form of *Taylor's Theorem*) *Let X and Y be finite-dimensional vector spaces and let the map-germ $f : X$,a*

$\rightarrowtail Y, b$ *be at least* C^m, *where* $1 \leqslant m$. *Then there is a map-germ* $\varphi : X, a \rightarrowtail Y, 0$ *such that*

$$f(x) = \sum_{k=0}^{m} \frac{1}{k!} f_k(a)(x-a)^k + \varphi(x),$$

where

$$\lim_{x \to a} \frac{|\varphi(x)|}{|x-a|^m} = 0. \qquad \square$$

The finite sum here is called the *m-jet* of the *germ* of f at a and the remainder φ the *m-tayl* (Poston and Stewart, 1978) of f at a.

Corollary 4.25 to Theorem 4.24 may be paraphrased by saying that any C^1 map of constant rank can have its 1-tayl at a point cut off by local diffeomorphisms of source and target. A somewhat analogous theorem in which the 2-tayl is cut off is that known as the *Morse Lemma*.

Theorem 4.32 (*The Morse Lemma*) *Let* $f : X \rightarrowtail \mathbb{R}$ *be a* C^2 *map*, X *being a finite-dimensional vector space, let* $a \in \operatorname{dom} f$ *and suppose that* $f_1(a) = 0$, *the linear map* $f_2(a) : X \to L(X, \mathbb{R})$ *being invertible* (*that is* f_2 *is* non-degenerate *at* a). *Then there is a local diffeomorphism* $h : X, a \to X, a$ *such that*, *for* x *near* a,

$$fh(x) = f(a) + \tfrac{1}{2} f_2(a)(x-a)^2.$$

For a simple proof see Poston and Stewart (1978), pp. 54–6. $\square$

Much of *Singularity Theory* is concerned with determining under what conditions the m-tayl of a map-germ at a point can be removed in some such way. See the ‘Further reading’ section for references.

The Morse Lemma has immediate application to the classification of *non-degenerate* critical points of smooth real-valued functions. In particular such a map if $f : \mathbb{R}^n \rightarrowtail \mathbb{R}$ has a local minimum or maximum at a point $a \in \operatorname{dom} f$ according as the quadratic form $f_2(a)\mathbf{u}^2$ is negative definite or positive definite.

4.3 The Faà de Bruno formula

The Leibniz formula for the nth derivative of a *product* of two smooth maps is well known and easy to prove. The formula for the nth derivative of the *composite* of two smooth maps is much less well

Table 4.1

$V\mathbf{a}$	$(V\mathbf{a})_1$	$(V\mathbf{a})_2$	$(V\mathbf{a})_3$	$(V\mathbf{a})_4$
	$V_1\mathbf{a}$	$(V_1\mathbf{a})_1$	$(V_1\mathbf{a})_2$	$(V_1\mathbf{a})_3$
		$V_2\mathbf{a}$	$(V_2\mathbf{a})_1$	$(V_2\mathbf{a})_2$
			$V_3\mathbf{a}$	$(V_3\mathbf{a})_1$
				$V_4\mathbf{a}$

explicitly

$V\mathbf{a}$	$V_1\mathbf{a}_1$	$V_2\mathbf{a}_1^2 + V_1\mathbf{a}_2$	$V_3\mathbf{a}_1^3 + 3V_2\mathbf{a}_1\mathbf{a}_2 + V_1\mathbf{a}_3$	$V_4\mathbf{a}_1^4 + 6V_3\mathbf{a}_1^2\mathbf{a}_2 + 4V_2\mathbf{a}_1\mathbf{a}_3 + 3V_2\mathbf{a}_2^2 + V_1\mathbf{a}_4$
	$V_1\mathbf{a}$	$V_2\mathbf{a}_1$	$V_3\mathbf{a}_1^2 + V_2\mathbf{a}_2$	$V_4\mathbf{a}_1^3 + 3V_3\mathbf{a}_1\mathbf{a}_2 + V_2\mathbf{a}_3$
		$V_2\mathbf{a}$	$V_3\mathbf{a}_1$	$V_4\mathbf{a}_1^2 + V_3\mathbf{a}_2$
			$V_3\mathbf{a}$	$V_4\mathbf{a}_1$
				$V_4\mathbf{a}$

where $V_1\mathbf{a}_1 = (V_1\mathbf{a})\mathbf{a}_1$, $V_2\mathbf{a}_1^2 + V_1\mathbf{a}_2 = (V_2\mathbf{a})\mathbf{a}_1^2 + (V_1\mathbf{a})\mathbf{a}_2$, etc, and where for example

$$\begin{aligned} V_4\mathbf{a}_1^4 + \text{etc} = (V_4\mathbf{a}_1^3 &+ 3V_3\mathbf{a}_1\mathbf{a}_2 + V_2\mathbf{a}_3)\mathbf{a}_1 \\ &+ 3(V_3\mathbf{a}_1^2 + V_2\mathbf{a}_2)\mathbf{a}_2 \\ &+ 3(V_2\mathbf{a}_1)\mathbf{a}_3 \\ &+ V_1\mathbf{a}_4, \end{aligned}$$

and so on, from which it follows that if the first few entries in any row are zero then each of the entries in the preceding row directly above these also is equal to zero.

known. In the section which follows, on the nth derivative of the composite $V\mathbf{a}$ of two smooth maps, $\mathbf{a} : W \rightarrowtail X$ and $V : X \rightarrowtail Y$, the crucial thing to note is that the map $\mathbf{a}$ is composable not only with the map V but with each of its derivatives, V_i. The somewhat bizarre choice of the letters V and $\mathbf{a}$ to denote these maps is influenced by the fact that in the main application that we shall make of the formula X will be $\mathbb{R}^2$ and both W and Y will be $\mathbb{R}$, $\mathbf{a}$ being a *regular* smooth curve on $\mathbb{R}^2$ and V a real-valued map from $\mathbb{R}^2$.

Table 4.1 shows the first few derivatives, not only of the composite map $V\mathbf{a}$ ($= V \circ \mathbf{a}$), but also, in successive rows, those of the composite maps $V_1\mathbf{a}$, $V_2\mathbf{a}$ and $V_3\mathbf{a}$. To save space and for clarity the formulas have all been slightly abbreviated. For example, the first derivative of $V\mathbf{a}$ should properly be written $(V_1 \circ \mathbf{a}) \cdot \mathbf{a}_1$ or $(V_1\mathbf{a})\mathbf{a}_1$ rather than $V_1\mathbf{a}_1$, and so on, throughout the table. In the complete form of the formula the symbol $\circ$ denotes *direct* composition of maps, while the dot $\cdot$ denotes the product, or composition, *of values*. In performing the differentiations it is essential to be clear at each stage which is which, for the Chain Rule applies to the one while the Leibniz formula applies to the other. What we have here is a delicate mixture of the two!

The explicit general formula here is

$$\text{for any } h \text{ and } k, \ (V_h\mathbf{a})_{k+1} = \sum_{j=0}^{k} \binom{k}{j} (V_{h+1}\mathbf{a})_{k-j}\mathbf{a}_{j+1}.$$

It has been rediscovered many times. An early attribution is to Faà de Bruno (1857). See also a paper by Felice Ronga (1983). He gives the following form for the mth derivative of a composite map gf:

$$(gf)_m(t) = \sum \frac{m!}{k_1! \ldots k_m!(1!)^{k_1} \ldots (m!)^{k_m}} g_n(f_1)^{k_1} \ldots (f_m)^{k_m},$$

where the suffices on f, g and gf denote differentiation, and the sum ranges over $n = 1, \ldots, m$ and all non-negative integers $k_1, \ldots, k_m$ such that $k_1 + 2k_2 + \ldots + mk_m = m$. All the derivatives of f are taken at t, those of g at $f(t)$.

Exercises

4.1 Let $f : \mathbb{R}^2 \to \mathbb{R}^2$ be defined by $f(x, y) = (x^2 - y^2, 2xy)$. For which (x, y) is $\mathrm{d}f(x, y)$ invertible? Is f itself invertible?

4.2 Define $f : \mathbb{R}^2 \rightarrowtail \mathbb{R}^3$ and $g : \mathbb{R}^3 \rightarrowtail \mathbb{R}^2$ by the formulas

$$f(x, y) = (x^2, xy, y^2),\ g(u, v, w) = (uw + v^2, uw - v^2),$$

and let $h = gf : \mathbb{R}^2 \rightarrowtail \mathbb{R}^2$ be the composite map.

Calculate $\mathrm{d}f(x, y)$, $\mathrm{d}f(1, 1)$, $\mathrm{d}g(u, v, w)$ and $\mathrm{d}g(f(1, 1))$.

Calculate also $\mathrm{d}h(1, 1)$ using

(i) an explicit formula for h,

(ii) the Chain Rule.

Is $\mathrm{d}h(1, 1)$ either injective or surjective?

4.3 Let $\mathbf{r} : \mathbb{R} \rightarrow \mathbb{R}^2$ be defined by $\mathbf{r}(t) = (t^2 - 1,\ t(t^2 - 1))$. For which $t \in \mathbb{R}$ is the derivative $\mathrm{d}\mathbf{r}t : \mathbb{R} \rightarrow \mathbb{R}^2$ injective? Is the map $\mathbf{r}$ itself injective?

4.4 Let $F : \mathbb{R}^2 \rightarrow \mathbb{R}$ be defined by $F(x, y) = y^2 - x^2(1 + x)$. For which $(x, y) \in \mathbb{R}^2$ is the derivative $\mathrm{d}F(x, y) : \mathbb{R}^2 \rightarrow \mathbb{R}$ surjective? Is the map F itself surjective?

4.5 Let $f : \mathbb{R}^3 \rightarrow \mathbb{R}^3;\ (x, y, z) \mapsto (u, v, w)$ be the map defined by the equations

$$\begin{aligned} u + v + w &= x, \\ v + w &= xy, \\ w &= xyz. \end{aligned}$$

Compute the Jacobian matrix of f at (x, y, z). For which (x, y, z) does it fail to be invertible?

4.6 Determine those points $(x, y, z) \in \mathbb{R}^3$ at which the derivative of the map $\mathbb{R}^3 \rightarrow \mathbb{R}^3;\ (x, y, z) \mapsto (x^2 + y^2,\ y^2 + z^2,\ xz)$ is invertible.

4.7 Let $u = x^2y$, $v = 4x^2 + xy + y^2$. For $(u, v) = (0, 4)$ verify that two solutions of these equations are $(x, y) = (1, 0)$ and $(x, y) = (0, -2)$. Show that the first but not the second of these extends to a C^1 solution $(x, y) = g(u, v)$ of the equations, the domain of definition of g being some open set containing $(0, 4)$.

4.8 Prove that the map $]-1, 1[\rightarrow \mathbb{R};\ x \rightarrow x/(1 - x^2)$, with domain the interval $]-1, 1[= \{x \in \mathbb{R} : -1 < x < 1\}$, has a C^1 inverse.

(Hint: Set about solving the equation $x/(1 - x^2) = y$ for x in terms of y, proving that for each y there is exactly one solution of the equation in the interval $]-1, 1[$. Then use the Inverse Function Theorem to establish the differentiability of the inverse at the origin.)

4.9 Let $f : \mathbb{R}^2 \rightarrowtail \mathbb{R}$ be a differentiable function, let $(a, b) \in \mathbb{R}^2$ and let $g : \mathbb{R}^2 \rightarrowtail \mathbb{R} : (r, \theta) \mapsto g(r, \theta)$ be defined by the formula

$$g(r, \theta) = f(a + r\cos\theta,\ b + r\sin\theta).$$

Express the partial derivative of g with respect to r at $(0, \theta)$ in terms of the partial derivatives of f at (a, b) and θ. (This partial derivative is called the *directional derivative* of f at (a, b) in the direction θ.)

Do this for the function f defined by the formula $f(x, y) = 2x + 3y$ at the point $(a, b) = (0, 0)$.

4.10 Attempt to carry out the procedure of Exercise 4.9 for the function $f : \mathbb{R}^2 \to \mathbb{R}$ defined by

$$f(x, y) = \frac{xy^2}{x^2 + y^2}, \text{ for } (x, y) \neq (0, 0), \; f(0, 0) = 0,$$

at $(a, b) = (0, 0)$. What do you deduce about the function f?

4.11 Let $f : \mathbb{R}^2 \to \mathbb{R}^2$ be defined by $f(x, y) = (\frac{1}{3}x^3 - yx, y)$. For which points of the domain of f does the derivative f_1 fail to be bijective? Sketch this set and its image by f in $\mathbb{R}^2$.

Show that there are vectors $\mathbf{a}_1 \neq 0$ and $\mathbf{a}_2 \in \mathbb{R}^2$ such that $f_1(0, 0)\mathbf{a}_1 = 0$ and $f_2(0, 0)\mathbf{a}_1{}^2 + f_1(0, 0)\mathbf{a}_2 = 0$.

(This example is known as the *Whitney cusp* or *pleat* (1955).)

4.12 Prove Proposition 4.10.

4.13 Let $\mathbf{r}$ be a non-singular point of the curve given by the equation $F(\mathbf{r}) = 0$, where $F : \mathbb{R}^2 \rightarrowtail \mathbb{R}$ is smooth, and let α be a tangent vector and $\mathbf{n}$ a normal vector to the curve at $\mathbf{r}$. Prove that if $\mathbf{r}$ is an inflection of the curve then $F_2(\mathbf{r})\alpha^2 = 0$, while if $\mathbf{r}$ is a vertex of the curve then

$$(F_1(\mathbf{r})\mathbf{n})(F_3(\mathbf{r})\alpha^3) - 3(F_2(\mathbf{r})\alpha\mathbf{n})(F_2(\mathbf{r})\alpha^2) = 0.$$

5
Curves on the unit sphere

5.0 Introduction

Before moving on to discuss arbitrary smooth curves in $\mathbb{R}^3$ we spend a moment considering the special case of curves on the unit sphere, that is $S^2 = \{(x, y, z) \in \mathbb{R}^3 : x^2 + y^2 + z^2 = 1\}$. Now each plane $\{\mathbf{r} \in \mathbb{R}^3 : \mathbf{c} \cdot \mathbf{r} = d\}$, where $\mathbf{c}$ is a unit vector and $-1 < d < 1$, cuts out a *circle* on S^2, and any circle on S^2 is so defined, the circle being said to be a *great circle* or *geodesic* of the sphere if $d = 0$. Any circle on S^2 that is not a great circle has an equation of the form $\mathbf{b} \cdot \mathbf{r} = 1$, where $\mathbf{b} \cdot \mathbf{b} > 1$. The point $\mathbf{b}$ is called the *geodesic centre* of the circle and the distance ρ of $\mathbf{b}$ from any point $\mathbf{r}$ of the circle is the *geodesic radius of curvature* of the circle. The *tangential* component of the curvature vector of the circle is its *geodesic curvature* κ, this being zero if the circle is great and otherwise is the reciprocal of the geodesic radius of curvature ρ (Figure 5.1).

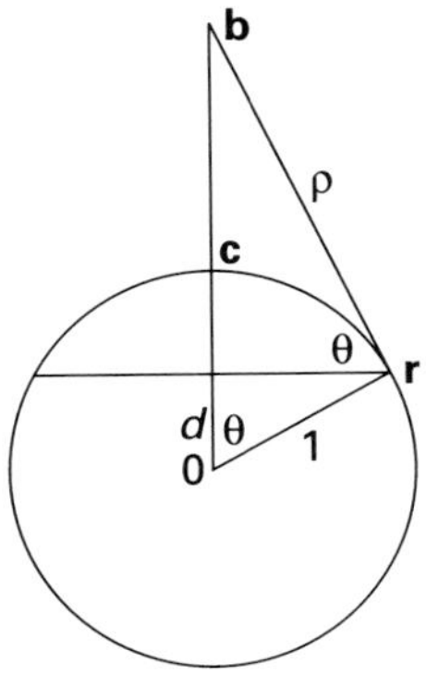

Figure 5.1

An alternative measure of the curvature of the circle is its *angular curvature* θ, half the angle subtended at the centre of the unit sphere by any diameter of the circle.

Proposition 5.1 Let θ be the angular curvature of a circle on the unit sphere. Then the geodesic curvature of the circle is equal to $\cot\theta$. *In particular the circle is great where $\theta = \frac{1}{2}\pi$ and $\cot\theta = 0$.* □

In discussing the curvature of a regular smooth curve lying on the unit sphere we are concerned at each point with the circle that most closely approximates the curve there. The details are closely analogous to the concepts defined and results obtained for regular smooth plane curves.

5.1 Geodesic curvature

Let $\mathbf{m}$ be a regular curve on the unit sphere S^2, that is a regular map $\mathbf{m}: \mathbb{R} \rightarrowtail \mathbb{R}^3$ with image a subset of S^2. Then $\mathbf{m}_1(t) \neq 0$, for all t, and $\mathbf{m}\cdot\mathbf{m} = 1$. Differentiating this last equation twice we get $\mathbf{m}\cdot\mathbf{m}_1 = 0$ and $\mathbf{m}\cdot\mathbf{m}_2 + \mathbf{m}_1\cdot\mathbf{m}_1 = 0$, from which it follows that at each $t \in \mathbb{R}$ the vector $\mathbf{m}_2(t)$ is not 0, nor is it a multiple of $\mathbf{m}_1(t)$. So the curve $\mathbf{m}$ is *not linear* anywhere. The *circle of curvature* of $\mathbf{m}$ at t is that passing through $\mathbf{m}(t)$ and cut out by the plane with equation $\mathbf{c}\cdot\mathbf{r} = \mathbf{c}\cdot\mathbf{m}(t)$, where $\mathbf{c}$ is a unit vector orthogonal both to $\mathbf{m}_1(t)$ and to $\mathbf{m}_2(t)$. Clearly this circle is a great circle if and only if $\mathbf{m}(t)$ depends linearly on $\mathbf{m}_1(t)$ and $\mathbf{m}_2(t)$. We say the curve $\mathbf{m}$ is *great* at t if its circle of curvature is a great circle on the sphere. It is *planar* at t if and only if $\mathbf{m}_3(t)$ depends linearly on $\mathbf{m}_1(t)$ and $\mathbf{m}_2(t)$, being A_k-planar at t, for $k \geqslant 3$, if and only if $\mathbf{m}_i(t)$ depends linearly on $\mathbf{m}_1(t)$ and $\mathbf{m}_2(t)$ for $1 \leqslant i \leqslant k$, but $\mathbf{m}_{k+1}(t)$ is linearly independent of them.

From our introductory remarks the appropriate notion of *curvature* for a regular curve $\mathbf{m}$ on the unit sphere is its *geodesic curvature*, measuring the extent to which the curve departs from being a great circle, this being the geodesic curvature $\kappa(t)$ of its circle of curvature at any point t. The *geodesic centre of curvature* of $\mathbf{m}$ at any point t where the geodesic curvature is non-zero is the geodesic centre $\mathbf{b}(t)$ of the circle of curvature (Figure 5.2).

The *focal centre* of $\mathbf{m}$ at any point t is either of the points $\mathbf{e}(t) = \pm\mathbf{b}(t)/|\mathbf{b}(t)|$ of the sphere. Normally one would choose the positive sign. However the focal centre exists also, but this time with essential ambiguity of sign, where the geodesic curvature is

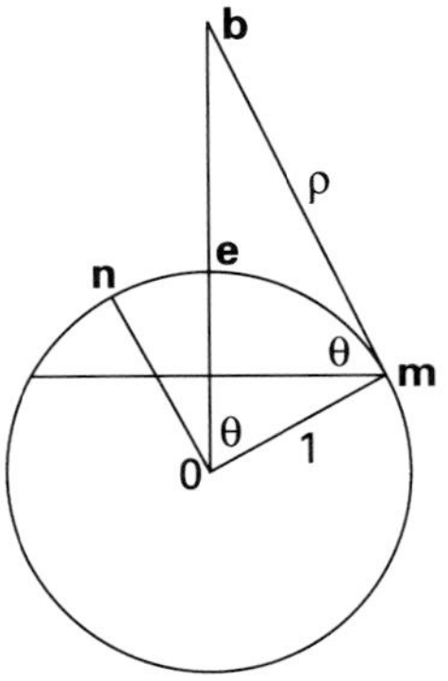

Figure 5.2

zero and the circle of curvature is a great circle of the sphere. One then defines the *unit normal* $\mathbf{n}(t)$ of $\mathbf{m}$ at t by the equation $\mathbf{e}(t) = \mathbf{m}(t)\cos\theta(t) + \mathbf{n}(t)\sin\theta(t)$, where not only $\mathbf{m}(t)\cdot\mathbf{n}(t) = 0$ but also $\mathbf{m}_1(t)\cdot\mathbf{n}(t) = 0$.

The following proposition is easily proved.

Proposition 5.2 Let $\mathbf{m}$ *be a regular curve on* S^2, *not great at* t. *Then*

(i) *the vectors* $\mathbf{m}(t)$, $\mathbf{m}_1(t)$ *and* $\mathbf{m}_2(t)$ *are linearly independent*;

(ii) *the focal centre* $\mathbf{e}(t)$ *of* $\mathbf{m}$ *at* t *is determined up to sign by the equations* $\mathbf{e}(t)\cdot\mathbf{e}(t) = 1$, $\mathbf{e}(t)\cdot\mathbf{m}_1(t) = 0$ *and* $\mathbf{e}(t)\cdot\mathbf{m}_2(t) = 0$;

(iii) *the vectors* $\mathbf{e}(t)$, $\mathbf{m}_1(t)$ *and* $\mathbf{m}_2(t)$ *are linearly independent*;

(iv) *the vectors* $\mathbf{m}(t)$, $\mathbf{m}_1(t)$ *and* $\mathbf{n}(t)$ *are linearly independent*;

(v) $\mathbf{m}_1(t)\cos\theta(t) + \mathbf{n}_1(t)\sin\theta(t) = 0$, *that is* $\kappa(t)\mathbf{m}_1(t) + \mathbf{n}_1(t) = 0$, *where* $\kappa(t) = \cot\theta$ *is the geodesic curvature of* $\mathbf{m}$ *at* t, *this equation being the analogue for the spherical curve* $\mathbf{m}$ *of the equation* $\kappa(t)\mathbf{r}_1(t) + n_1(t) = 0$ *for a regular plane curve* $\mathbf{r}$;

(vi) $\mathbf{e}_1(t) = (-\mathbf{m}(t)\sin\theta(t) + \mathbf{n}(t)\sin\theta(t))\theta_1(t)$, *this being the analogue for* $\mathbf{m}$ *of the equation* $\mathbf{e}_1(t) = \rho_1(t)\mathbf{n}(t)$ *for a regular plane curve* $\mathbf{r}$. □

The curve $\mathbf{e}$ of focal centres will be called the *evolute* of m.

By analogy with the case of plane curves a point $\mathbf{c}$ of S^2 will be said to be an A_k *centre* of the curve $\mathbf{m}$ at t if, for all i such that $1 \leqslant i \leqslant k$, $\mathbf{c}\cdot\mathbf{m}_i(t) = 0$, but $\mathbf{c}\cdot\mathbf{m}_{k+1}(t) \neq 0$. The focal centre $\mathbf{e}(t)$ is then at least an A_2 centre of $\mathbf{m}$ at t. Again by analogy with the case of plane curves the curve $\mathbf{m}$ will be said to have an *ordinary vertex* at t if its focal centre $\mathbf{e}(t)$ is A_3, that is if also $\mathbf{e}(t)\cdot\mathbf{m}_3(t) = 0$, but $\mathbf{e}(t)\cdot\mathbf{m}_4(t) \neq 0$.

Proposition 5.3 Let $\mathbf{m}$ *be a regular curve on* S^2, *not great at* t, *with evolute* $\mathbf{e}$. *Then*

(i) $\mathbf{e}$ *is not great at* t;
(ii) $\mathbf{m}$ *has an ordinary vertex at* t *if and only if* $\mathbf{e}_1(t) = 0$ *but* $\mathbf{e}_2(t) \neq 0$, *if and only if the geodesic curvature* κ *has an ordinary critical point, if and only if the angular curvature* θ *has an ordinary critical point*;
(iii) *at an ordinary vertex* t *of* $\mathbf{m}$ *the vectors* $\mathbf{e}(t)$, $\mathbf{e}_1(t)$ *and* $\mathbf{e}_2(t)$ *are linearly independent.* □

The theories of parallels and of involutes all have their analogues for spherical curves also, part (vi) of Proposition 5.2 playing a vital role in the latter theory.

5.2 Spherical kinematics

The theory that follows has application to plate tectonics, the modern theory of the structure of the surface of the Earth.

Suppose that the unit sphere (or some part of it – a *spherical lamina*) moves rigidly over S^2. This is the same as $\mathbb{R}^3$ moving rigidly over $\mathbb{R}^3$, with the origin remaining fixed. Any subsequent position $\mathbf{r}(t)$ of a vector $\mathbf{r}(0)$ is then obtainable from $\mathbf{r}(0)$ by the action of some element $g(t)$ of the rotation group $SO(3)$ of the Euclidean space $\mathbb{R}^3$. That is $\mathbf{r}(t) = g(t)\mathbf{r}(0)$, where $g(0) = 1$, and of course $g(t)^\tau g(t) = 1$. Differentiating these equations at 0 we find that $\mathbf{r}_1(0) = g_1(0)\mathbf{r}(0)$, while $g_1(0)^\tau + g_1(0) = 0$. Now a 3×3 skew-symmetric matrix has determinant zero and so has a non-zero kernel vector. So there exists $\mathbf{k} \in \mathbb{R}^3$ such that $g_1(0)\mathbf{k} = 0$, $\mathbf{k}$ being uniquely determined provided that $g_1(0) \neq 0$. What this means is that at 0 there is an *instantaneous*, or *polar*, *axis* through the origin, at rest at that moment, and intersecting the sphere in the *poles* of the motion at $t = 0$.

Differentiating the equation $g(t)^\tau g(t) = 1$ several more times at the origin we find that

$$g_2(0) + 2g_1(0)^\tau g_1(0) + g_2(0)^\tau = 0,$$

$$g_3(0) + 3g_2(0)^\tau g_1(0) + 3g_1(0)^\tau g_2(0) + g_3(0)^\tau = 0,$$

and so on. Thus for each $k > 1$ the symmetric part of $g_k(0)$ is determined by the $g_i(0)$ for $1 \leq i < k$.

Now consider the spherical curve $t \mapsto \mathbf{r}(t)$. This will be great at 0 if and only if $\det[\mathbf{r}(0), \mathbf{r}_1(0), \mathbf{r}_2(0)] = \det[\mathbf{r}(0), g_1(0)\mathbf{r}(0), g_2(0)\mathbf{r}(0)] = 0$. This is clearly the intersection of the sphere with a cubic cone with vertex at the origin. Since $g_1(0)\mathbf{k} = 0$ the determinant is zero for $\mathbf{r}(0) = \mathbf{k}$. So this curve passes through the poles of the instantaneous axis.

To get more information about this curve suppose that $\mathbf{k} = (0, 0, 1)$. Then $g_1(0)$ has matrix of the form

$$\begin{bmatrix} 0 & -c & 0 \\ c & 0 & 0 \\ 0 & 0 & 0 \end{bmatrix}, \text{ for some } c \in \mathbb{R}.$$

Reparametrising the motion, and ignoring the possibility that $c = 0$, we may without loss of generality assume that $c = 1$. Now

$$\begin{bmatrix} 0 & 1 & 0 \\ -1 & 0 & 0 \\ 0 & 0 & 0 \end{bmatrix}\begin{bmatrix} 0 & -1 & 0 \\ 1 & 0 & 0 \\ 0 & 0 & 0 \end{bmatrix} = \begin{bmatrix} 1 & 0 & 0 \\ 0 & 1 & 0 \\ 0 & 0 & 0 \end{bmatrix},$$

from which it follows from the equation

$$g_2(0) + 2g_1(0)^\tau g_1(0) + g_2(0)^\tau = 0$$

that the matrix of $g_2(0)$ is of the form

$$\begin{bmatrix} -1 & \gamma & -\beta \\ -\gamma & -1 & \alpha \\ \beta & -\alpha & 0 \end{bmatrix}, \text{ for some } \alpha, \beta, \gamma \in \mathbb{R}.$$

Accordingly the equation of the inflection cubic is

$$\det\begin{bmatrix} x & -y & -x + \gamma y - \beta z \\ y & x & -\gamma x - y + \alpha z \\ z & 0 & \beta x - \alpha y \end{bmatrix} = \det\begin{bmatrix} x & -y & -\beta z \\ y & x & \alpha z \\ z & 0 & \beta x - \alpha y + z \end{bmatrix}$$

$$= 0,$$

that is $(\beta x - \alpha y)(x^2 + y^2 + z^2) + z(x^2 + y^2) = 0$. The tangent plane to the cone at the pole then has equation $\beta x - \alpha y = 0$. Generically α and β are not both zero. Without loss of generality we may suppose that $\alpha = 0$ and that $\beta > 0$. This cubic cone may then most conveniently be studied as a cubic curve in $\mathbb{R}P^3$ or, on taking the line $z = 0$ as line at infinity, as the cubic curve in $\mathbb{R}^3$ with equation

$$y^2 = -\frac{x(x^2 + \beta' x + 1)}{x + \beta'}, \text{ where } \beta' = \beta^{-1}.$$

For $\beta' = 2$ this curve is a crunodal cubic with node at $(x, y) = (-1, 0)$, that is at $[x, y, z] = [-1, 0, 1]$. Otherwise it has no singularities. For $0 < \beta' < 2$ it has a single component, while for $\beta' > 2$ it has two.

The curve $\mathbf{r}$ will have a point of higher circularity at 0 provided that $\det[\mathbf{r}_1(0), \mathbf{r}_2(0), \mathbf{r}_3(0)] = \det[g_1(0)\mathbf{r}(0), g_2(0)\mathbf{r}(0), g_3(0)\mathbf{r}(0)] = 0$. This set also is the intersection of the sphere with a cubic cone, vertex at 0. The derivative of the map $\mathbf{r} \mapsto \det[g_1(0)\mathbf{r}, g_2(0)\mathbf{r}, g_3(0)\mathbf{r}]$ is the linear map

$$\begin{aligned}\mathbf{s} \mapsto {} & \det[g_1(0)\mathbf{s}, g_2(0)\mathbf{r}, g_3(0)\mathbf{r}] \\ & + \det[g_1(0)\mathbf{r}, g_2(0)\mathbf{s}, g_3(0)\mathbf{r}] + \det[g_1(0)\mathbf{r}, g_2(0)\mathbf{r}, g_3(0)\mathbf{s}],\end{aligned}$$

which is the zero linear map for $\mathbf{r} = \mathbf{k}$ again since $g_1(0)\mathbf{k} = 0$ and since

$$\mathbf{k}^\tau[g_1(0)\mathbf{s}, g_2(0)\mathbf{k}, g_3(0)\mathbf{k}] = [0, \mathbf{k}^\tau g_2(0)\mathbf{k}, \mathbf{k}^\tau g_2(0)\mathbf{k}] = 0.$$

Accordingly this cubic curve on the sphere also passes through the poles of the instantaneous axis, at which points it is singular.

The points of intersection of the two cubic curves away from the poles are termed the *Ball points* of the motion, these being analogues of the Ball point of the motion of a plane lamina. These are points of the motion that are *greater* than any others! Obviously they come in antipodal pairs. These are points $\mathbf{r}(0)$ where the rank of the matrix $[\mathbf{r}(0), \mathbf{r}_1(0), \mathbf{r}_2(0), \mathbf{r}_3(0)]$ is less than 3. In that case not only $\det[\mathbf{r}_1(0), \mathbf{r}_2(0), \mathbf{r}_3(0)] = 0$ but also $\det[\mathbf{r}(0), \mathbf{r}_1(0), \mathbf{r}_3(0)] = 0$. Now from the equation

$$g_3(0) + 3g_2(0)^\tau g_1(0) + 3g_1(0)^\tau g_2(0) + g_3(0)^\tau = 0$$

it follows from the choices already made for $g_1(0)$ and $g_2(0)$ that

$$g_3(0) = \begin{bmatrix} 3\gamma & \zeta & -\eta \\ -\zeta & 3\gamma & -\frac{3}{2}\beta + \xi \\ \eta & -\frac{3}{2}\beta - \xi & 0 \end{bmatrix}, \text{ where } \xi, \eta, \zeta \in \mathbb{R}.$$

The two cubic equations then become

$$\beta x(x^2 + y^2 + z^2) + z(x^2 + y^2) = 0$$

and

$$(-\tfrac{3}{2}\beta + \xi)y(x^2 + y^2 + z^2) + (3\beta y + (3\gamma + \eta/\beta)z)(x^2 + y^2) = 0.$$

Away from the poles, where $x^2 + y^2 = 0$, we have

$$3\beta^2 xy + (3\beta\gamma + \eta)xz + (\tfrac{3}{2}\beta - \xi)yz = 0,$$

a quadric cone that passes through the polar axis, with tangent plane there

$$(3\beta\gamma + \eta)x + (\tfrac{3}{2}\beta + \xi)y = 0.$$

This quadric cone and the inflectional cubic cone intersect in general in six common axes, one of which is the polar axis, generated by the vector $(0, 0, 1)$, a simple root unless $\frac{3}{2}\beta + \xi = 0$. Another is the axis generated by the vector $(0, 1, 0)$, not relevant since it is a solution of the first cubic equation but not the second. The remaining four axes are the Ball axes of the motion. All may be real.

This proves:

Theorem 5.4 For a generic rigid motion of S^2 over S^2 there may be as many as four Ball axes. □

For further details of spherical kinematics, with applications to mechanism theory, see Hunt (1978), Section 14.4, *et seq*.

Exercises

5.1 Prove Proposition 5.2.

5.2 Prove that at any point the geodesic curvature of a regular curve on the unit sphere is the component tangential to the sphere of the principal curvature vector there.

5.3 Discuss the geometry of smooth curves on a sphere of radius R and investigate what happens as $R \to \infty$.

5.4 Discuss rigid motions of a sphere of radius R over a sphere of radius R. Investigate what happens to the cubic cone of inflectional points as $R \to \infty$.

5.5 Prove that, under either an inversion or stereographic projection that maps a sphere to a plane, any vertex of a curve that lies on the sphere maps to a vertex of the image curve that lies in the plane.

6
Space curves

6.0 Introduction

Our approach to space curves is modelled on our treatment of plane curves, the central features being the analogues of the evolute of a plane curve and its cusps. We lay much less emphasis than is usual on the Serret–Frenet equations of a space curve, though these naturally have their part to play in appropriate places.

6.1 Space curves

As with curves in the plane so here with curves in space we begin by concentrating almost entirely on curves presented parametrically.

A *smooth parametric curve* in $\mathbb{R}^3$ is a smooth map

$$\mathbf{r} : \mathbb{R} \rightarrowtail \mathbb{R}^3;\ t \mapsto \mathbf{r}(t).$$

It is *regular* (or *immersive*) at t if its first derivative $\mathbf{r}_1(t)$ is non-zero. The vector $\mathbf{r}_1(t)$, which may be regarded as the *velocity* of the curve $\mathbf{r}$ at time t, generates, at a regular point t, the *tangent vector line* to $\mathbf{r}$ at t. The *tangent line* to $\mathbf{r}$ at t (or at $\mathbf{r}(t)$) is then the line

$$u \mapsto \mathbf{r}(t) + u\mathbf{r}_1(t).$$

Likewise a *smooth parametric surface* in $\mathbb{R}^3$ is a smooth map

$$\mathbf{s} : \mathbb{R}^2 \rightarrowtail \mathbb{R}^3;\ w = (u, v) \mapsto \mathbf{s}(w).$$

It is *regular* (or *immersive*) at w if its first derivative $\mathbf{s}_1(w)$ is injective. Here $\mathbf{s}_1(w)$ is the linear map $\mathbb{R}^2 \to \mathbb{R}^3$ which (up to an additive constant) best approximates $\mathbf{s}$ at w, its matrix being the 3×2 Jacobian matrix of the partial derivatives of the components of $\mathbf{s}$ at w. This is in direct analogy with the definition of the regularity of a space curve above if its derivative $\mathbf{r}_1(t)$ at t is thought of, as it may be, not as a vector but rather as the linear map $\mathbb{R} \to \mathbb{R}^3$ which (up to an additive constant) best approximates $\mathbf{r}$ at t, its matrix being the 3×1 column vector of the derivatives of the components of $\mathbf{r}$ at t. Where $\mathbf{s}_1(w)$ is injective, its image is a two-dimensional linear subspace of $\mathbb{R}^3$, the *tangent vector plane* to $\mathbf{s}$ at w. Its translate by $\mathbf{s}(w)$ is then the *tangent plane* to $\mathbf{s}$ at w.

Given a smooth space curve $\mathbf{r}$ its *tangent bundle* or *tangent developable* is the surface $\mathbf{s}$ formed from its tangent lines, given parametrically by

$$(t, u) \mapsto \mathbf{s}(t, u) = \mathbf{r}(t) + u\mathbf{r}_1(t).$$

This being so

$$\mathbf{s}_1(t, u) = [\mathbf{r}_1(t) + u\mathbf{r}_2(t) \quad \mathbf{r}_1(t)].$$

This is a 3×2 matrix whose first column is the vector $\mathbf{r}_1(t) + u\mathbf{r}_2(t)$ and whose second is the vector $\mathbf{r}_1(t)$. These are linearly independent if and only if $u\mathbf{r}_2(t)$ is linearly independent of $\mathbf{r}_1(t)$, which is so unless either $u = 0$, as at the point of contact of the tangent line with the curve, or $\mathbf{r}_2(t)$ is a multiple of $\mathbf{r}_1(t)$, that is $\mathbf{r}$ is *linear* at t, in which case the tangent developable fails to be regular all along the tangent line to $\mathbf{r}$ at t. In the generic case, where $\mathbf{r}_2$ is nowhere a multiple of $\mathbf{r}_1$ the developable has an *ordinary cuspidal edge*.

It is easy with a couple of sheets of reasonably stiff paper, a pair of scissors and a mini-stapler, or, last century, muslin clamps(!) (Thomson and Tait, 1879), to make a model of a space curve and its tangent developable as in Figure 6.1. Some people have a little bit of trouble learning to twist the model correctly. Properly done it should be able to stand on its own on the table, for all to admire!

Note how each tangent transfers from one sheet of the model to the other at its point of tangency with the curve. We shall be making explicit reference to this model from time to time throughout this chapter.

There is a sense in which a *generic* space curve has no linear points. Roughly speaking such a point, if it is isolated on the curve, may be got rid of by imparting a slight twist to the curve at the point. This is in contrast to the case of a plane curve with an ordinary inflection, for in that case any slight perturbation of the curve will just move the inflection slightly. It will not remove it. For the time being we shall generally assume that the space curves under study are nowhere linear.

At a non-linear point of the space curve $\mathbf{r}$ the vectors $\mathbf{r}_1(t)$ and $\mathbf{r}_2(t)$

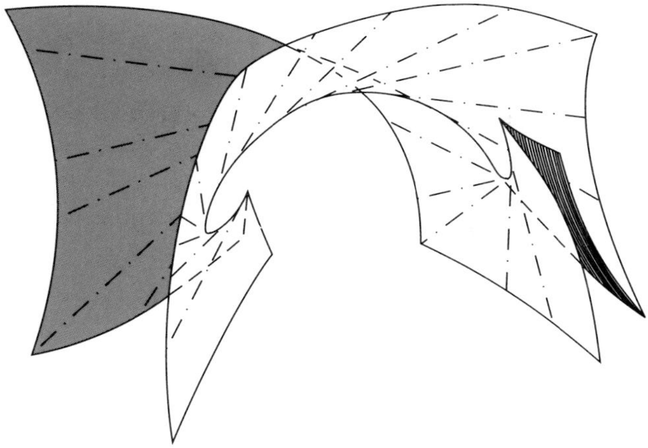

Figure 6.1

span a plane, the *osculating plane* to $\mathbf{r}$ at t, the plane that hugs the curve most closely at the point of contact. This plane coincides with the tangent plane to the tangent developable of $\mathbf{r}$ at each point of the tangent line to $\mathbf{r}$ at t, each being spanned by the vectors $\mathbf{r}_1(t)$, $\mathbf{r}_2(t)$.

We say that a (nowhere linear) smooth curve $\mathbf{r}$ is *planar* at t if $\mathbf{r}_3(t)$ depends linearly on the linearly independent vectors $\mathbf{r}_1(t)$ and $\mathbf{r}_2(t)$, an *ordinary* planar point being one where the fourth derivative of $\mathbf{r}$ is linearly independent of the first and second derivatives. Generically a space curve is non-planar except at isolated points.

At an ordinary planar point the tangent developable has a *cuspidal pinch point*, as was proved by John Cleave only a few years ago (1980). The pinch point is at the end of a double line of the developable, a curve along which one leaf of the developable intersects the other transversally. Figure 6.2 from Cleave's paper illustrates the tangent developable of a curve on a circular cylinder, planar where it is simply tangent to one of the generators of the cylinder (cf. Exercise 6.18). We shall have more to say about cuspidal pinch points in Chapter 14.

Non-regular points of smooth space curves also have to be considered. The smooth space curve $\mathbf{r}$ will be said to have an *ordinary cusp* at t if $\mathbf{r}_1(t) = 0$ but $\mathbf{r}_2(t) \neq 0$, with $\mathbf{r}_3(t)$ linearly independent of $\mathbf{r}_2(t)$, and to have an *ordinary non-planar cusp* at t if also $\mathbf{r}_4(t)$ is linearly independent of $\mathbf{r}_2(t)$ and $\mathbf{r}_3(t)$. If $\mathbf{r}_1(t) = 0$, $\mathbf{r}_2(t) \neq 0$ and $\mathbf{r}_3(t)$ is a multiple of $\mathbf{r}_2(t)$ then the cusp is *rhamphoid*.

The tangent developable of a smooth space curve in the neighbourhood of an ordinary non-planar cusp forms a *swallow-tail*, with a curve

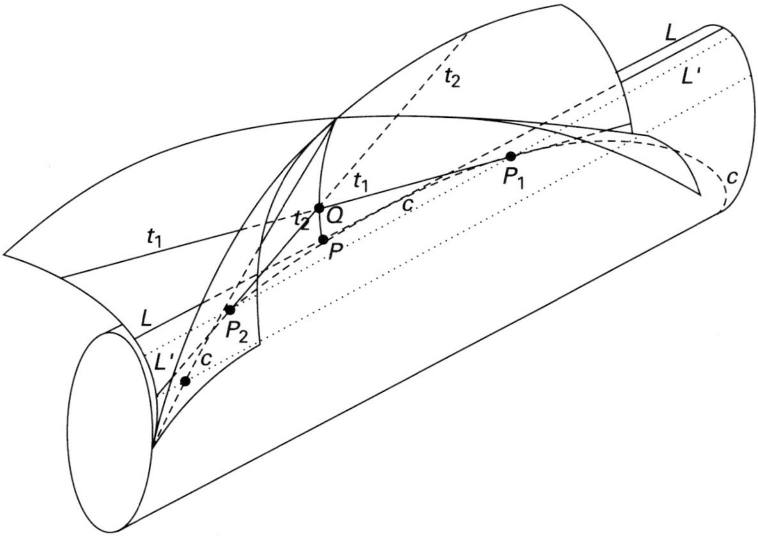

Figure 6.2

of points of self intersection terminating at the cusp. The standard swallow-tail is the discriminant of the vector space of quartic polynomials of the form $x^4 + ax^2 + bx + c$ consisting of all those points $(a, b, c) \in \mathbb{R}^3$ for which the polynomial has a repeated real root. This discriminant is the tangent developable of the curve consisting of all such polynomials having a triple real root for, since the sum of the roots is zero, such are of the form

$$(x - t)^3(x + 3t) = x^4 - 6t^2x^2 + 8t^3x - 3t^4.$$

The space curve that represents these is $t \mapsto (-6t^2, 8t^3, -3t^4)$ and its tangent developable represents the polynomials

$$x^4 + (-6t^2 - 12tu)x^2 + (8t^3 + 24t^2u)x + (-3t^4 - 12t^3u)$$
$$= (x - t)^3(x + 3t) - 12tu(x - t)^2 = (x - t)^2(x^2 + 2tx - 3t^2 - 12tu).$$

The double line of the swallow-tail represents the quartic forms

$$(x - t)^2(x + t)^2 = x^4 - 2t^2x^2 + t^4,$$

the point $(-2t^2, 0, t^4)$ being the point of intersection of the tangents to the space curve at the points of parameter t and $-t$, where $3u = -t$.

Some kind of paper model can be made of such a swallow-tail, three sheets of paper now being necessary, and account having to be taken of the fact that the surface unavoidably cuts itself near the cusp of the edge. See Exercise 6.2 and Figure 14.4.

We shall have rather more to say about cuspidal edges, and swallow-tails and cuspidal pinch points in particular, in later chapters, in particular in Chapters 13 and 14. It will not then normally be the case that the surface with the edge is developable.

It is less likely for a space curve than for a plane one that it intersects itself, two or more distinct values of the parameter having the same image point in $\mathbb{R}^3$. When this does occur the common image point is said to be a *singularity*, but not a point of non-regularity, of the curve.

When one looks at a space curve from a distance one sees a plane curve, which typically may have points of non-regularity at points where the original curve is regular. We formalise this as follows.

Proposition 6.1 Let $\mathbf{r}$ *be a regular space curve, let* π *be the orthogonal projection of the ambient space* $\mathbb{R}^3$ *to a plane in* $\mathbb{R}^3$ *and let* t *be a point where* $\mathbf{r}$ *is not planar. Then the plane curve* $\pi\mathbf{r}$ *is regular and non-linear at* t *if and only if the kernel of* π*, the direction of projection, fails to lie in the osculating plane to* $\mathbf{r}$ *at* t*, has an ordinary inflection at* t *if and only if the kernel of* π *lies in the osculating plane to* $\mathbf{r}$ *at* t *but is not tangent at* t*, and has an ordinary cusp at* t *if and only if the kernel of* π *is tangent to* $\mathbf{r}$ *at* t.

Proof We prove things in the reverse order.

First we remark that, since π is linear, $(\pi\mathbf{r})_i = \pi\mathbf{r}_i$ for all i. Moreover, π is surjective, of rank 2 and kernel rank 1. Now the kernel is tangent to $\mathbf{r}$ at t exactly when $(\pi\mathbf{r})_1 = 0$, when necessarily $(\pi\mathbf{r})_2$ and $(\pi\mathbf{r})_3$ are linearly independent, that is when $\pi\mathbf{r}$ has an ordinary cusp at t. The kernel of π lies in the osculating plane to $\mathbf{r}$ at t but is not tangent there exactly when some linear combination of $\mathbf{r}_1(t)$ and $\mathbf{r}_2(t)$, say $\lambda\mathbf{r}_1(t) + \mu\mathbf{r}_2(t)$, with $\mu \neq 0$, generates the kernel. But then $(\pi\mathbf{r})_1(t) \neq 0$ while $\lambda(\pi\mathbf{r})_1(t) + \mu(\pi\mathbf{r})_2(t) = 0$, so that $\pi\mathbf{r}$ has an ordinary inflection at t. The remaining case is when $\ker\pi$ does not lie in the osculating plane to $\mathbf{r}$ at t, in which case $(\pi\mathbf{r})_1$ and $(\pi\mathbf{r})_2$ are linearly independent; that is $\pi\mathbf{r}$ is regular and not linear at t. □

We assumed in the above proposition that the curve $\mathbf{r}$ was not planar at t.

Proposition 6.2 With $\mathbf{r}$ *and* π *as above, suppose that* $\mathbf{r}$ *has an ordinary planar point at* t *and suppose that the kernel of* π *lies in the osculating plane to* $\mathbf{r}$ *at* t *but is not tangent to* $\mathbf{r}$ *at* t. *Then the plane curve* $\pi\mathbf{r}$ *has an ordinary undulation at* t. *On the other hand if the kernel of* π *is tangent to* $\mathbf{r}$ *at* t *then the curve* $\pi\mathbf{r}$ *has a rhamphoid cusp at* t, *in general ordinary*. □

Both these propositions are concerned with distant views of curves. Close-up views are considered in the final section of this chapter.

6.2 The focal surface and space evolute

Our first approach to the curvature of a regular smooth space curve $\mathbf{r}$ is to study how closely it approximates at each point to one lying on a *sphere*. Now the sphere with centre $\mathbf{c}$ and radius P (Capital rho!) consists of all points $\mathbf{r}$ in $\mathbb{R}^3$ such that $(\mathbf{r}-\mathbf{c})\cdot(\mathbf{r}-\mathbf{c}) = P^2$, where the dot $\cdot$ now indicates the standard Euclidean scalar product on $\mathbb{R}^3$. As with the analogous equation in the plane this is equivalent to

$$\mathbf{c}\cdot\mathbf{r} - \tfrac{1}{2}\mathbf{r}\cdot\mathbf{r} = \tfrac{1}{2}(\mathbf{c}\cdot\mathbf{c} - P^2),$$

the right-hand side of this equation being constant. So for a smooth parametric curve $\mathbf{r}$ with image lying on the sphere the function $V(\mathbf{c})$ defined by the equation $V(\mathbf{c})(t) = \mathbf{c}\cdot\mathbf{r}(t) - \tfrac{1}{2}\mathbf{r}(t)\cdot\mathbf{r}(t)$ is constant, so that all its derivatives with respect to t are everywhere zero, namely

$$V(\mathbf{c})_1 = (\mathbf{c}-\mathbf{r})\cdot\mathbf{r}_1 = 0,$$

$$V(\mathbf{c})_2 = (\mathbf{c}-\mathbf{r})\cdot\mathbf{r}_2 - \mathbf{r}_1\cdot\mathbf{r}_1 = 0,$$

$$V(\mathbf{c})_3 = (\mathbf{c}-\mathbf{r})\cdot\mathbf{r}_3 - 3\mathbf{r}_1\cdot\mathbf{r}_2 = 0,$$

$$V(\mathbf{c})_4 = (\mathbf{c}-\mathbf{r})\cdot\mathbf{r}_4 - 4\mathbf{r}_1\cdot\mathbf{r}_3 - 3\mathbf{r}_2\cdot\mathbf{r}_2 = 0,$$

and so on, just as before. If such a curve is regular at t then it is not linear at t, since $\mathbf{r}_1(t)\cdot\mathbf{r}_1(t) \neq 0$ if $\mathbf{r}_1(t) \neq 0$. The *circle of curvature* of $\mathbf{r}$ at t is the circle passing through $\mathbf{r}(t)$ and cut out on the sphere by the osculating plane to $\mathbf{r}$ at t. Clearly this circle is a *great* circle, that is it has centre at $\mathbf{c}$, the centre of the sphere, if and only if $\mathbf{r}(t)$ depends linearly on $\mathbf{r}_1(t)$ and $\mathbf{r}_2(t)$. We say the curve $\mathbf{r}$ is *great* at t if its circle of curvature is a great circle on the sphere.

The curve $\mathbf{r}$ may be planar at t. Indeed it can clearly be planar at every point, as it will be if its image is the circle of intersection of the sphere with some plane.

Now suppose that $\mathbf{r}$ is a regular space curve not necessarily lying on a sphere. Clearly $V(\mathbf{c})_1(t) = \mathbf{r}_1(t) \cdot (\mathbf{c} - \mathbf{r}(t)) = 0$ whenever $\mathbf{c}$ happens to lie on the *normal plane* to $\mathbf{r}$ at t, the plane through $\mathbf{r}(t)$ orthogonal to the tangent line there. Provided only that $\mathbf{r}$ is not linear at t there will then be a unique line of points $\mathbf{c}$ of this plane at which also $V(\mathbf{c})_2(t) = 0$, for this latter equation, namely $\mathbf{r}_2(t) \cdot (\mathbf{c} - \mathbf{r}(t)) = \mathbf{r}_1(t) \cdot \mathbf{r}_1(t)$, is that of a plane normal to the vector $\mathbf{r}_2(t)$ and therefore one that is not parallel to the normal plane. The *focal line* of $\mathbf{r}$ at t, the line of intersection of the two planes, is normal both to $\mathbf{r}_1(t)$ and to $\mathbf{r}_2(t)$, that is it is normal to the osculating plane to $\mathbf{r}$ at t. Since $\mathbf{r}$ is regular at t it does not pass through $\mathbf{r}(t)$. See Figure 6.3 below.

The next stage is more complicated than before. Provided only that $\mathbf{r}$ is not planar at t there will be a unique point $\mathbf{c} = \mathbf{e}(t)$ on the focal line at which $V(\mathbf{c})_3(t) = 0$, the *centre of spherical curvature* of $\mathbf{r}$ at t. For a nowhere planar curve on a sphere with centre $\mathbf{c}$ the spherical centre of curvature of the curve is everywhere at $\mathbf{c}$. In general, however, the point $\mathbf{e}(t)$ is not constant, but lies on a space curve, the *space evolute* of $\mathbf{r}$, $\mathbf{e} : t \mapsto \mathbf{e}(t)$. At a planar point there are two possibilities, as is usual when one has three linear equations in three unknowns and the matrix of the left-hand side is singular. In fact the three equations for $\mathbf{c}$, namely $V(\mathbf{c})_1 = 0$, $V(\mathbf{c})_2 = 0$ and $V(\mathbf{c})_3 = 0$ can be written together in the matrix form

$$\begin{bmatrix} \mathbf{r}_1(t)\cdot \\ \mathbf{r}_2(t)\cdot \\ \mathbf{r}_3(t)\cdot \end{bmatrix} \mathbf{c} = \begin{bmatrix} \mathbf{r}(t)\cdot\mathbf{r}_1(t) \\ \mathbf{r}(t)\cdot\mathbf{r}_2(t) + \mathbf{r}_1(t)\cdot\mathbf{r}_1(t) \\ \mathbf{r}(t)\cdot\mathbf{r}_3(t) + 3\mathbf{r}_1(t)\cdot\mathbf{r}_2(t) \end{bmatrix}.$$

Either there is no solution at all (when $\mathbf{e}(t)$ has to be thought of as lying at infinity) or less probably the whole focal line consists of points $\mathbf{c}$ at which $V(\mathbf{c})_3(t) = 0$. The latter possibility does not occur for a generic curve. In the case that it does occur and provided that $\mathbf{r}_4(t)$ is linearly independent of $\mathbf{r}_1(t)$ and $\mathbf{r}_2(t)$ there is then a unique point of the focal line where $V(\mathbf{c})_4(t) = 0$, the centre of spherical curvature in this case.

At a point t where the centre of spherical curvature is uniquely determined the distance P of $\mathbf{e}(t)$ from $\mathbf{r}(t)$ is called the *radius of spherical curvature* of $\mathbf{r}$ at t. Its reciprocal is the *spherical curvature* of $\mathbf{r}$ at t, this being taken to be zero at planar points.

By analogy with the circularity of plane curves a regular space curve $\mathbf{r}$ will be said to be *spherical* at t if it is non-planar at t and

$V_4(\mathbf{e})(t) = 0$, where as before we write $V_k(\mathbf{e})$ for $V(\mathbf{c})_k$, with $\mathbf{c}$ afterwards put equal to $\mathbf{e}$, and to have an *ordinary vertex* at t if it is spherical at t and $V_5(\mathbf{e})(t) \neq 0$.

The following proposition lists some elementary properties of the space evolute of a regular space curve.

Proposition 6.3 Let $\mathbf{r}$ *be a regular space curve, nowhere linear nor planar, with space evolute* $\mathbf{e}$. *Then*

(*a*) *at a point t where* $\mathbf{e}$ *is regular* $\mathbf{e}$ *is not only not linear but also is not planar;*
(*b*) *at a point t where* $\mathbf{e}$ *is regular the tangent line to* $\mathbf{e}$ *at t coincides with the focal line of* $\mathbf{r}$ *at t;*
(*c*) *at a point t where* $\mathbf{e}$ *is regular the osculating plane to* $\mathbf{e}$ *at t coincides with the normal plane to* $\mathbf{r}$ *at t;*
(*d*) *at a point t where* $\mathbf{e}_1 = 0$ *but* $\mathbf{e}_2 \neq 0$ *the evolute* $\mathbf{e}$ *has an ordinary non-planar cusp;*
(*e*) *the curve* $\mathbf{e}$ *is regular at t if and only if* $V_4(\mathbf{e})(t) \neq 0$, *and has an ordinary non-planar cusp at t if and only if* $V_4(\mathbf{e})(t) = 0$ *but* $V_5(\mathbf{e})(t) \neq 0$, *that is if and only if* $\mathbf{r}$ *has an ordinary vertex at t.*

Proof The defining equations of the space evolute $\mathbf{e}$ are

$$(\mathbf{e} - \mathbf{r}) \cdot \mathbf{r}_1 = 0, \ (\mathbf{e} - \mathbf{r}) \cdot \mathbf{r}_2 = \mathbf{r}_1 \cdot \mathbf{r}_1 \text{ and } (\mathbf{e} - \mathbf{r}) \cdot \mathbf{r}_3 = 3\mathbf{r}_1 \cdot \mathbf{r}_2,$$

and by differentiating the first two and then using the second and third one deduces at once that everywhere $\mathbf{e}_1 \cdot \mathbf{r}_1 = 0$ and $\mathbf{e}_1 \cdot \mathbf{r}_2 = 0$. Since the first of these implies that $\mathbf{e}_2 \cdot \mathbf{r}_1 + \mathbf{e}_1 \cdot \mathbf{r}_2 = 0$ we have also $\mathbf{e}_2 \cdot \mathbf{r}_1 = 0$ everywhere. Differentiating these equations again we get

$$\mathbf{e}_1 \cdot \mathbf{r}_3 + \mathbf{e}_2 \cdot \mathbf{r}_2 = 0 \text{ and } \mathbf{e}_2 \cdot \mathbf{r}_2 + \mathbf{e}_3 \cdot \mathbf{r}_1 = 0.$$

Suppose now that, at t, $\mathbf{e}_1 \neq 0$. Since $\mathbf{r}$ is not planar at t it follows that, at t, $\mathbf{e}_1 \cdot \mathbf{r}_3 \neq 0$ and so also $\mathbf{e}_2 \cdot \mathbf{r}_2 \neq 0$, implying at once that $\mathbf{e}_2$ is not a multiple of $\mathbf{e}_1$ at t. That is $\mathbf{e}$ is not linear at t. But then also $\mathbf{e}_3 \cdot \mathbf{r}_1 \neq 0$, implying at once that $\mathbf{e}_3$ is linearly independent of $\mathbf{e}_1$ and $\mathbf{e}_2$ at t. That is $\mathbf{e}$ is not planar at t. This is assertion (a).

Since $\mathbf{e}_1 \cdot \mathbf{r}_1 = 0$ and $\mathbf{e}_1 \cdot \mathbf{r}_2 = 0$ not only $\mathbf{c} = \mathbf{e}(t)$ but every point $\mathbf{c} = \mathbf{e}(t) + u\mathbf{e}_1(t)$ of the tangent line to $\mathbf{e}$ at t satisfies the equations $(\mathbf{c} - \mathbf{r}) \cdot \mathbf{r}_1 = 0$ and $(\mathbf{c} - \mathbf{r}) \cdot \mathbf{r}_2 = \mathbf{r}_1 \cdot \mathbf{r}_1$, and so coincides with the focal line of $\mathbf{r}$ at t. This is assertion (b).

Since $\mathbf{e}_1 \cdot \mathbf{r}_1 = 0$ and $\mathbf{e}_2 \cdot \mathbf{r}_1 = 0$ not only $\mathbf{c} = \mathbf{e}(t)$ but every point $\mathbf{c} = \mathbf{e}(t) + u\mathbf{e}_1(t) + v\mathbf{e}_2(t)$ of the osculating plane to $\mathbf{e}$ at t satisfies the equation $(\mathbf{c}-\mathbf{r})\cdot\mathbf{r}_1=0$, and so this plane coincides with the normal plane to $\mathbf{r}$ at t. This is assertion (c).

Differentiating the equations $\mathbf{e}_1\cdot\mathbf{r}_2=0$ and $\mathbf{e}_2\cdot\mathbf{r}_1=0$ twice and then setting $\mathbf{e}_1=0$ we have that in that case $\mathbf{e}_2\cdot\mathbf{r}_2=0$, $\mathbf{e}_3\cdot\mathbf{r}_1=0$, $2\mathbf{e}_2\cdot\mathbf{r}_3+\mathbf{e}_3\cdot\mathbf{r}_2=0$ and $\mathbf{e}_2\cdot\mathbf{r}_3+2\mathbf{e}_3\cdot\mathbf{r}_2+\mathbf{e}_4\cdot\mathbf{r}_1=0$. Suppose now that, at t, $\mathbf{e}_2\neq 0$. Since $\mathbf{r}$ is not planar at t it follows that, at t, $\mathbf{e}_2\cdot\mathbf{r}_3\neq 0$ and so also $\mathbf{e}_3\cdot\mathbf{r}_2\neq 0$, implying that $\mathbf{e}_3$ is not a multiple of $\mathbf{e}_2$ at t. But then also $\mathbf{e}_4\cdot\mathbf{r}_1\neq 0$, for otherwise both $\mathbf{e}_2\cdot\mathbf{r}_3$ and $\mathbf{e}_3\cdot\mathbf{r}_2$ would be equal to zero, implying that $\mathbf{e}_4$ is linearly independent of $\mathbf{e}_2$ and $\mathbf{e}_3$ at t. This is assertion (d).

Finally, differentiating twice the equation $(\mathbf{e}-\mathbf{r})\cdot\mathbf{r}_3=3\mathbf{r}_1\cdot\mathbf{r}_2$ we find that $\mathbf{e}_1\cdot\mathbf{r}_3=0$ if and only if $V(\mathbf{c})_4=0$, where $\mathbf{c}=\mathbf{e}(t)$, and $\mathbf{e}_2\cdot\mathbf{r}_3+2\mathbf{e}_1\cdot\mathbf{r}_4\neq 0$ if and only if $V(\mathbf{c})_5\neq 0$, where $\mathbf{c}=\mathbf{e}(t)$. From these and (d) assertion (e) follows at once. □

Proposition 6.4 Let $\mathbf{r}$ *be a nowhere linear or planar regular space curve* $\mathbf{r}$ *whose space evolute* $\mathbf{e}$ *is everywhere regular. Then the tangent developable* $\mathbf{s}$ *of* $\mathbf{e}$, *representable as a smooth parametric surface, regular except along the curve* $\mathbf{e}$, *is the union of the focal lines of* $\mathbf{r}$. *The tangent line to* $\mathbf{e}$ *at* t *coincides with the focal line of* $\mathbf{r}$ *at* t *and the osculating plane to* $\mathbf{e}$ *at* t *coincides with the tangent plane to* $\mathbf{s}$ *at any other point of the focal line.* □

It is natural to call the tangent developable $\mathbf{s}$ of $\mathbf{e}$ the *focal surface* of the space curve $\mathbf{r}$. Our earlier paper model (see Figure 6.1) gives an idea of what it looks like. In the sequel the family of focal lines of $\mathbf{r}$ will be regarded as forming the focal surface of $\mathbf{r}$, regardless of whether $\mathbf{r}$ is planar or whether $\mathbf{e}$ is regular. In the case that $\mathbf{e}$ has a non-planar ordinary cusp at t the focal line of $\mathbf{r}$ at t is the limiting tangent line to $\mathbf{e}$ at t, having the vector $\mathbf{e}_2(t)$ lying along it (Exercise 6.6).

In contrast to the case of plane curves the non-regular points on the space evolute of a regular space curve do not quite correspond to critical points of the spherical radius of curvature.

Proposition 6.5 Let $\mathbf{r}$ *be a regular nowhere linear or planar space curve. Then its spherical radius of curvature* P *is critical at a point* t *if and only if either*

(a) $\mathbf{e}_1(t)=0$; or
(b) $\mathbf{e}(t)-\mathbf{r}(t)$ *is normal to the focal line of* $\mathbf{r}$ *at* t.

Proof Consider $(\mathbf{e}-\mathbf{r})\cdot(\mathbf{e}-\mathbf{r}) = P^2$. Then P^2 is critical, and so also is the radius P, at a point t if and only if $(\mathbf{e}-\mathbf{r})\cdot(\mathbf{e}_1-\mathbf{r}_1) = 0$ there, that is if and only if $(\mathbf{e}-\mathbf{r})\cdot\mathbf{e}_1 = 0$, for we already have $(\mathbf{e}-\mathbf{r})\cdot\mathbf{r}_1 = 0$. So there are two possibilities. Either $\mathbf{e}_1 = 0$ or the vector $\mathbf{e}-\mathbf{r}$ is orthogonal to the focal line. □

One's first intuition in the second case probably is that there must there be a local *minimum* of the spherical radius of curvature, but this need not be so, as we shall shortly see.

The closest point to $\mathbf{r}(t)$ on the focal line of $\mathbf{r}$ at t is the *principal centre of curvature* of $\mathbf{r}$ at t, this being the centre of the *circle* that most closely approximates the curve $\mathbf{r}$ at t. The radius $\rho(t)$ of this circle is what is called the *principal radius of curvature* of $\mathbf{r}$ at t. It is the reciprocal of $\rho(t)$ that is defined to be the *curvature* $\kappa(t)$ of $\mathbf{r}$ at t. The *principal normal line* to $\mathbf{r}$ at t is the line through $\mathbf{r}(t)$ and the principal centre of curvature. Clearly it intersects the focal line of $\mathbf{r}$ at t at right angles.

Figure 6.3 depicts the normal plane to the curve $\mathbf{r}$ at a point t. The tangent line to $\mathbf{r}$ at t is normal to the page, which is why the projection of that curve on the plane is cuspidal. By contrast the projection of the space evolute $\mathbf{e}$ is shown as being ordinarily tangent to the focal line, and not inflectional there, for since the osculating plane to $\mathbf{e}$ at t is the

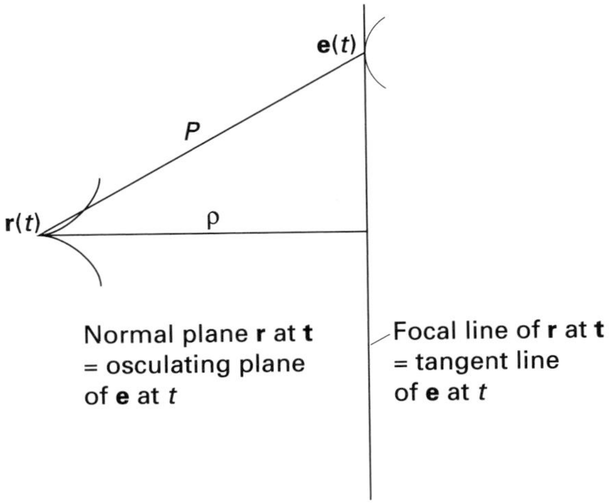

Figure 6.3

page of the diagram it cannot contain the kernel of the projection. The osculating plane to $\mathbf{r}$ at t is normal to the page and to the focal line.

In answer to the question raised above what we shall shortly show, as a corollary of Proposition 6.7, is that the spherical centre of curvature $\mathbf{e}(t)$ of $\mathbf{r}$ at t coincides with the principal centre of curvature there if and only if $\rho_1(t) = 0$, regardless of whether this is a local maximum or local minimum of $\rho(t)$.

6.3 The Serret–Frenet equations

The classical approach of Serret and Frenet to the curvature and torsion of a regular space curve $\mathbf{r}$ with connected domain is first of all to take *arc-length* s from some suitable base point as parameter. This can theoretically be done, just as for plane curves. The curve may then be regarded as a *unit-speed* curve, the *velocity* vector $\mathbf{t}(s) = \mathbf{r}_1(s)$ being everywhere of unit length.

Let $\mathbf{r}$ be such a curve. As in the plane case the vector $\mathbf{t}_1(s) = \mathbf{r}_2(s)$ is normal to the curve. We set $\mathbf{t}_1(s) = \kappa(s)\mathbf{n}(s)$, where $\mathbf{n}(s)$ is the (*unit*) *principal normal* to $\mathbf{r}$ at s, defined up to sign, provided that $\kappa(s) \neq 0$, the number $\kappa(s)$ being defined to be the *curvature* of $\mathbf{r}$ at s. We suppose in what follows that κ is everywhere non-zero in which case we may, without loss of generality, suppose that κ is everywhere positive.

This being so, the equations for the focal line of $\mathbf{r}$ at s are

$$(\mathbf{c} - \mathbf{r}(s)) \cdot \mathbf{t}(s) = 0 \text{ and } (\mathbf{c} - \mathbf{r}(s)) \cdot \kappa(s)\mathbf{n}(s) = 1.$$

It follows at once that the principal centre of curvature of $\mathbf{r}$ at s is the point $\mathbf{r}(s) + \rho(s)\mathbf{n}(s)$, where $\rho(s) = 1/\kappa(s)$, in accord with our earlier remarks, the principal normal vector $\mathbf{n}(s)$ being the unit normal vector orthogonal to the focal line and pointing towards it. Finally the (*unit*) *binormal* $\mathbf{b}(s)$ is defined to be the vector $\mathbf{t}(s) \times \mathbf{n}(s)$, the triad of unit vectors $\mathbf{t}(s)$, $\mathbf{n}(s)$, $\mathbf{b}(s)$ forming a right-handed orthonormal basis for the tangent vector space to $\mathbf{r}$ at s. The direction of the vector $\mathbf{b}(s)$ coincides with that of the focal line of $\mathbf{r}$ at s.

Clearly $\mathbf{t}$, $\mathbf{n}$ and $\mathbf{b}$ are everywhere unit spherical smooth curves associated to $\mathbf{r}$. Each of the derivative vectors $\mathbf{t}_1(s)$, $\mathbf{n}_1(s)$, $\mathbf{b}_1(s)$ is linearly dependent on $\mathbf{t}(s)$, $\mathbf{n}(s)$, $\mathbf{b}(s)$, the matrix of coefficients being skew-symmetric by virtue of the six equations that we obtain on differentiating the orthonormality relations, namely $\mathbf{t}_1 \cdot \mathbf{t} = 0$, etc., and $\mathbf{t}_1 \cdot \mathbf{n} + \mathbf{n}_1 \cdot \mathbf{t} = 0$, etc. The first of these we already know, namely

$\mathbf{t}_1 = \kappa\mathbf{n}$. Accordingly the full set is as follows:

$$\begin{aligned} \mathbf{t}_1 &= \qquad\ \ \kappa\mathbf{n} \\ \mathbf{n}_1 &= -\kappa\mathbf{t} \qquad\qquad +\tau\mathbf{b} \\ \mathbf{b}_1 &= \qquad\quad -\tau\mathbf{n}. \end{aligned}$$

The coefficient τ making its appearance here is defined to be the *torsion* of the curve $\mathbf{r}$. Like the curvature κ it depends smoothly on s.

These equations are the *equations of Serret and Frenet*.

Proposition 6.6 Let $\mathbf{r}$ *be a nowhere linear regular space curve. Then* $\mathbf{r}$ *has at t an ordinary planar point if and only if the torsion is zero and non-critical there.*

Proof We suppose $\mathbf{r}$ parametrised by arc-length. Then by the Serret–Frenet equations we have

$$\begin{aligned} \mathbf{r}_1 &= \mathbf{t} \\ \mathbf{r}_2 &= \kappa\mathbf{n} \\ \mathbf{r}_3 &= \kappa_1\mathbf{n} - \kappa^2\mathbf{t} + \kappa\tau\mathbf{b} \\ \mathbf{r}_4 &= \kappa_2\mathbf{n} - 3\kappa\kappa_1\mathbf{t} + 2\kappa_1\tau\mathbf{b} - \kappa^3\mathbf{n} + \kappa\tau_1\mathbf{b} - \kappa\tau^2\mathbf{n}. \end{aligned}$$

Since the coefficient of $\mathbf{b}$ in the expression for $\mathbf{r}_3$ is $\kappa\tau$ and since by hypothesis $\kappa \neq 0$ it follows that $\mathbf{r}$ is planar if and only if $\tau = 0$. The curve will at such a point have an ordinary planar point if and only if then the coefficient of $\mathbf{b}$ in the expression for $\mathbf{r}_4$, namely $\kappa\tau_1$, is non-zero, this being so if and only if $\tau_1 \neq 0$. □

Planar points of a regular space curve $\mathbf{r}$ are frequently referred to as *points of zero torsion*, just as linear points of the curve are frequently referred to as *points of zero curvature*.

Proposition 6.7 Let $\mathbf{r}$ *be a nowhere linear smooth space curve. Then at a point t where* $\mathbf{r}$ *is not planar the evolute* $\mathbf{e}$ *is given by*

$$\mathbf{e} = \mathbf{r} + \rho\mathbf{n} + \frac{\rho_1}{\tau}\mathbf{b},$$

where ρ_1 *denotes the derivative of* ρ *with respect to* arc-length.

Proof Clearly we again suppose the curve parametrised by arc-length. Then all we have to do is to verify that $V(\mathbf{c})_i = 0$, with $\mathbf{c} = \mathbf{e}$, for $i = 1$, 2, 3. This is an easy exercise (Exercise 6.5). □

It is clear from the formula that as one approaches a planar point or point of zero torsion s on the curve $\mathbf{r}$ the centre of curvature moves off to infinity unless $\rho_1(s) = 0$. One can verify that in the latter case the three equations for the centre of spherical curvature are satisfied by the entire focal line. From the formula also it is clear that the spherical centre of curvature at a non-planar point of $\mathbf{r}$ coincides with the principal centre of curvature of $\mathbf{r}$ at the point if and only if the principal radius of curvature ρ, regardless of parametrisation, is critical there.

We must stress the necessity of using a unit-speed parametrisation in applying the Serret–Frenet equations. In our earlier discussion of the space evolute of $\mathbf{r}$ there was no need at all to take arc-length as parameter.

6.4 Parallels

In our earlier study of the evolute $\mathbf{e}$ of a nowhere linear regular plane curve $\mathbf{r}$ we discussed how we can recover the original curve $\mathbf{r}$ as well as its parallels or offsets $\mathbf{r} + \delta\mathbf{n}$ by rolling a straight line along the evolute so that it coincides at any moment t with the tangent line to $\mathbf{e}$ at t. As it rolls its points describe the involutes of $\mathbf{e}$, one being the curve $\mathbf{r}$ and each other point a parallel or offset to $\mathbf{r}$. This rolling is succinctly summed up in the equation

$$\mathbf{e}_1 = \rho_1\mathbf{n}$$

which, regardless of the parameter chosen, equates the speed of travel of the point of tangency along the evolute with the speed of travel of the corresponding point on the rolling line.

A similar trick can be played with the space evolute $\mathbf{e}$ of a nowhere linear regular space curve $\mathbf{r}$. Let E denote the tangent developable of $\mathbf{e}$, the focal surface of $\mathbf{e}$, let $L(t)$ denote the tangent line to $\mathbf{e}$ at t, the focal line of $\mathbf{r}$ at t and let $P(t)$ denote the osculating plane to $\mathbf{e}$ at t, this plane being both tangent to E at any point of $L(t)$ and normal to $\mathbf{r}$ at $\mathbf{r}(t)$. Then imagine a plane rolling on E so that at any moment t it coincides with $P(t)$. As the plane rolls, its points describe the *involutes* of $\mathbf{e}$, one of these describing the curve $\mathbf{r}$ and each other point a *parallel* to $\mathbf{r}$. In the remainder of this section we validate this intuition.

If all this is correct then every line on the rolling plane rolls on E, the point of contact describing a smooth curve on E. Our strategy for verifying this will be to prove first that each normal line to $\mathbf{r}$ is instantaneously rolling on E and that, in this rolling, each point of each normal describes a parallel to $\mathbf{r}$ in $\mathbb{R}^3$. Moreover, each such parallel curve has the same focal surface and space evolute as $\mathbf{r}$. It then follows that every line in the normal plane may be thought of as rolling on E and not just those which happen to pass through $\mathbf{r}$.

A less obvious feature of this rolling is that with increasing t those normal lines intersecting the focal line on one side of the point of contact with E are rolling in one direction while those intersecting it on the opposite side are rolling in the opposite direction.

Yet another point to notice is that in the rolling no line remains permanently orthogonal to the focal line. Such orthogonality, if it occurs, is momentary only.

As an aid to intuition at this point it may be helpful to roll into a cylinder or cone a sheet of paper on which several lines have been drawn, all passing through the same point of the paper. As one then unrolls the paper observe how the various lines on it unroll. Notice, in particular, that no line remains permanently orthogonal to the line along which the unrolling occurs, except in the cylinder case.

With this in mind we introduce a family of smooth curves on the focal surface E, curves which we call the *focal curves* of $\mathbf{r}$, not to be confused with the *focal lines* of $\mathbf{r}$. We characterise a focal curve of $\mathbf{r}$ as a smooth space curve $\mathbf{f}$ such that $\mathbf{f} - \mathbf{r} = \sigma\mathbf{m}$ and $\mathbf{f}_1 = \mu\mathbf{m}$, where $\mathbf{m}$ is a smooth curve on the unit sphere such that $\mathbf{m} \cdot \mathbf{r}_1 = 0$. At a regular point of $\mathbf{f}$ this is equivalent to requiring that the tangent line to $\mathbf{f}$ there is normal to the curve $\mathbf{r}$.

Proposition 6.8 Let $\mathbf{f} = \mathbf{r} + \sigma\mathbf{m}$ *be a focal curve of a nowhere linear or planar regular space curve* $\mathbf{r}$, $\mathbf{m}$ *being a smooth curve on the unit sphere such that* $\mathbf{m} \cdot \mathbf{r}_1 = 0$. *Then* $\mathbf{f}$ *lies on the focal surface E of* $\mathbf{r}$. *Moreover*, $\mathbf{f}_1 = \sigma_1\mathbf{m}$ *and* $\mathbf{r}_1 + \sigma\mathbf{m}_1 = 0$.

Proof We remark first that $(\mathbf{f} - \mathbf{r}) \cdot \mathbf{r}_1 = \sigma\mathbf{m} \cdot \mathbf{r}_1 = 0$. Differentiating this we get $\mathbf{f}_1 \cdot \mathbf{r}_1 + (\mathbf{f} - \mathbf{r}) \cdot \mathbf{r}_2 - \mathbf{r}_1 \cdot \mathbf{r}_1 = 0$. But $\mathbf{f}_1 \cdot \mathbf{r}_1 = 0$, since, by hypothesis, $\mathbf{f}_1 = \mu\mathbf{m}$, for some smooth function μ. So $\mathbf{f}$ lies on the focal surface of $\mathbf{r}$. Moreover,

$$\mu = \mathbf{m} \cdot \mathbf{f}_1 = \mathbf{m} \cdot (\mathbf{r}_1 + \sigma_1 \mathbf{m} + \sigma \mathbf{m}_1) = \sigma_1.$$

That is $\mathbf{f}_1 = \sigma_1 \mathbf{m}$. By differentiating the equation $\mathbf{f} = \mathbf{r} + \sigma \mathbf{m}$ we deduce that $\mathbf{r}_1 + \sigma \mathbf{m}_1 = 0$. □

All this begs the question of the existence of such focal curves. Later, in Chapter 16, these will be shown to coincide with one family of focal curves of any tube of small fixed radius with core the space curve. Their existence essentially follows also from the unrolling construction discussed at the end of this section. In the meantime we determine some properties of these curves and the other curves related to them.

Proposition 6.9 *Let* $\mathbf{r}$ *be a nowhere linear or planar regular space curve and* $\mathbf{m}$ *a smooth curve on the unit sphere such that* $\mathbf{r}_1 + \sigma \mathbf{m}_1 = 0$ *for some smooth function* σ. *Then the curve* $\mathbf{m}$ *is regular and nowhere planar, being great at* t *if and only if* $\mathbf{m}(t) = \mathbf{n}(t)$, *up to sign.*

Proof We have not only $\mathbf{r}_1 + \sigma \mathbf{m}_1 = 0$ but also $\mathbf{r}_2 + \sigma \mathbf{m}_2 + \sigma_1 \mathbf{m}_1 = 0$ and $\mathbf{r}_3 + \sigma \mathbf{m}_3 + 2\sigma_1 \mathbf{m}_2 + \sigma_2 \mathbf{m}_1 = 0$. The independence of $\mathbf{m}_1$, $\mathbf{m}_2$ and $\mathbf{m}_3$ then follows at once from the independence of $\mathbf{r}_1$, $\mathbf{r}_2$ and $\mathbf{r}_3$. Moreover, $\mathbf{m}(t)$ depends linearly on $\mathbf{m}_1(t)$ and $\mathbf{m}_2(t)$ if and only if $\mathbf{m}(t)$ depends linearly on $\mathbf{r}_1(t)$ and $\mathbf{r}_2(t)$, that is $\mathbf{m}(t) = \mathbf{n}(t)$, up to sign. □

Proposition 6.10 *Let* $\mathbf{f} = \mathbf{r} + \sigma \mathbf{m}$ *be a focal curve of a nowhere linear or planar regular space curve* $\mathbf{r}$, *with* $\mathbf{f}_1 = \sigma_1 \mathbf{m}$, *where* $\mathbf{m}$ *is a smooth curve on the unit sphere. Then*

(*a*) $\mathbf{f}$ *is regular at* $t \Leftrightarrow \sigma_1(t) \neq 0 \Leftrightarrow \mathbf{f}(t) \neq \mathbf{e}(t)$;
(*b*) *where* $\mathbf{f}$ *is regular* $\mathbf{f}$ *is not linear*;
(*c*) $\mathbf{f}$ *is regular and not planar at* $t \Leftrightarrow \mathbf{f}(t) \neq \mathbf{e}(t)$ *and* $\mathbf{m}(t) \neq \pm\mathbf{n}(t)$;
(*d*) $\mathbf{f}$ *has an ordinary cusp at* $t \Leftrightarrow \sigma_1(t) = 0$ *but* $\sigma_2(t) \neq 0 \Leftrightarrow \mathbf{f}(t) = \mathbf{e}(t)$ *and* $\mathbf{e}_1(t) \neq 0$;
(*e*) $\mathbf{f}$ *has an ordinary non-planar cusp at* $t \Leftrightarrow$ *also* $\mathbf{m}(t) \neq \pm\mathbf{n}(t)$.

Proof Since $\mathbf{f}_1 = \sigma_1 \mathbf{m}$ it is clear that $\mathbf{f}$ is regular at t if and only if $\sigma_1(t) \neq 0$. Moreover, $\mathbf{f} = \mathbf{e}$ at t if and only if not only $\mathbf{f}_1 \cdot \mathbf{r}_1 = 0$ but also $\mathbf{f}_1 \cdot \mathbf{r}_2 =$ at t. Now

$$\mathbf{f}_1 \cdot \mathbf{r}_2 = \sigma_1 \mathbf{m} \cdot (-\sigma \mathbf{m}_2 - \sigma_1 \mathbf{m}_1) = \sigma_1 \sigma \mathbf{m}_1 \cdot \mathbf{m}_1 = 0 \Leftrightarrow \sigma_1 = 0.$$

Assertion (a) follows, since, by the previous proposition, $\mathbf{m}$ is regular.

Since $\mathbf{f}_1 = \sigma_1 \mathbf{m}$ we also have $\mathbf{f}_2 = \sigma_2 \mathbf{m} + \sigma_1 \mathbf{m}_1$. Since $\mathbf{m}$ is regular $\mathbf{m}_1$ is linearly independent of $\mathbf{m}$. The linear independence of $\mathbf{f}_1$ and $\mathbf{f}_2$ follows whenever $\sigma_1 \neq 0$. This is assertion (b).

Differentiating the same equation a second time we obtain

$$\mathbf{f}_3 = \sigma_3 \mathbf{m} + 2\sigma_2 \mathbf{m}_1 + \sigma_1 \mathbf{m}_2 = 0.$$

Since $\mathbf{m}$ is linearly independent of $\mathbf{m}_1$ and $\mathbf{m}_2$ if and only if $\mathbf{m} \neq \pm \mathbf{n}$, assertion ($c$) follows.

Suppose now that $\sigma_1 = 0$. Then $\mathbf{f}_1 = 0$, $\mathbf{f}_2 = \sigma_2 \mathbf{m}$ and $\mathbf{f}_3 = \sigma_3 \mathbf{m} + 2\sigma_2 \mathbf{m}_1$. The first part of assertion (d) follows at once. Moreover, on differentiating $(\mathbf{f} - \mathbf{r}) \cdot \mathbf{r}_2 = \mathbf{r}_1 \cdot \mathbf{r}_1$ twice we deduce at once that $\mathbf{f}(t) = \mathbf{e}(t)$ and $\mathbf{e}_1(t) \neq 0$ if and only if $\mathbf{f}_2 \cdot \mathbf{r}_2 \neq 0$ at t. But, with $\sigma_1 = 0$, we have $\mathbf{f}_2 \cdot \mathbf{r}_2 = \sigma_2 \mathbf{m} \cdot (-\sigma \mathbf{m}_2) = \sigma_2 \sigma \mathbf{m}_1 \cdot \mathbf{m}_1$. The second part then follows.

A third differentiation gives $\mathbf{f}_4 = \sigma_4 \mathbf{m} + 3\sigma_3 \mathbf{m}_1 + 3\sigma_2 \mathbf{m}_2 = 0$, when $\sigma_1 = 0$. Assertion (e) then follows as (d) did above. □

The next propostion is perhaps of greater interest and will be seen in a wider context later on, for what it tells us is that the focal curves of $\mathbf{r}$ are *geodesics* on the focal surface of $\mathbf{r}$.

Proposition 6.11 Let $\mathbf{r}$ *be a nowhere linear or planar regular space curve and let* $\mathbf{f}$ *be a focal curve of* $\mathbf{r}$, *lying necessarily on the focal surface of* $\mathbf{r}$. *Then at each t for which* $\mathbf{f}$ *is regular the principal normal of* $\mathbf{f}$ *is normal to the focal surface of* $\mathbf{r}$.

Proof Let $\mathbf{f} = \mathbf{r} + \sigma \mathbf{m}$, with $\mathbf{f}_1 = \sigma_1 \mathbf{m}$, $\mathbf{m}$ being a smooth curve on the unit sphere. Then $\mathbf{f}_2 = \sigma_2 \mathbf{m} + \sigma_1 \mathbf{m}_1$. Moreover, $\mathbf{r}_1 = -\sigma \mathbf{m}_1$. So when $\sigma_1 \neq 0$, that is when $\mathbf{f}$ is regular, the vector $\mathbf{r}_1$ depends linearly on $\mathbf{f}_1$ and $\mathbf{f}_2$. Moreover, since $\mathbf{m} \cdot \mathbf{r}_1 = 0$, $\mathbf{f}_1 \cdot \mathbf{r}_1 = 0$, from which it follows that the vector $\mathbf{r}_1$ is a principal normal vector to $\mathbf{f}$. But since $\mathbf{e}_1 \cdot \mathbf{r}_1 = \mathbf{e}_2 \cdot \mathbf{r}_1 = 0$ all along the focal line on which the point $\mathbf{f}$ lies the vector $\mathbf{r}_1$ also is normal to the focal surface of $\mathbf{r}$. □

We are now in a position at last to describe the parallels to a regular space curve $\mathbf{r}$. Let $\mathbf{m}$ be any smooth curve on the unit sphere such that

$\mathbf{r}_1 = \lambda \mathbf{m}_1$, for some smooth function λ. Then, for each $\delta \in \mathbb{R}$, the curve $\mathbf{r} + \delta\mathbf{m}$ is parallel to $\mathbf{r}$. To each focal curve of $\mathbf{r}$, $\mathbf{f} = \mathbf{r} + \sigma\mathbf{m}$, there is associated a set of parallels $\mathbf{r} + \delta\mathbf{m}$ to $\mathbf{r}$, each of which is described by some point of a line rolling on the focal curve in such a way that at each moment it coincides with a tangent line to the focal curve, all this being guaranteed by the equation $\mathbf{f}_1 = \sigma_1 \mathbf{m}$.

Proposition 6.12 Let $\mathbf{p} = \mathbf{r} + \delta\mathbf{m}$ *be parallel to a given regular nowhere linear or planar space curve* $\mathbf{r}$. *Then, except where* $\delta = \sigma$, $\mathbf{p}$ *is a regular nowhere linear or planar space curve, but, where* $\delta = \sigma$, *but* $\sigma_1 \neq 0$, *that is where* $\mathbf{p}$ *lies on the focal line to* $\mathbf{r}$ *but not at* $\mathbf{e}$, $\mathbf{p}$ *has an ordinary non-planar cusp, while, if* $\delta = \sigma$ *and* $\sigma_1 = 0$ *but* $\sigma_2 \neq 0$, *that is where* $\mathbf{p} = \mathbf{e}$ *but* $\mathbf{e}_1 \neq 0$, $\mathbf{p}$ *has its first two derivatives zero, but its next three linearly independent.*

Proof All this follows from the fact that, since $\mathbf{r}_1 + \sigma\mathbf{m}_1 = 0$,

$$(\mathbf{r} + \delta\mathbf{m})_1 = (\delta - \sigma)\mathbf{m}_1,$$

so that

$$(\mathbf{r} + \delta\mathbf{m})_2 = (\delta - \sigma)\mathbf{m}_2 - \sigma_1\mathbf{m}_1,$$

$$(\mathbf{r} + \delta\mathbf{m})_3 = (\delta - \sigma)\mathbf{m}_3 - 2\sigma_1\mathbf{m}_2 - \sigma_2\mathbf{m}_1,$$

and so on. Now, by Proposition 6.9, $\mathbf{m}_1$, $\mathbf{m}_2$ and $\mathbf{m}_3$ are everywhere linearly independent. It follows at once that, provided $\delta \neq \sigma$, $(\mathbf{r} + \delta\mathbf{m})_1$, $(\mathbf{r} + \delta\mathbf{m})_2$ and $(\mathbf{r} + \delta\mathbf{m})_3$ are linearly independent. The other assertions follow in like manner. □

For a regular smooth space curve $\mathbf{e}$, parametrised by arc-length s, one can give an alternative construct of the involutes of $\mathbf{e}$ as follows. First, by Theorem 1.21, one constructs a plane curve $\mathbf{e}^*$, also parametrised by s, having the same curvature κ as $\mathbf{e}$. In what follows $\mathbf{t}$, $\mathbf{n}$ and $\mathbf{b}$ denote the unit tangent, principal normal and binormal of $\mathbf{e}$ and $\mathbf{t}^*$ and $\mathbf{n}^*$ the unit tangent and unit normal of $\mathbf{e}^*$. We select first of all a fixed point $\mathbf{r}^*$ of the plane in which $\mathbf{e}^*$ lies. Then there exist smooth functions α, β of s such that, for all s, $\mathbf{r}^* = \mathbf{e}^*(s) + \alpha(s)\mathbf{t}^*(s) + \beta(s)\mathbf{n}^*(s)$, with

$$(\mathbf{e}^* + \alpha\mathbf{t}^* + \beta\mathbf{n}^*)_1 = (1 + \alpha_1 - \kappa\beta)\mathbf{t}^* + (\alpha\kappa + \beta_1)\mathbf{n}^* = 0,$$

implying that $1 + \alpha_1 - \kappa\beta = 0$ and that $\alpha\kappa + \beta_1 = 0$.

We now 'roll' this plane over the tangent developable of $\mathbf{e}$ in $\mathbb{R}^3$, the curve swept out by $\mathbf{r}^*$ then being the curve $\mathbf{r} = \mathbf{e} + \alpha\mathbf{t} + \beta\mathbf{n}$.

The next proposition provides a check on the accuracy of this.

Proposition 6.13 The evolute of the curve $\mathbf{r}$ *constructed above is the curve* $\mathbf{e}$.

Proof What has to be proved is that

$$(\mathbf{e} - \mathbf{r}) \cdot \mathbf{r}_1 = 0,$$

$$(\mathbf{e} - \mathbf{r}) \cdot \mathbf{r}_2 = \mathbf{r}_1 \cdot \mathbf{r}_1,$$

and

$$(\mathbf{e} - \mathbf{r}) \cdot \mathbf{r}_3 = 3\mathbf{r}_1 \cdot \mathbf{r}_2.$$

Now, since

$$\mathbf{r} = \mathbf{e} + \alpha\mathbf{t} + \beta\mathbf{n},$$

$$\mathbf{r}_1 = \beta\tau\mathbf{b},$$

since $(\mathbf{r}^*)_1 = 0$, from which it follows that

$$\mathbf{r}_2 = -\beta\tau^2\mathbf{n} + (\beta\tau)_1\mathbf{b}$$

and

$$\mathbf{r}_3 = -\beta\tau^2(-\kappa\mathbf{t} + \tau\mathbf{b}) + (\beta\tau)_1(-\tau\mathbf{n}) - (\beta\tau^2)_1\mathbf{n} + (\beta\tau)_2\mathbf{b}.$$

Accordingly

$$(\mathbf{e} - \mathbf{r}) \cdot \mathbf{r}_1 = (-\alpha\mathbf{t} - \beta\mathbf{n}) \cdot \beta\tau\mathbf{b} = 0,$$

$$(\mathbf{e} - \mathbf{r}) \cdot \mathbf{r}_2 = (-\alpha\mathbf{t} - \beta\mathbf{n}) \cdot (-\beta\tau^2\mathbf{n} + (\beta\tau)_1\mathbf{b})$$

$$= \beta^2\tau^2 = \mathbf{r}_1 \cdot \mathbf{r}_1,$$

while

$$(\mathbf{e} - \mathbf{r}) \cdot \mathbf{r}_3 = -\alpha\beta\kappa\tau^2 + \beta\tau(\beta\tau)_1 + \beta(\beta\tau^2)_1$$

$$= 3\beta\tau(\beta\tau)_1 = 3\mathbf{r}_1 \cdot \mathbf{r}_2,$$

since $\alpha\kappa + \beta_1 = 0$. □

Note that if, at an isolated point of $\mathbf{e}$, $\kappa = 0$ but $\tau \neq 0$, $\mathbf{n}$ at that point being defined as its limiting value there, then, provided that $\beta \neq 0$, the curve $\mathbf{r}$ is regular and planar at that point, $\mathbf{r}_1$, $\mathbf{r}_2$ and $\mathbf{r}_3$ all lying in the plane spanned by $\mathbf{b}$ and $\mathbf{n}$. Then each point of the tangent to $\mathbf{e}$ at s also satisfies the equation $(\mathbf{e}-\mathbf{r})\cdot\mathbf{r}_3 = 3\mathbf{r}_1\cdot\mathbf{r}_2$, as is readily checked, $\mathbf{e}$ itself satisfying the equation

$$(\mathbf{e}-\mathbf{r})\cdot\mathbf{r}_4 = 4\mathbf{r}_1\cdot\mathbf{r}_3 + 3\mathbf{r}_2\cdot\mathbf{r}_2.$$

On the other hand, at a point of $\mathbf{e}$ where $\kappa \neq 0$ but $\tau = 0$ each involute $\mathbf{r}$ is non-regular. In the case that $\tau_1 \neq 0$ and $\beta \neq 0$ one has in fact $\mathbf{r}_1 = 0$, $\mathbf{r}_2 = \beta\tau_1\mathbf{b} \neq 0$ and $\mathbf{r}_3 = (2\beta_1\tau_1 + \beta\tau_2)\mathbf{b}$, so that the cusp is rhamphoidal, further work being required to establish the generic relation then to be expected between $\mathbf{r}_4$, $\mathbf{r}_5$ and higher derivatives.

It should be noted in all this that our model (Figure 6.1) of a space curve and its tangent developable is relevant here. The model for the rolling plane is the plane sheet of paper, with the original curve drawn on it, before the paper is split in two and the two sheets are twisted apart. The construction shows vividly exactly how the plane curve and space curve are related, and especially how they share the same curvature function. As the sheets of the model are twisted apart the curvature remains the same but the torsion, originally zero, changes.

6.5 Close up views

If one looks at a regular space curve, not from infinity in some direction (a distant view), but from some finite point of space P, then it can be shown that if in the distant view in that direction one sees a non-linear regular point, an ordinary inflection or an ordinary cusp, then the view from P of the image of the projection of the curve from one's point of view on to the same plane is of the same type. What is new, however, is that from a point on the tangent to the curve at an ordinary planar point or point of zero torsion, the view in general is of an ordinary rhamphoid cusp, where the radius of curvature is not critical, but for exactly one point of the tangent the view is of a rhamphoid cusp for which the radius of curvature is critical, this property being not only an affine invariant but, more surprisingly, a diffeomorphism invariant of the viewed curve. That one expects it to be an affine invariant follows from the fact that if the projection point is finite then there is some ambiguity as to the plane in which the viewed curve may be supposed to be lying, any one such plane curve being an affine transform of any other.

The argument for the case of a point of zero torsion goes as follows. Let $t \mapsto \mathbf{r}(t) = (x(t), y(t), z(t))$ be a regular space curve passing through the origin, and let the curve be viewed from the point $(0, 0, c)$ on the z-axis, with $c \neq 0$. Then the curve projects to the curve $\mathbf{p}$ in the (x, y)-plane, where $\mathbf{q}(t) = (x(t), y(t))$ and $\mathbf{p}(t) = f(t)\mathbf{q}(t)$, with $f(0) = 1$, $(c - z(t))f(t) = c$ and, since $(c - z(t))f_1(t) - z_1(t)f(t) = 0$, $cf_1(0) = z_1(0)$. Then we have

$$\mathbf{p}_1 = f\mathbf{q}_1 + f_1\mathbf{q},$$

$$\mathbf{p}_2 = f\mathbf{q}_2 + 2f_1\mathbf{q}_1 + f_2\mathbf{q},$$

$$\mathbf{p}_3 = f\mathbf{q}_3 + 3f_1\mathbf{q}_2 + 3f_2\mathbf{q}_1 + f_3\mathbf{q},$$

$$\mathbf{p}_4 = f\mathbf{q}_4 + 4f_1\mathbf{q}_3 + 6f_2\mathbf{q}_2 + 4f_3\mathbf{q}_1 + f_4\mathbf{q},$$

$$\mathbf{p}_5 = f\mathbf{q}_5 + 5f_1\mathbf{q}_4 + 10f_2\mathbf{q}_3 + 10f_3\mathbf{q}_2 + 5f_4\mathbf{q}_1 + f_5\mathbf{q}.$$

At the origin, where $\mathbf{q} = 0$ and $f = 1$ these reduce to

$$\mathbf{p}_1 = \mathbf{q}_1,$$

$$\mathbf{p}_2 = \mathbf{q}_2 + 2f_1\mathbf{q}_1,$$

$$\mathbf{p}_3 = \mathbf{q}_3 + 3f_1\mathbf{q}_2 + 3f_2\mathbf{q}_1,$$

$$\mathbf{p}_4 = \mathbf{q}_4 + 4f_1\mathbf{q}_3 + 6f_2\mathbf{q}_2 + 4f_3\mathbf{q}_1,$$

$$\mathbf{p}_5 = \mathbf{q}_5 + 5f_1\mathbf{q}_4 + 10f_2\mathbf{q}_3 + 10f_3\mathbf{q}_2 + 5f_4\mathbf{q}_1.$$

Suppose now that $\mathbf{q}_1(0) = 0$, that is that the viewing point lies on the tangent to the curve at the origin. Then the equations reduce at the origin to

$$\mathbf{p}_1 = 0,$$

$$\mathbf{p}_2 = \mathbf{q}_2,$$

$$\mathbf{p}_3 = \mathbf{q}_3 + 3f_1\mathbf{q}_2,$$

$$\mathbf{p}_4 = \mathbf{q}_4 + 4f_1\mathbf{q}_3 + 6f_2\mathbf{q}_2,$$

$$\mathbf{p}_5 = \mathbf{q}_5 + 5f_1\mathbf{q}_4 + 10f_2\mathbf{q}_3 + 10f_3\mathbf{q}_2.$$

Now if $\mathbf{r}$ is not planar at the origin then $\mathbf{q}_2$ and $\mathbf{q}_3$ are linearly independent and accordingly $\mathbf{p}_2$ and $\mathbf{p}_3$ are linearly independent, that is the curve $\mathbf{p}$ has an ordinary cusp at the origin. On the other hand if $\mathbf{r}$ has an ordinary planar point there then $\mathbf{q}_2 \neq 0$ but $\mathbf{q}_3$ depends linearly on $\mathbf{q}_2$ though $\mathbf{q}_4$ does not, implying that $\mathbf{p}_2 \neq 0$ but $\mathbf{p}_3$ depends linearly on $\mathbf{p}_2$ though $\mathbf{p}_4$ does not; that is the curve $\mathbf{p}$ has a rhamphoid cusp at the origin.

It remains to check whether or not this rhamphoid cusp is ordinary. Suppose that at the origin $\mathbf{q}_3 = 3\lambda\mathbf{q}_2$. Then there also

$$\mathbf{p}_3 = 3\lambda\mathbf{q}_2 + 3f_1\mathbf{q}_2 = 3(\lambda + f_1)\mathbf{p}_2 = 3(\lambda + c^{-1}z_1)\mathbf{p}_2.$$

By Theorem 1.23 the question is whether $\mathbf{p}_5 - 10(\lambda + f_1)\mathbf{p}_4$ is linearly independent of $\mathbf{p}_2$. But

$$\begin{aligned}\mathbf{p}_5 - 10(\lambda + f_1)\mathbf{p}_4 \\ &= \mathbf{q}_5 + 5f_1\mathbf{q}_4 + 10f_2\mathbf{q}_3 + 10f_3\mathbf{q}_2 - 10(\lambda + f_1)(\mathbf{q}_4 + 4f_1\mathbf{q}_3 + 6f_2\mathbf{q}_2), \\ &= \mathbf{q}_5 + 5f_1\mathbf{q}_4 - 10(\lambda + f_1)\mathbf{q}_4 + \text{a multiple of } \mathbf{q}_2, \\ &= \mathbf{q}_5 - 10\lambda\mathbf{q}_4 - 5c^{-1}z_1\mathbf{q}_4 + \text{a multiple of } \mathbf{q}_2.\end{aligned}$$

Now $\mathbf{q}_5 - 10\lambda\mathbf{q}_4 - 5c^{-1}z_1\mathbf{q}_4$ is a multiple of $\mathbf{q}_2$ for at most one value of c; for since $\mathbf{q}_2$ and $\mathbf{q}_4$ are linearly independent at 0 there are unique numbers α and β such that $\mathbf{q}_5 - 10\lambda\mathbf{q}_4 = \alpha\mathbf{q}_2 + \beta\mathbf{q}_4$, β being equal to $5c^{-1}z_1$ only for $c = 5\beta^{-1}z_1$. Finally $\mathbf{q}_2 = \mathbf{p}_2$ at the origin. It follows that, provided $\beta \neq 0$, there is just one point of view on the tangent to $\mathbf{r}$ at an ordinary planar point from which the observed rhamphoid cusp fails to be ordinary.

What if $\beta = 0$? Then since at the origin $\mathbf{q}_3 = 3\lambda\mathbf{q}_2$ and $\mathbf{q}_5 = 10\lambda\mathbf{q}_4 + \alpha\mathbf{q}_2$ it follows, again from Theorem 1.23, that the curve $\mathbf{q}$ in the (x, y)-plane, which is the projection of the space curve $\mathbf{r}$ from the point at infinity on the z-axis, then has a rhamphoid cusp at the origin that fails to be ordinary.

The reference for the unique special point on the tangent to a regular space curve at a point of zero torsion is to a paper by David (1983).

It is also of interest to consider what the space evolute looks like at a point where the space curve $\mathbf{r}$ has an ordinary non-planar point in the particular case that the entire focal line satisfies the equation

$$\mathbf{r}_3 \cdot (\mathbf{e} - \mathbf{r}) = 3\mathbf{r}_1 \cdot \mathbf{r}_2,$$

that is in the particular case that not only $\tau = 0$ but also $\rho_1 = 0$. The point $\mathbf{e}(t)$ is then the unique solution at t of the three equations

$$\begin{aligned}\mathbf{r}_1 \cdot (\mathbf{e} - \mathbf{r}) &= 0 \\ \mathbf{r}_2 \cdot (\mathbf{e} - \mathbf{r}) &= \mathbf{r}_1 \cdot \mathbf{r}_1 \\ \mathbf{r}_4 \cdot (\mathbf{e} - \mathbf{r}) &= 4\mathbf{r}_1 \cdot \mathbf{r}_3 + 3\mathbf{r}_2 \cdot \mathbf{r}_2.\end{aligned}$$

What one can show is that the space evolute is then regular and linear at t, that is the vector $\mathbf{e}_1(t)$ is non-zero and the vector $\mathbf{e}_2(t)$ is a multiple of it. Moreover, the three vectors $\mathbf{e}_1(t)$, $\mathbf{e}_3(t)$, and $\mathbf{e}_4(t)$ are linearly independent. The projection of the space evolute on the normal plane to $\mathbf{r}$ at t has an ordinary inflection at t, the inflectional line, being the actual tangent line to $\mathbf{e}$ at t, coinciding with the focal line to $\mathbf{r}$ as one would expect. By what we have just been saying the projection of the curve $\mathbf{r}$ itself on the normal plane to $\mathbf{r}$ at t has a rhamphoid cusp at t, the cuspidal tangent coinciding with the principal normal line. The details are left as Exercise 6.16.

Also left for investigation (Exercise 6.17) are the involutes of a space curve $\mathbf{e}$ with an ordinary planar point, and what happens to the involutes at special planar points.

6.6 Historical note

Space curves were originally known as *curves of double curvature*, first studied by Alexis-Claude Clairaut (1731) when in his teens. The date of the papers of Serret (1851) and Frenet (1852) is surprisingly late. It seems, however, that they were the last to discover the equations attributed to them, for these are to be found in a book by Carl Eduard Senff (1831) of the University of Dorpat (now Tartu in Estonia), who attributed them to his teacher Martin Bartels. Indeed the concept of torsion is already explicit in a paper of Michel-Ange Lancret (1806), a pupil of Monge, of whom we have more to say in a later note (10.5).

Exercises

6.1 Construct a space curve with two sheets of firm A4 paper, a pair of scissors and a mini-stapler (or muslin clamps(!)).

(To do this you should cut out a roughly elliptical curve from both sheets held together, and then staple the cut edges together all the way round. Finally you should make a cut from the outside to the elliptical hole, and you are then free to give what was originally a plane curve some torsion. Nearly everyone gets the twisting wrong at the first try! (See Figure 6.1.))

6.2 Construct a swallow-tail cusp with three sheets of firm A4 paper, a pair of scissors and a mini-stapler.

(This is harder and less satisfactory in view of the unavoidable presence of the double curve; but worth a try nevertheless.)

6.3 Verify that the space evolute of the space curve $t \mapsto (t, t^2, t^3)$ is everywhere regular, but that there is at least one point where the radius of spherical curvature is critical.

6.4 Prove Proposition 6.2.

6.5 Let $\mathbf{r}$ be a nowhere linear smooth space curve. Prove that at a point t where $\mathbf{r}$ is not planar the evolute $\mathbf{e}$ is given by

$$\mathbf{e} = \mathbf{r} + \rho\mathbf{n} + \frac{\rho_1}{\tau}\mathbf{b},$$

where ρ_1 denotes the derivative of ρ with respect to *arc-length*.

(We know to start with that the three familiar equations that determine $\mathbf{e}$ have a unique solution since at a non-planar point the vectors $\mathbf{r}_1$, $\mathbf{r}_2$, $\mathbf{r}_3$ are linearly independent. All you have to do is to 'interview' the above 'candidate' for the job!)

6.6 Let $\mathbf{r}$ be a regular smooth space curve whose space evolute $\mathbf{e}$ has a non-planar ordinary cusp at a point t. Prove that the focal line of $\mathbf{r}$ at t is the limiting tangent line to $\mathbf{e}$ at t, having the vector $\mathbf{e}_2(t)$ lying along it.

6.7 Determine the curvature and torsion of the *helix*

$$t \mapsto (\cos t, \sin t, kt).$$

6.8 By considering the determinant of the matrix

$$\begin{bmatrix} \mathbf{r}_1 \cdot & 0 \\ \mathbf{r}_2 \cdot & \mathbf{r}_1 \cdot \mathbf{r}_1 \\ \mathbf{r}_3 \cdot & 3\mathbf{r}_1 \cdot \mathbf{r}_2 \\ \mathbf{r}_4 \cdot & 4\mathbf{r}_1 \cdot \mathbf{r}_3 + 3\mathbf{r}_2 \cdot \mathbf{r}_2 \end{bmatrix}$$

or otherwise find all the ordinary vertices of the space curve

$$t \mapsto (t, \tfrac{1}{2}t^2, \tfrac{1}{6}t^3 - 2t).$$

(The definition of an *ordinary vertex* is given just before Proposition 6.3.)

6.9 Find the vertices, if any, of each of the following space curves:

$$t \mapsto (t, t^2, t^3),$$

$$t \mapsto (t, t^2, t^4) \text{ (a tricky case)},$$

$$t \mapsto (t, \tfrac{1}{2}t^2, t^2 + \tfrac{1}{2}t^3).$$

6.10 Let $\mathbf{r}$ be a regular nowhere linear or planar space curve parametrised by arc-length from some reference point $s = 0$ where $\rho_1 = 0$. Prove that $P_1(0) = 0$ and that at 0

$$PP_2 = \rho\rho_2 + \frac{\rho_2^{\,2}}{\tau^2}.$$

Deduce that if the principal radius of curvature ρ has a local minimum at 0 then the radius of spherical curvature P has a local minimum at 0 but that if ρ has a local maximum at 0 then P does not necessarily have a local maximum at 0.

6.11 Let $\mathbf{r}$ be a nowhere linear regular space curve. Prove that the three equations for the centre of spherical curvature $\mathbf{e}(t)$ at a point t are satisfied by the entire focal line at t if and only if the torsion is zero at t and also the radius of spherical curvature is critical there.

6.12 Let $\mathbf{r}$ be a unit-speed regular curve lying on a sphere of radius P, with $\rho_1 = 0$ only at isolated points. Prove that $\rho\tau^3 + \rho_2\tau = \rho_1\tau_1$ along the curve. Deduce that if $\tau = 0$ but $\tau_1 \neq 0$ then $\rho_1 = 0$, but that $\rho_1 = 0$ does not necessarily imply that $\tau = 0$. Show, however, that if $\rho_1 = 0$ but $\tau \neq 0$ then ρ necessarily has a local maximum, being there equal to P.

6.13 Perform the exercise recommended on p. 108 with a sheet of paper rolled into a cylinder or cone on which several lines have been drawn, all passing through the same point of the paper.

6.14 Prove parts (*d*) and (*e*) of Proposition 6.10.

6.15 Complete the proof of Proposition 6.12.

6.16 Let $\mathbf{r}$ be a nowhere linear or planar regular space curve and let $\mathbf{m}$ be a unit normal with $\mathbf{r}_1 + \sigma\mathbf{m}_1 = 0$. Let π denote the projection of $\mathbb{R}^3$ to the normal plane to $\mathbf{r}$ at some point t, with non-zero kernel vector $\mathbf{r}_1(t)$, and suppose that the space evolute $\mathbf{e}$ is regular at t. Let $\mathbf{f} = \mathbf{r} + \sigma\mathbf{m}$. Show that in general the curve $\pi\mathbf{f}$ has an ordinary inflection at t but that if $\mathbf{m}(t) = \pm\mathbf{n}(t)$ then $\pi\mathbf{f}$ has an ordinary undulation at t while if $\mathbf{f}(t) = \mathbf{e}(t)$, with $\mathbf{m}(t) \neq \pm\mathbf{n}(t)$, then $\pi\mathbf{f}$ has a rhamphoid cusp at t. Sketch the various possibilities on the normal plane to $\mathbf{r}$ at t.

6.17 Investigate the involutes of a space curve $\mathbf{e}$ with an ordinary planar point, and what happens to the involutes at special planar points.

6.18 (Cleave (1980) – see Figure 6.2) Let $\mathbf{r} : \mathbb{R} \to \mathbb{R}^3$; $t \mapsto (\cos\theta(t), \sin\theta(t), z(t))$ be a regular curve on the cylinder with equation $x^2 + y^2 = 1$, with $\theta(0) = z(0) = 0$, that is simply tangent to the generator $x = 1$, $y = 0$ of the cylinder at $(1, 0, 0)$. Prove that $\mathbf{r}$ is planar at $(1, 0, 0)$. Show also, by considering the tangents to $\mathbf{r}$ at the points of intersection with the curve of generators close to the generator $x = 1$, $y = 0$, that the tangent developable of $\mathbf{r}$ has a double line that terminates at the origin.

7
k-times linear forms

7.0 Introduction

In this chapter we study *several-times* or *multiply linear forms*. In doing so the logic behind our earlier non-standard use of subscripts to denote derivatives may begin to become clear. Here as there a subscript will indicate how many slots there are available to be filled following the symbol to which the subscript is attached. This is particularly helpful to have clearly signed when not all the available slots are filled, as will frequently be the case. The use of subscripts in this way will not be restricted to derivatives, but will be employed whenever we have occasion to discuss multiply linear forms.

The long section on real cubic forms on $\mathbb{R}^2$ is an essential preliminary for our later study of the geometry of regular surfaces in $\mathbb{R}^3$ and especially their ridges and umbilics.

7.1 k-times linear forms

Let X be a real vector space. A *two-times linear* or *twice linear form* V_2 on X is by definition a linear map

$$V_2 : X \to L(X, \mathbb{R}),$$

$L(X, \mathbb{R})$ being the real linear space of linear maps $X \to \mathbb{R}$. Such a form induces a *bilinear* form

$$X \times X \to \mathbb{R};\ (\mathbf{u}, \mathbf{v}) \mapsto V_2\mathbf{uv} = (V_2(\mathbf{u}))(\mathbf{v}),$$

said to be *symmetric* if, for all $\mathbf{u}, \mathbf{v} \in X$, $V_2\mathbf{vu} = V_2\mathbf{uv}$. It also induces a *quadratic form*

$$X \to \mathbb{R};\ \mathbf{u} \mapsto V_2\mathbf{u}^2 = V_2\mathbf{uu},$$

this being a homogenous polynomial map of degree two in the case that $X = \mathbb{R}^n$. Since, for any symmetric twice-linear form V_2 and any $\mathbf{u}, \mathbf{v}$,

$$V_2\mathbf{uv} = V_2(\tfrac{1}{2}(\mathbf{u}+\mathbf{v}))^2 - V_2(\tfrac{1}{2}(\mathbf{u}-\mathbf{v}))^2,$$

the form V_2 is uniquely determined by the quadratic form that it induces.

Example 7.1

$$I_2 : \mathbb{R}^n \longrightarrow L(\mathbb{R}^n, \mathbb{R});\ \mathbf{u} \mapsto \mathbf{u}\cdot\ = \mathbf{u}^\tau,$$

with $I_2\mathbf{uv} = \mathbf{u}\cdot\mathbf{v} = \mathbf{v}\cdot\mathbf{u} = I_2\mathbf{vu}$ and $I_2\mathbf{u}^2 = \mathbf{u}\cdot\mathbf{u}$, where $\cdot$ denotes the standard scalar product on $\mathbb{R}^n$. □

Clearly in the case that $X = \mathbb{R}^n$ matrix notations are available. Suppose that we denote by B the matrix of the composite linear map ${I_2}^{-1}V_2 : \mathbb{R}^2 \mapsto \mathbb{R}^2$. Then, for any $\mathbf{u}, \mathbf{v} \in \mathbb{R}^n$,

$$V_2\mathbf{uv} = I_2{I_2}^{-1}V_2\mathbf{uv} = I_2(B\mathbf{u})\mathbf{v} = (B\mathbf{u})^\tau\mathbf{v} = \mathbf{u}^\tau B^\tau\mathbf{v},$$

and moreover, since $V_2\mathbf{vu} = V_2\mathbf{uv}$, $\mathbf{v}^\tau B^\tau\mathbf{u} = \mathbf{u}^\tau B\mathbf{v} = \mathbf{u}^\tau B^\tau\mathbf{v}$, from which it follows that $B^\tau = B$, that is B is *symmetric*.

Example 7.2

$$\mathbb{R}^2 \longrightarrow L(\mathbb{R}^2, \mathbb{R})$$
$$\begin{bmatrix} x \\ y \end{bmatrix} \mapsto [x \quad y]\begin{bmatrix} a & b \\ b & c \end{bmatrix}$$

is the twice linear form inducing the quadratic form

$$\mathbb{R}^2 \longrightarrow \mathbb{R};\begin{bmatrix} x \\ y \end{bmatrix} \mapsto [x \quad y]\begin{bmatrix} a & b \\ b & c \end{bmatrix}\begin{bmatrix} x \\ y \end{bmatrix} = ax^2 + 2bxy + cy^2. \qquad \square$$

It is a standard theorem of linear algebra that the eigenvalues of a real symmetric matrix are all real and that the dimension of any eigenspace of such a matrix is equal to the multiplicity of the eigenvalue as a root of the characteristic polynomial. Moreover, the eigenspaces of any two distinct eigenvalues are mutually orthogonal. Accordingly for any $n \times n$ symmetric matrix B there is an orthonormal basis for $\mathbb{R}^n$ consisting of eigenvectors of B. If P is the orthogonal matrix whose columns are the vectors of such an orthonormal basis then the matrix $P^{-1}BP = P^\tau BP$ is a diagonal matrix Λ, whose diagonal entries are the eigenvalues of B.

It follows at once that the quadratic form $\mathbf{x}^\tau B\mathbf{x}$ is reducible by the orthogonal change of coordinates $\mathbf{x} = P\mathbf{x}'$ to the form $\mathbf{x}'^\tau \Lambda \mathbf{x}'$, a 'sum of squares' with coefficients the eigenvalues of the matrix B.

The rank of the matrix B, equal to the number of non-zero entries in the diagonal matrix Λ, is called the *rank* of the quadratic form $\mathbf{x}^\tau B\mathbf{x}$, the form being said to be *non-degenerate* if its rank is equal to the dimension n of the vector space X, that is if the kernel of B is zero.

So much for twice linear forms. A *three-times* or *thrice linear form* V_3 on a real vector space X is by definition a linear map

$$V_3 : X \to L(X, L(X, \mathbb{R})).$$

Such a form induces a *trilinear form*

$$X \times X \times X \to \mathbb{R}; (\mathbf{u}, \mathbf{v}, \mathbf{w}) \mapsto V_3\mathbf{uvw} = ((V_3(\mathbf{u}))(\mathbf{v}))(\mathbf{w}),$$

said to be *symmetric* if, for all $\mathbf{u} \in X$, $V_3\mathbf{u}$ is symmetric and, for all $\mathbf{u}$, $\mathbf{v} \in X$, $V_3\mathbf{vu} = V_3\mathbf{uv}$, this implying that $V_3\pi(\mathbf{uvw}) = V_3\mathbf{uvw}$, for any permutation π of $\mathbf{u}$, $\mathbf{v}$, $\mathbf{w}$. It also induces a *cubic form*

$$X \to \mathbb{R}; \mathbf{u} \mapsto V_3\mathbf{u}^3 = V_3\mathbf{uuu},$$

this being a homogenous polynomial map of degree three in the case that $X = \mathbb{R}^n$. A symmetric thrice linear form is uniquely determined by the cubic form that it induces.

Higher symmetric k-times linear forms and their associated k-ic forms, or forms of *order* k, are defined in similar manner.

Matrix notations are not appropriate for the descriptions of k-times linear forms on $\mathbb{R}^n$, for $k \geqslant 3$. In the sequel we shall mainly be concerned with forms on $\mathbb{R}^2$ and in this case a convenient alternative notation is available for *symmetric* k-times linear forms.

For any $a_0, a_1, \ldots, a_k \in \mathbb{R}$, where $k \geqslant 2$, $[a_0, a_1, \ldots, a_k]_k$ is defined to be the k-times linear map on $\mathbb{R}^2$ determined by

$$[a_0, a_1, \ldots, a_k]_k \begin{bmatrix} x \\ y \end{bmatrix} = [a_0 x + a_1 y, a_1 x + a_2 y, \ldots, a_{k-1}x + a_k y]_{k-1}.$$

This is readily verified to be symmetric.

In the case $k = 1$, $[a, b]_1 = [a \quad b] : \mathbb{R}^2 \to \mathbb{R}$.

As an example, when $k = 3$,

$$[a, b, c, d]_3 \begin{bmatrix} x \\ y \end{bmatrix}^3 = ax^3 + 3bx^2y + 3cxy^2 + dy^3.$$

7.2 Quadratic forms on $\mathbb{R}^2$

The following remarks concerning quadratic forms on $\mathbb{R}^2$ are preparatory to a detailed study of cubic forms on $\mathbb{R}^2$, essential, as we have indicated already, for an understanding of the finer points of the geometry of surfaces in $\mathbb{R}^3$.

A non-zero symmetric twice linear form V_2 on $\mathbb{R}^2$ and its associated quadratic form are each said to be *elliptic*, *parabolic* or *hyperbolic* according as the determinant of its matrix, which is the product of the eigenvalues of the matrix, is positive, zero or negative. In the hyperbolic or parabolic case the quadratic polynomial form admits a factorisation as the product of two real linear forms, which, in the parabolic, degenerate case, coincide (up to a real multiple). The kernels of these linear forms are called the *root lines* of the original form. In the elliptic case the quadratic polynomial admits factorisation as the product of two complex conjugate linear forms. The kernels of these forms will be regarded as the root lines of the forms in this case also, though they lie in $\mathbb{C}^2$ and not in $\mathbb{R}^2$. Either of these root lines is the complex conjugate of the other. The word 'conjugate' is also employed in another way, vectors $\mathbf{u}, \mathbf{v} \in \mathbb{R}^2$ being said to be *conjugate* with respect to a symmetric twice linear form V_2 on $\mathbb{R}^2$ if $V_2\mathbf{u}\mathbf{v} = 0$.

The characterisation of parabolic quadratic forms on $\mathbb{R}^2$ provided by the next theorem will be useful in the sequel.

Theorem 7.3 Let V_2 be a non-zero symmetric twice real linear form on $\mathbb{R}^2$. Then V_2 is parabolic, that is the quadratic form is, up to sign, a perfect square, if and only if there is a non-zero vector $\mathbf{u} \in \mathbb{R}^2$ such that $V_2\mathbf{u} = 0$.

Proof This is simply a restatement of the fact that the form V_2 is parabolic if and only if one of the eigenvalues of its matrix is zero.

The following alternative proofs may be of some interest.

(1) Let the matrix of V_2 be

$$\begin{bmatrix} a & b \\ b & c \end{bmatrix}$$

and let $\mathbf{u} = (\xi, \eta)$ be a non-zero vector in $\mathbb{R}^2$. Then $V_2\mathbf{u} = 0$ implies that $a\xi + b\eta = 0$ and that $b\xi + c\eta = 0$. Suppose that $\xi \neq 0$ and let $t = -\eta\xi^{-1}$. Then $b = ct$ and $a = ct^2$. That is

$$ax^2 + 2bxy + cy^2 = c(t^2x^2 + 2txy + y^2) = c(tx + y)^2.$$

Conversely, if $V_2(x, y)^2 = (lx + my)^2 = l^2x^2 + 2lmxy + m^2y^2$, then

$$V_2(m, -l) = \begin{bmatrix} l^2 & lm \\ lm & m^2 \end{bmatrix} \begin{bmatrix} m \\ -l \end{bmatrix} = [l^2m - lm \quad lm^2 - m^2l]$$
$$= [0 \quad 0].$$

(2) Suppose that $V_2\mathbf{u} = 0$, with $\mathbf{u} \neq 0$ and let $\mathbf{v}$ be any non-zero vector such that $V_2\mathbf{v}^2 = 0$. Then not only $V_2\mathbf{u}^2 = 0$ and $V_2\mathbf{v}^2 = 0$ but also $V_2(\mathbf{u} + \mathbf{v})^2 = V_2\mathbf{u}(\mathbf{u} + 2\mathbf{v}) + V_2\mathbf{v}^2 = 0$. Now the vectors $\mathbf{u}$, $\mathbf{v}$ and $\mathbf{u} + \mathbf{v}$ cannot all be root vectors unless two are linearly dependent. But then all three are. It follows that V_2 is parabolic.

The proof of the reverse implication is as before.

□

The next proposition will be of frequent application later.

Proposition 7.4 Let B_2 be a parabolic twice linear form on $\mathbb{R}^2$ with $B_2\mathbf{u} = 0$, where $\mathbf{u} \neq 0$, and let $A_1 \in L(\mathbb{R}^2, \mathbb{R})$ be such that $A_1\mathbf{u} = 0$. Then there exists $\mathbf{v} \in \mathbb{R}^2$ such that $A_1 = B_2\mathbf{v}$.

Proof Since the linear map $B_2 : \mathbb{R}^2 \to L(\mathbb{R}^2, \mathbb{R})$ has rank 1 its image is one-dimensional. So, for any vector $\mathbf{w} \in \mathbb{R}^2$ that is not in the kernel of B_2, $B_2\mathbf{w}$ spans the image of B_2, with $B_2\mathbf{w}\mathbf{u} = 0$. Accordingly $A_1 = \mu B_2\mathbf{w}$, for some $\mu \in \mathbb{R}$. Choose $\mathbf{v} = \mu\mathbf{w}$. Clearly the vector $\mathbf{v}$ is not unique. The vector $\mathbf{v} + \lambda\mathbf{u}$ would do equally well, for any $\lambda \in \mathbb{R}$. □

A non-zero quadratic form on $\mathbb{R}^2$ will be said to be *right-angled* if it is hyperbolic and if its root directions are mutually orthogonal with respect to the standard scalar product on $\mathbb{R}^2$.

Proposition 7.5 The non-zero quadratic form $ax^2 + 2bxy + cy^2$ is right-angled if and only if $a + c = 0$. □

A pair of real quadratic forms will generally be regarded as *equivalent* if either is a non-zero real multiple of the other.

The real quadratic forms on $\mathbb{R}^2$ form a vector space of dimension three. The set of equivalence classes then form a real projective space of dimension two. If we represent the non-zero form $ax^2 + 2bxy + cy^2$ by the point $[a, b, c]$ of $\mathbb{R}P^2$ then the parabolic forms correspond to points of the conic with equation $ac = b^2$. The tangent line to this conic at a point then consists of all the quadratic forms one of whose factors is the double factor of the form represented by the point of contact.

Given two twice linear forms V_2 and W_2 on $\mathbb{R}^2$ a vector $\mathbf{u}$ is said to be a *Jacobian* vector of the pair if the linear forms $V_2\mathbf{u}$ and $W_2\mathbf{u}$ are equivalent; the same as asking that the determinant of the 2×2 matrix

$$\begin{bmatrix} V_2\mathbf{u} \\ W_2\mathbf{u} \end{bmatrix}$$

should be zero. Up to a factor 2 this matrix is just the Jacobian matrix of the map $\mathbb{R}^2 \rightarrowtail \mathbb{R}^2 : \mathbf{u} \mapsto (V_2\mathbf{u}^2, W_2\mathbf{u}^2)$ – hence the name.

For the quadratic forms $ax^2 + 2bxy + cy^2$ and $a'x^2 + 2b'xy + c'y^2$ the Jacobian vectors are root vectors of the quadratic form

$$\begin{bmatrix} ax + by & bx + cy \\ a'x + b'y & b'x + c'y \end{bmatrix} = (ab' - a'b)x^2 + (ac' - a'c)xy$$
$$+ (bc' - b'c)y^2$$

and accordingly there are three cases depending on whether this quadratic form is elliptic, parabolic or hyperbolic. In the elliptic case the Jacobian vectors must be interpreted as vectors of $\mathbb{C}^2$ not of $\mathbb{R}^2$.

7.3 Cubic forms on $\mathbb{R}^2$

The classification of a symmetric thrice linear form on $\mathbb{R}^2$ and its associated cubic form is reducible to the classification of an associated quadratic form.

Consider the symmetric thrice linear form $V_3 = [a, b, c, d]_3$ on $\mathbb{R}^2$. A non-zero element $\mathbf{u} = (x, y) \in \mathbb{R}^2$ is said to be a *Hessian vector* of the form if and only if the twice linear form $V_3\mathbf{u} = [a, b, c, d]_3(x, y)$ is parabolic, the condition for this being that there is a non-zero vector $\mathbf{v} \in \mathbb{R}^2$ such that $V_3\mathbf{uv} = 0$, or equivalently such that the matrix of the quadratic form

$$\begin{bmatrix} ax + by & bx + cy \\ bx + cy & cx + dy \end{bmatrix},$$

has determinant zero, this being a quadratic condition on x and y. Up to a factor 6 this matrix is just the matrix of second derivatives of the map $\mathbb{R}^2 \rightarrowtail \mathbb{R}^2 : \mathbf{u} \mapsto V_3\mathbf{u}^3$. The quadratic form which is the determinant of the matrix, namely

$$(ac - b^2)x^2 + (ad - bc)xy + (bd - c^2)y^2,$$

is called the *Hessian* of the thrice linear form. The two root lines of this form are the *Hessian lines* of the form V_3. Since V_3 is symmetric $V_3\mathbf{vu} = V_3\mathbf{uv}$, for any $\mathbf{u}, \mathbf{v} \in \mathbb{R}^2$, implying that if $\mathbf{u}$ and $\mathbf{v}$ are non-zero vectors such that $V_3\mathbf{uv} = 0$ then $\mathbf{u}$ and $\mathbf{v}$ generate the Hessian lines of V_3.

Proposition 7.6 The condition for a vector (x, y) to be a Hessian vector of the thrice linear form $V_3 = [a, b, c, d]_3$ on $\mathbb{R}^2$ is independent of the choice of basis for $\mathbb{R}^2$.

Proof This is a direct consequence of the fact that the definition of Hessian lines does not involve the basis for $\mathbb{R}^2$ at all.

Alternatively, direct computation shows that the actual Hessian quadratic forms for different choices of basis differ by a factor which is the square of the (necessarily non-zero) determinant of the change of basis matrix. □

Clearly the Hessian of a thrice linear form may be elliptic, parabolic or hyperbolic. In the elliptic case the Hessian lines are not real but complex conjugate and in the parabolic case they coincide. Another possibility is that the Hessian may vanish, in which case the Hessian lines are not defined.

The following theorem classifies cubic forms on $\mathbb{R}^2$ in terms of their Hessian forms.

Theorem 7.7 Let V_3 be a thrice linear form on $\mathbb{R}^2$. Then the three root directions of its cubic form are real and distinct, are real with two directions coincident, are real with all three directions coincident, or are such that only one direction is real, the other two being complex conjugate, according as the Hessian directions are not real, but complex conjugate, are real and coincident, are not defined, since the Hessian vanishes, or are real and distinct, the four cases being known as the elliptic, parabolic, perfect *and* hyperbolic cases, *respectively.*

Sketch of proof The proof proceeds by case examination, it being enough to determine those cubic forms which have Hessians non-zero real multiples of $x^2 + y^2$, x^2, 0 and xy, since this exhausts all possible types and the concept of Hessian does not depend on choice of basis.

As an example, those cubic forms with Hessian a non-zero real multiple of xy are easily seen to be of the form $ax^3 + dy^3$, where a and

d are not both zero, that is of the form $(\alpha x)^3 + (\delta y)^3$ where α and δ are the real cube roots of a and d respectively. But this factorises into one real linear factor and an elliptic quadratic factor.

The details are left as an exercise. □

Proposition 7.8 The Hessian of the thrice linear form V_3 *on* $\mathbb{R}^2$ *is parabolic if and only if there is a non-zero vector* $\mathbf{u} \in \mathbb{R}^2$ *such that* $V_3\mathbf{u}^2 = 0$.

Proof Clearly if the Hessian of V_3 is parabolic and if $\mathbf{u}$ is a Hessian vector then $V_3\mathbf{u}^2 = 0$, for the kernel of $V_3\mathbf{u}$ consisting itself of Hessian vectors can only consist of multiples of $\mathbf{u}$. Conversely, suppose that $V_3\mathbf{u}^2 = 0$ but the Hessian is not parabolic. Then there must exist a Hessian vector $\mathbf{v}$, not a multiple of $\mathbf{u}$, such that $V_3\mathbf{v}^2 = 0$. But $V_3\mathbf{u}^2 = V_3\mathbf{v}^2$ implies that $V_3(\mathbf{u}+\mathbf{v})(\mathbf{u}-\mathbf{v}) = 0$, implying that $\mathbf{u}+\mathbf{v}$ and $\mathbf{u}-\mathbf{v}$ are also Hessian vectors of V_3, clearly not the case. That is the Hessian is parabolic. □

Until now, apart from our definition of a right-angled quadratic form, we have only considered the underlying space $\mathbb{R}^2$ as a vector space, without using in any essential way its scalar product and the derived concept of orthogonality. In the applications of cubic forms that we shall make to the theory of smooth surfaces in $\mathbb{R}^3$ the Euclidean structure of the ambient space plays an important role. The following property of the Hessian of a cubic form does involve this extra structure.

Proposition 7.9 Let $ax^3 + 3bx^2y + 3cxy^2 + dy^3$ *be a cubic form on* $\mathbb{R}^2$ *with coefficients* a, b, c, *and* d. *Then there exists a non-zero vector*

$$\begin{bmatrix} x \\ y \end{bmatrix},$$

such that the matrix

$$\begin{bmatrix} ax+by & bx+cy \\ bx+cy & cx+dy \end{bmatrix}$$

is a real multiple of the identity matrix, if and only if the Hessian lines of the cubic, the root lines of its Hessian, are mutually orthogonal, or the Hessian vanishes.

Proof There exist real numbers x, y, λ, with x, y not both zero, such that $ax + by = \lambda$, $bx + cy = 0$ and $cx + dy = \lambda$ if and only if the matrix

$$\begin{bmatrix} a & b & 1 \\ b & c & 0 \\ c & d & 1 \end{bmatrix}$$

has a non-zero kernel vector, but this is so if and only if its determinant $(ac - b^2) + (bd - c^2)$ is zero. But then the sum of the coefficients of the terms x^2 and y^2 in the Hessian of the cubic form is zero, implying that the Hessian lines are mutually orthogonal.

An exceptional case, where $\lambda = 0$ and $ac - b^2 = bd - c^2 = 0$, is when the cubic is a perfect cube and the Hessian vanishes, $ad - bc$ also being zero. □

We shall say that a cubic form on $\mathbb{R}^2$, furnished with its standard scalar product, is *right-angled* if the root directions *of its Hessian* are mutually orthogonal. Such a cubic form necessarily is hyperbolic. A right-angled cubic form is not to be confused with an elliptic form, two of whose root lines happen to be mutually orthogonal.

Put more succinctly, what Proposition 7.9 says is that a thrice linear form V_3 is right-angled if and only if there is a non-zero vector $\mathbf{w} \in \mathbb{R}^2$ such that the twice linear form $V_3\mathbf{w}$ is a non-zero scalar multiple of the form I_2.

There is an alternative geometrical version of Proposition 7.9 that will be useful in Chapter 11. First we prove the following Lemma.

Lemma 7.10 The necessarily real eigenvalues of a symmetric 2×2 real matrix

$$\begin{bmatrix} a & b \\ b & c \end{bmatrix}$$

are equal if and only if the matrix is a multiple of the identity matrix.

Proof The quadratic equation for the eigenvalues λ, namely

$$(a - \lambda)(c - \lambda) - b^2 = 0,$$

or equivalently

$$\lambda^2 - (a + c)\lambda + ac - b^2 = 0$$

has equal roots if and only if

$$(a + c)^2 = 4(ac - b^2)$$

that is

$$(a - c)^2 + 4b^2 = 0$$

that is $a = c$ and $b = 0$. □

Proposition 7.11 The set of points $(x, y, z) \in \mathbb{R}^3$ such that

$$\begin{vmatrix} px + qy + \kappa z & qx + ry \\ qx + ry & rx + sy + \kappa z \end{vmatrix} = 0$$

is a non-degenerate quadric cone with vertex the origin, that is it is not the union of a pair of planes through the origin, if and only if the thrice linear form $[p, q, r, s]_3$ is not right-angled.

Proof By the symmetry of the matrix

$$\begin{bmatrix} px + qy & qx + ry \\ qx + ry & rx + sy \end{bmatrix}$$

there are, for each $(x, y) \in \mathbb{R}^2$, two real values of z such that (x, y, z) belongs to the cone. By Proposition 7.9, these coincide for $(x, y) \neq (0, 0)$ if and only if the cubic form is right-angled, in which case they form the line of intersection of the two planes into which the quadric degenerates. □

The sections $z = \text{const}$ of the quadric cone of Proposition 7.11 are conics of various type according to the type of the cubic form. These are stated in the next proposition.

Proposition 7.12 The sections $z = \text{const}$ of the quadric cone associated to a cubic form are ellipses if the cubic is elliptic, parabolas if the cubic is parabolic, and hyperbolas if the cubic is hyperbolic, except that in the case that the cubic is right-angled (cf. Proposition 7.11) the sections are orthogonal line-pairs, the cone in this case degenerating to a plane-pair, while in the case that the cubic is perfect the cone degenerates to a plane-pair, one factor being the plane $z = 0$. □

In the hyperbolic case one can even go further.

Proposition 7.13 Let $[p, q, r, s]_3$ be a real thrice linear form on $\mathbb{R}^2$ that is hyperbolic but not right-angled. Then either each of the hyperbolic horizontal sections of the cone of Proposition 7.11 lies within the acute angle formed by its asymptotes or each lies within the obtuse angle formed by its asymptotes. □

A hyperbolic cubic form on $\mathbb{R}^2$ furnished with its usual scalar product will be said to be *acute-angled* or *obtuse-angled* according as any horizontal section of its quadric cone lies within the acute angle or the obtuse angle formed by its asymptotes.

Certain applications of cubic forms to the study of smooth surfaces involve the Jacobian of a pair of forms one of which is a positive-definite form, which by a suitable change of basis may be taken to be the twice linear form I_2, associated with the standard scalar product. We shall consider such a Jacobian shortly, after looking at cubic forms on $\mathbb{R}^2$ in quite another way.

7.4 Use of complex numbers

When the scalar product on $\mathbb{R}^2$ is taken to be the standard one it is sometimes convenient to identify $\mathbb{R}^2$ with the complex numbers $\mathbb{C}$ and to rewrite forms in x and y as forms in $z = x + iy$ and $\bar{z} = x - iy$, the quadratic form $x^2 + y^2$, for example, becoming the form $z\bar{z}$.

It is easily verified that the real quadratic form $ax^2 + 2bxy + cy^2$ may be written in the form $\gamma z^2 + 2rz\bar{z} + \bar{\gamma}\bar{z}^2$, where $4\gamma = a - c - 2bi$ and $2r = a + c$. Conversely, any such form is a real quadratic form in x and y for any $\gamma \in \mathbb{C}$, $r \in \mathbb{R}$.

Proposition 7.14 The quadratic form $\gamma z^2 + 2rz\bar{z} + \bar{\gamma}\bar{z}^2$ is parabolic if and only if $\gamma\bar{\gamma} = |\gamma|^2 = r^2$ and has mutually orthogonal root lines if and only if $r = 0$.

Proof Easy exercise. □

Similarly the real cubic form $ax^3 + 3bx^2y + 3cxy^2 + dy^3$ can be written as $\alpha z^3 + 3\bar{\beta}z^2\bar{z} + 3\beta z\bar{z}^2 + \bar{\alpha}\bar{z}^3$ for suitable $\alpha, \beta \in \mathbb{C}$. Conversely any such form is real.

The real cubic forms on $\mathbb{R}^2$ form a vector space of dimension four. Two such forms, neither of them the zero form, are commonly regarded as *equivalent* if either is a real multiple of the other. The set of equivalence classes then forms a three-dimensional real projective space. We can, however, go further and consider cubic forms to be equivalent if one is obtainable from another by a rotation of $\mathbb{R}^2$. To be explicit, consider the form $\alpha z^3 + 3\bar{\beta}z^2\bar{z} + 3\beta z\bar{z}^2 + \bar{\alpha}\bar{z}^3$ and suppose that $\alpha \neq 0$. Then by setting $\alpha^{1/3}z = \zeta$, where $\alpha^{1/3}$ denotes one of the three cube roots of α, the form reduces to one for which the first and last coefficients are 1, without losing any of the essential features of the form, such as the number of real linear factors. Suppose this done. The cubic form $z^3 + 3\bar{\beta}z^2\bar{z} + 3\beta z\bar{z}^2 + \bar{z}^3$ may then be represented by the complex number β. In this way the complex plane may be taken to be a map of almost all cubic forms. Those left out of the picture are those of the form $3\bar{\beta}z^2\bar{z} + 3\beta z\bar{z}^2 = 3z\bar{z}(\bar{\beta}z + \beta\bar{z})$, namely those that have $z\bar{z} = x^2 + y^2$ as a factor, and these may be thought of as lying *at infinity*.

Proposition 7.15 The cubic form $z^3 + 3\bar{\beta}z2\bar{z} + 3\beta z\bar{z}^2 + \bar{z}^3$ *is parabolic if and only if* $3\beta = 2\gamma + \bar{\gamma}^2$ *for some* $\gamma = e^{i\theta}$ *on the unit circle, where* 3θ *is not an integral multiple of* 2π.

Proof Such a form is parabolic if and only if it is not a perfect cube but has as a factor a parabolic quadratic form which, by Proposition 7.9, may be taken to be of the form $\gamma z^2 + 2z\bar{z} + \bar{\gamma}\bar{z}^2$, where $\gamma = e^{i\theta}$. The third factor must then be, up to a multiple, $\bar{\gamma}z + \gamma\bar{z}$. Equating the coefficients of $z\bar{z}^2$, we have $3\beta = 2\gamma + \bar{\gamma}^2$. Finally

$$\gamma z^2 + 2z\bar{z} + \bar{\gamma}\bar{z}^2 = (\bar{\gamma}z + \gamma z)^2 = \bar{\gamma}^2z^2 + 2z\bar{z} + \gamma^2\bar{z}^2$$

if and only if $\gamma = \bar{\gamma}^2$, that is if and only if $\gamma^3 = 1$. But this is the case if and only if 3θ is an integral multiple of 2π. □

Proposition 7.16 The cubic form $z^3 + 3\bar{\beta}z^2\bar{z} + 3\beta z\bar{z}^2 + \bar{z}^3$ *is right-angled if and only if* $|\beta| = 1$ $(\beta^3 \neq 1)$.

Proof Easy exercise. □

By Proposition 7.15 parabolic cubic forms are represented in the 'β-plane' by the *deltoid*, the image Δ of the curve $\theta \mapsto \frac{1}{3}(2e^{i\theta} + e^{-2i\theta})$, which has ordinary cusps wherever 3θ is a multiple of 2π, as is readily verified. Moreover, the images of these cusps lie at the cube roots of 1, namely 1, ω and ω^2. The curve, sketched in Figure 7.1, is that traced by a point on the circumference of a circle of radius $\frac{1}{3}$ rolling on the

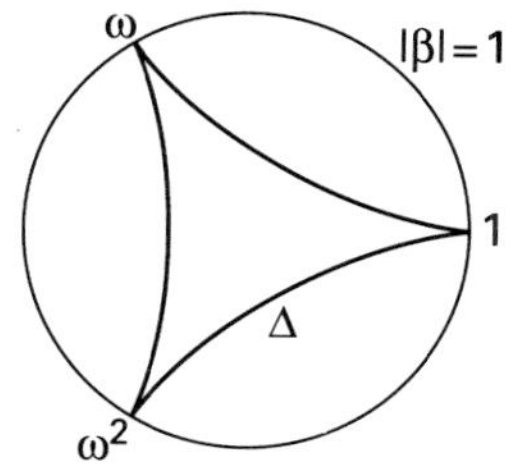

Figure 7.1

inside of the unit circle, touching the circle three times for each circuit of the unit circle, at 1, ω and ω^2. The points of the β-plane inside the deltoid represent elliptic forms, with in particular the origin representing the form $z^3 + \bar{z}^3 = 2(x^3 - 3xy^2)$. Those outside the deltoid represent hyperbolic forms. By Proposition 7.16 the points of the unit circle (other than the cusps of the deltoid) represent right-angled cubic forms. It is a nice exercise (Exercise 7.12) to determine which points of the plane represent acute-angled hyperbolic cubic forms and which represent obtuse ones!

Given homogeneous polynomials $p(x, y)$ and $q(x, y)$ in x and y of degrees n and m then the *Jacobian* of p and q, being the determinant of the Jacobian matrix of the map $\mathbb{R}^2 \to \mathbb{R}^2$; $(x, y) \mapsto (p(x, y), q(x, y))$, is a homogeneous polynomial in x and y of degree $(n-1)(m-1)$. In particular the Jacobian of a quadratic form and a cubic form is necessarily another cubic form.

In our description of the geometry round an umbilic of a surface we shall encounter two real cubic forms the second of which is the Jacobian of the quadratic form $x^2 + y^2 = z\bar{z}$ and the first cubic form.

Proposition 7.17 Let $P(r, \theta) = p(r\cos\theta, r\sin\theta)$, *where* $p(x, y)$ *is a homogeneous polynomial in* x, y *of degree* n. *Then, up to a multiple, the Jacobian of the polynomials* $x^2 + y^2$ *and* $p(x, y)$ *is, as a function of* r *and* θ, *equal to* $\partial P/\partial\theta$.

Proof Exercise. □

Corollary 7.18 The Jacobian of $z\bar{z}$ *and* $z^3 + 3\bar{\beta}z^2\bar{z} + 3\beta z\bar{z}^2 + \bar{z}^3$ *is, up to a multiple, equal to* $(iz)^3 - \bar{\beta}(iz)^2(\overline{iz}) - \beta(iz)(\overline{iz})^3 + (\overline{iz})^3$, *parabolic if and only if* $-(\frac{1}{3})\beta$ *lies on the tricuspidal curve* Δ, *that is if and only if* β *lies on the reflection in the origin of a tricuspidal curve three times the size of* Δ. *(See Figure 7.2.)* □

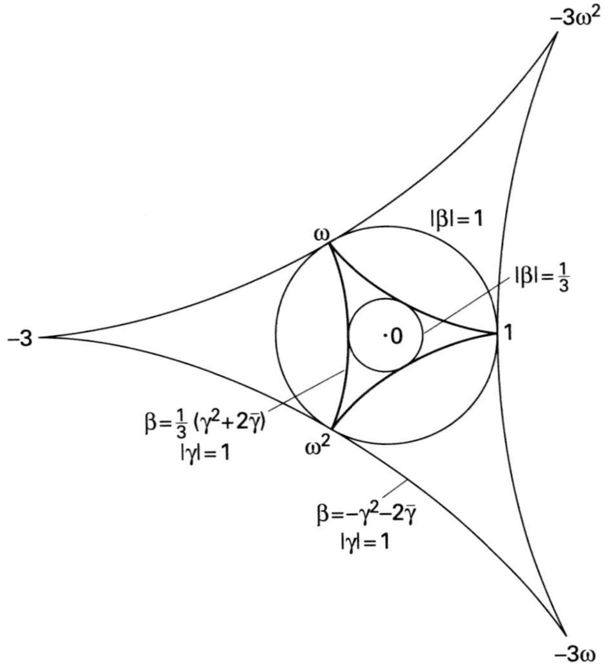

Figure 7.2

In Figure 7.2 two circles appear, not only the unit circle $|z| = 1$ representing the right-angled cubic forms but also the circle $|z| = \frac{1}{3}$. The geometrical significance of the smaller circle is indicated in the next proposition and the theorem which follows it.

Proposition 7.19 The cubic form $z^3 + 3\bar{\beta}z^2\bar{z} + 3\beta z\bar{z}^2 + \bar{z}^3$ *has two of its root lines mutually orthogonal if and only if* $|\beta| = \frac{1}{3}$. □

Theorem 7.20 Let $V_3(x, y)^3$ *be a cubic form on* $\mathbb{R}^2$ *and let* $J_3(x, y)^3$ *denote the Jacobian of* $V_3(x, y)^3$ *and the quadratic form* $x^2 + y^2$. *Then* V_3 *is right-angled if and only if two of the root-lines of* J_3 *are mutually orthogonal, these root lines being the root lines of the Hessian of* V_3. □

Yet another application of cubic forms to the geometry of a regular surface in $\mathbb{R}^3$ in the neighbourhood of an umbilic involves the concept of the *harmonic part* of a cubic form.

The *harmonic part* of a real quadratic form $\gamma z^2 + rz\bar{z} + \bar{\gamma}\bar{z}^2$ is defined to be the quadratic form $\gamma z^2 + \bar{\gamma}\bar{z}^2$. This is of course zero whenever $\gamma = 0$, that is whenever the form is a multiple of $z\bar{z} = x^2 + y^2$. A quadratic form equal to its harmonic part is what we have previously called an *orthogonal* quadratic form, its root lines being mutually orthogonal. We could just as well term such a form a *harmonic* quadratic form.

In similar fashion the *harmonic part* of a real cubic form $\alpha z^3 + 3\bar{\beta}z^2\bar{z} + 3\beta z\bar{z}^2 + \bar{\alpha}\bar{z}^3$ is defined to be the real cubic form $\alpha z^3 + \bar{\alpha}\bar{z}^3$. This is 0 whenever $\alpha = 0$, again whenever the form is a multiple of $z\bar{z}$. For $\alpha \neq 0$ the acute angle between any two of the root lines of the harmonic part is equal to $\pi/3$.

A cubic equal to its harmonic part is said to be *harmonic*.

Proposition 7.21 Let C_3 be any thrice linear form on $\mathbb{R}^2$ and let $\mathbf{u}$ *and* $\mathbf{v}$ *be vectors in $\mathbb{R}^2$ such that* $\mathbf{u} \cdot \mathbf{v} = 0$ *and* $\mathbf{v} \cdot \mathbf{v} = \mathbf{u} \cdot \mathbf{u}$. *Then $C_3\mathbf{u}^3 - 3C_3\mathbf{u}\mathbf{v}^2$ is equal to four times the harmonic part of $C_3\mathbf{u}^3$.*

Proof Using the complex numbers for vectors in $\mathbb{R}^2$ let $\mathbf{u}$ be denoted by z and $\mathbf{v}$ by $\mathrm{i}z$. Then, with $C_3\mathbf{u}^3 = \alpha z^3 + 3\bar{\beta}z^2\bar{z} + 3\beta z\bar{z}^2 + \bar{\alpha}\bar{z}^3$,

$$\begin{aligned} C_3\mathbf{u}^3 - 3C_3\mathbf{u}\mathbf{v}^2 &= \alpha z^3 + 3\bar{\beta}z^2\bar{z} + 3\beta z\bar{z}^2 + \bar{\alpha}\bar{z}^3 - 3(\alpha z(\mathrm{i}z)^2 \\ &\quad + 2\bar{\beta}z(\mathrm{i}z)(-\mathrm{i}\bar{z}) + \bar{\beta}(\mathrm{i}z)^2\bar{z} + 2\beta(\mathrm{i}z)\bar{z}(-\mathrm{i}\bar{z}) \\ &\quad + \beta z(-\mathrm{i}\bar{z})^2 + \bar{\alpha}\bar{z}(-\mathrm{i}\bar{z})^2) \\ &= 4\alpha z^3 + 4\bar{\alpha}\bar{z}^3. \end{aligned}$$ □

The harmonic, or orthogonal, quadratic forms form a two-dimensional linear subspace of the three-dimensional vector space of all quadratic forms on $\mathbb{R}^2$, a basis for the subspace consisting of the real and imaginary parts of z^2, namely $x^2 - y^2$ and $2xy$.

The harmonic cubic forms likewise form a two-dimensional linear subspace of the four-dimensional vector space of all cubic forms on $\mathbb{R}^2$, a basis for the subspace consisting of the real and imaginary parts of z^3, namely $x^3 - 3xy^2$ and $3x^2y - y^3$.

The terminology derives in either case from the fact that the forms are harmonic functions of x and y in the usual sense of analysis, namely that they satisfy the two-dimensional Laplace equation, the sum of the second derivatives with respect to x and with respect to y being 0.

As we have remarked, the real cubic forms on $\mathbb{R}^2$ form a vector space of dimension four, two such forms, neither of them the zero form, being regarded as *equivalent* if either is a real multiple of the other. The set of equivalence classes then forms a three-dimensional real projective space. Christopher Zeeman (1976) dubbed the set of equivalence classes of parabolic real cubic forms in this projective space the *umbilic* bracelet, the name deriving from the role that cubic forms play in the classification of umbilics, as we shall later see (Porteous, 1971). The tricuspidal deltoid Δ is just a section of this bracelet. As one travels once along the projective line of harmonic cubics the corresponding deltoid section twists through one third of a full rotation.

One of John Robinson's remarkable 'symbolic sculptures', *Eternity* (Figure 7.3), first shown to the mathematical community at the Pop Maths Roadshow at the University of Leeds in September 1989, almost models the umbilic bracelet – the 'fibre' in his sculpture, however, being an equilateral triangle, rather than a deltoid.

As a final comment it is worth remark that in applications of the material of this chapter the role of the form I_2, where $I_2(x, y)^2 = x^2 + y^2$, is often played by some other positive-definite twice linear form on $\mathbb{R}^2$. A fresh choice of basis orthonormal with respect to such a form immediately reduces the situation to the one considered here.

Exercises

7.1 Prove that a symmetric thrice linear form on $\mathbb{R}^2$ is uniquely determined by the cubic form that it induces.

7.2 Prove that the non-zero quadratic form $ax^2 + 2bxy + cy^2$ is right-angled if and only if $a + c = 0$.

7.3 Suppose that we represent the non-zero form $ax^2 + 2bxy + cy^2$ by the point $[a, b, c]$ of $\mathbb{R}P^2$. Prove that the parabolic forms correspond to points of the conic with equation $ac = b^2$ and that the tangent line to this conic at a point then consists of all the quadratic forms one of whose factors is the double factor of the form represented by the point of contact.

7.4 Prove that the Hessian quadratic forms of a thrice linear form on a two-dimensional real vector space X for different choices of basis differ by a factor which is the square of the (necessarily non-zero) determinant of the change of basis matrix.

7.5 Determine all those cubic forms whose Hessian is

(i) a non-zero multiple of xy,
(ii) a non-zero multiple of x^2,

(iii) a non-zero multiple of $x^2 + y^2$,
(iv) zero.

7.6 Prove that the Hessian quadratic forms of a thrice linear form on a two-dimensional real vector space X for different choices of

Figure 7.3

basis differ by a factor which is the square of the (necessarily non-zero) determinant of the change of basis matrix.

7.7 Verify that the quadratic form $\gamma z^2 + 2rz\bar{z} + \bar{\gamma}\bar{z}^2$, with $z = x + \mathrm{i}y$, $\gamma = \alpha + \mathrm{i}\beta \in \mathbb{C}$ and $r \in \mathbb{R}$, is a real quadratic form on $\mathbb{R}^2$. Verify that the form is parabolic if and only if $\gamma\bar{\gamma} = |\gamma|^2 = r^2$ and has mutually orthogonal root lines if and only if $r = 0$.

7.8 Prove Proposition 7.12.

7.9 Prove Proposition 7.13.

7.10 Prove Proposition 7.14.

7.11 Prove Proposition 7.16.

7.12 Determine which points of the 'β-plane' represent acute-angled cubic forms and which represent obtuse ones.

7.13 By considering when $\gamma z^2 + \bar{\gamma}\bar{z}^2$, with $|\gamma| = 1$, is a factor of $z^3 + 3\bar{\beta}z^2\bar{z} + 3\beta z\bar{z}^2 + \bar{z}^3$, prove that that cubic has two of its root lines mutually orthogonal if and only if $|\beta| = \frac{1}{3}$.

7.14 Prove Proposition 7.17.

7.15 Prove Corollary 7.18.

7.16 Show that the eigenspaces of the matrix

$$\begin{bmatrix} a & b \\ b & c \end{bmatrix},$$

where a, b and c are real numbers such that $(a - c)^2 + b^2 \neq 0$, are the root lines of the Jacobian of the quadratic forms $ax^2 + 2bxy + cy^2$ and $x^2 + y^2$, and verify that these root lines are mutually orthogonal.

7.17 Let V_2 be a (non-zero) symmetric twice linear form on $\mathbb{R}^2$ and V_3 a symmetric thrice linear form on $\mathbb{R}^2$ and suppose that there is a non-zero vector $\mathbf{a}_1$ in $\mathbb{R}^2$ such that $V_2\mathbf{a}_1 = 0$ and $V_3{\mathbf{a}_1}^3 = 0$. Prove that there exists a (possibly zero) vector $\mathbf{a}_2$ in $\mathbb{R}^2$ such that $V_3{\mathbf{a}_1}^2 + V_2\mathbf{a}_2 = 0$. What freedom is there in the choice of $\mathbf{a}_2$? (See Proposition 7.4.)

7.18 Let $ax^3 + 3bx^2y + 3cxy^2 + dy^3$ be a cubic form on $\mathbb{R}^2$ with real coefficients a, b, c, and d. Prove that there exists a non-zero vector

$$\begin{bmatrix} x \\ y \end{bmatrix},$$

such that the matrix

$$\begin{bmatrix} ax + by & bx + cy \\ bx + cy & cx + dy \end{bmatrix}$$

is a real multiple of the identity matrix, if and only if the Hessian lines of the cubic, the root lines of its Hessian, are mutually orthogonal.

7.19 Prove Proposition 7.19.

7.20 Prove Theorem 7.20.

8
Probes

8.0 Introduction

Much of this chapter should probably be skipped on a first reading, as it is rather more technical than what has gone before. Yet its logical place is here and its ideas do underlie much of what follows. It is here that some of the subtleties of the higher derivatives of a map begin to be unravelled, but it is only as these ideas find application in the geometry of the chapters that follow that they can really be appreciated. The final section of the chapter is given for its intrinsic interest, but will not be applied later.

8.1 Probes of smooth map-germs

Let $f : \mathbb{R}^n,a \rightarrowtail \mathbb{R}^p,f(a)$ be a smooth map-germ. Then a *k-probe* of f or *probe of index* k, is an immersive smooth map-germ $\sigma : \mathbb{R}^k,0 \rightarrowtail \mathbb{R}^n,a$, such that $(f\sigma)_1(0)$ $[= \mathrm{d}(f\sigma)0] = 0$. Asserting the existence of a probe of index k is equivalent to asserting that the dimension of the kernel of the first differential of f at a is greater than or equal to k. The map-germ f is said to be Σ^k if it has a k-probe but no $(k+1)$-probe. Necessarily $k \leqslant n$. Clearly f is Σ^k if and only if the dimension of the kernel of $f_1(a)$ is equal to k. Moreover, the probe σ may be taken simply to be the inclusion of the kernel of $f_1(a)$ in $\mathbb{R}^n$.

Suppose now that $\sigma : \mathbb{R}^k,0 \rightarrowtail \mathbb{R}^n,a$ is a k-probe of $f : \mathbb{R}^n,a \rightarrowtail \mathbb{R}^p,f(a)$ and that φ is a j-probe of $(f\sigma)_1 : \mathbb{R}^k,0 \rightarrowtail L(\mathbb{R}^k, \mathbb{R}^p)$, 0, that is an immersive map-germ $\varphi : \mathbb{R}^j,0 \rightarrowtail \mathbb{R}^k,0$ such that

$$((f\sigma)_1\varphi)_1(0)\ [= \mathrm{d}(\mathrm{d}(f\sigma)\varphi)0] = 0.$$

Then (σ, φ) is said to be a (k, j)-*probe* of f, or (*multi-*)*probe* of *multi-index* (k, j), f being said to be $\Sigma^{k,j}$ if it has a (k, j)-probe, but no $(k+1)$-probe and no $(k, j+1)$-probe. The probes σ and φ are the *terms* of the (multi-)probe (σ, φ).

For example the map-germ $f : \mathbb{R}^n,a \rightarrowtail \mathbb{R}^p,f(a)$ is $\Sigma^{k,k}$ if it has a k-probe $\sigma : \mathbb{R}^k,0 \rightarrowtail \mathbb{R}^n,a$ such that

$$(f\sigma)_1(0) = 0 \text{ and } (f\sigma)_2(0) = 0,$$

but no $(k+1)$-probe (and necessarily no $(k, k+1)$-probe), the j-probe $\varphi : \mathbb{R}^k,0 \rightarrowtail \mathbb{R}^k,0$ of $(f\sigma)_1$ in this case being the identity.

Example 8.1 The map-germ

$$f : \mathbb{R}^2, 0 \rightarrowtail \mathbb{R}^2, 0; (x, y), (0, 0) \mapsto (x^3 - 3xy, y), (0, 0)$$

is $\Sigma^{1,1}$, the map-germ $\sigma : \mathbb{R},0 \rightarrowtail \mathbb{R}^2,0;\ t,0 \mapsto (t, t^2), (0, 0)$, paired with the identity map-germ on $\mathbb{R}$ at 0, being an appropriate $(1, 1)$-probe. □

It is not difficult to see that whenever successive indices are equal, as is the case here, the linking term in the probe may without loss of generality be taken to be the identity.

The definitions extend indefinitely in the obvious way. For example, a (k, j, i)-probe of $f : \mathbb{R}^n,a \rightarrowtail \mathbb{R}^p,f(a)$, (σ, φ, χ), consists of a k-probe σ of f, a j-probe φ of $(f\sigma)_1$ and an i-probe χ of $((f\sigma)_1\varphi)_1$, the map-germ f being said to be $\Sigma^{k,j,i}$ if there is a (k, j, i)-probe but no $(k+1)$-probe, no $(k, j+1)$-probe and no $(k, j, i+1)$-probe.

It remains the case that where successive indices are equal the linking term in the probe may without loss of generality be taken to be the identity.

Note that in the special case that f is linear, with a (k, j, i)-probe, (σ, φ, χ) say, then $(f\sigma)_1(0) = f\sigma_1(0) = 0$, $((f\sigma)_1\varphi)_1(0) = f(\sigma_1\varphi)_1(0) = 0$ and $f((\sigma_1\varphi)_1\chi)_1(0) = 0$, so that the image of each relevant construct of the probe lies in the kernel of f. A similar conclusion clearly holds for any multi-index probe of a linear map f. This remark will be of relevance in the discussion of contact in Chapter 9.

A map $f : \mathbb{R}^n \rightarrowtail \mathbb{R}^p$ is said to be Σ^I *at* a, where $I = k, j, i, \ldots$ is a multi-index, if and only if its germ at a is Σ^I. The probe classification just given of a smooth map f at a point a of its domain, involving a single multi-probe only, is equivalent to the early classification of Boardman (1967). More detailed analysis of f at a is possible using more than one probe at a time, for example taking a multi-index probe together with those obtained from it by *elision*, the running together of successive terms of the probe. For example, the probes $(\sigma\varphi, \chi)$, $(\sigma, \varphi\chi)$ and $(\sigma\varphi\chi)$ are elisions of the probe (σ, φ, χ).

Table 8.1

$V\mathbf{a}$	$(V\mathbf{a})_1$	$(V\mathbf{a})_2$	$(V\mathbf{a})_3$	$(V\mathbf{a})_4$	$(V\mathbf{a})_5$
	$V_1\mathbf{a}$	$(V_1\mathbf{a})_1$	$(V_1\mathbf{a})_2$	$(V_1\mathbf{a})_3$	$(V_1\mathbf{a})_4$
		$V_2\mathbf{a}$	$(V_2\mathbf{a})_1$	$(V_2\mathbf{a})_2$	$(V_2\mathbf{a})_3$
			$V_3\mathbf{a}$	$(V_3\mathbf{a})_1$	$(V_3\mathbf{a})_2$
				$V_4\mathbf{a}$	$(V_4\mathbf{a})_1$
					$V_5\mathbf{a}$,

explicitly:

0	Aa	$Ba^2 + Ab$	$Ca^3 + 3Bab + Ac$	$Da^4 + 6Ca^2b + 3Bb^2 + 4Bac + Ad$	$Ea^5 + 10Da^3b + 15Cab^2 + 10Ca^2c + 10Bbc + 5Bad + Ae$
	A	Ba	$Ca^2 + Bb$	$Da^3 + 3Cab + Bc$	$Ea^4 + 6Da^2b + 3Cb^2 + 4Cac + Bd$
		B	Ca	$Da^2 + Cb$	$Ea^3 + 3Dab + Cc$
			C	Da	$Ea^2 + Db$
				D	Ea
					E

the next entry to the right in the first row being

$$Fa^6 + 15Ea^4b + 45Da^2b^2 + 20Da^3c + 15Cb^3 + 60Cabc + 15Ca^2d + 10Bc^2 + 15Bbd + 6Bae + Af$$

8.2 Probing a map-germ $V:\mathbb{R}^2 \rightarrowtail \mathbb{R}$

Some of the many possibilities are exemplified in the following detailed study of a map-germ $V:\mathbb{R}^2,0 \rightarrowtail \mathbb{R},0$. Since any probe between vector spaces of the same dimension may be taken to be the identity it follows that any single multi-index probe of V may be assumed to be of the form $({}^21, {}^21, \ldots, {}^21, \mathbf{a}, {}^11, \ldots, {}^11)$, where $\mathbf{a}:\mathbb{R},0 \rightarrowtail \mathbb{R}^2,0$ is immersive at 0 and ${}^21:\mathbb{R}^2,0 \to \mathbb{R}^2,0$ and ${}^11:\mathbb{R},0 \to \mathbb{R},0$ are the identity map-germs. In order to avoid a proliferation of subscripts it is a convenient shorthand in this particular case to employ the upper case letters A, B, C, ... to denote $V_1\mathbf{a}$, $V_2\mathbf{a}$, $V_3\mathbf{a}$, ... at 0 and the lower case letters a, b, c, ... to denote $\mathbf{a}_1(0)$, $\mathbf{a}_2(0)$, $\mathbf{a}_3(0)$, ..., it being permissible to think of the latter simply as vectors in $\mathbb{R}^2$, $\mathbf{a}$ being a map-germ $\mathbb{R},0 \rightarrowtail \mathbb{R}^2,0$. Since $\mathbf{a}$ is immersive at 0, $a = \mathbf{a}_1(0)$ is non-zero, but there is nothing to prevent any of the higher derivatives of $\mathbf{a}$ at 0 being zero.

From such a probe $({}^21, {}^21, \ldots, {}^21, \mathbf{a}, {}^11, \ldots, {}^11)$ many others may be formed by elision. For example from $({}^21, \mathbf{a}, {}^11)$ one may also form $({}^21, \mathbf{a})$, $(\mathbf{a}, {}^11)$ and $\mathbf{a}$ itself. Each of these will be associated to a particular entry in the 'Faà de Bruno Table', which we repeat here as Table 8.1. The game for a given map-germ V is to find those probes $\mathbf{a}$ which make zero the greatest number of entries in that table. As we previously remarked in Table 4.1, if the first few entries in any row are zero then each of the entries in the preceding row directly above these also is equal to zero.

It will be convenient to *code* a probe $\mathbf{a}$ by a *word* of 'non-decreasing' letters, where the last letter contains information about the first row in the table, the second last letter information about the second row in the table, the third last letter information about the third row of the table and so on. If the word contains say only three letters then that means that no significant entry in the fourth row is zero for $\mathbf{a}$. For example, the code CDE, of three letters only, describes a probe $\mathbf{a}$ for which the first entry in the fourth row of the table is non-zero but for which the entries up to and including $\underline{C}a$ in the third row, up to and including $\underline{D}a^3 + 3Cab + Bc$ $(= Da^3$, when $B = 0$ and $Ca = 0)$ in the second row, and up to and including $\underline{E}a^5 + \ldots$ $(= Ea^5$, when $A = 0$, $B = 0$, $Ca = 0$ and $Da^3 = 0)$ in the first row, are all zero, but the next entry to the right in each case is non-zero, and any subsequent zeros that might occur are of no consequence.

On occasion a row can be traversed *ad infinitum*, in which case we employ the letter Ω as an upper-case ∞. For example the code $C\Omega$ describes a probe $\mathbf{a}$ for which the first entry in the third row of the

Table 8.2

$A_1 : AA$ or $2A\Omega$				
$A_2 : BB$				
$A_3 : CC$; 0 or $2B\Omega$				
$A_4 : DD$	D_4 : 1 or $3BB\Omega$			
$A_5 : EE$; 0 or $2C\Omega$	$D_5 : BCC;\ BB\Omega$			
$A_6 : FF$	$D_6 : BDD$; 0 or $2BC\Omega;\ BB\Omega$	$E_6 : CCC$		
$A_7 : GG$; 0 or $2D\Omega$	$D_7 : BEE;\ BB\Omega$	$E_7 : CC\Omega$		
$A_8 : HH$	$D_8 : BFF$; 0 or $2BD\Omega;\ BB\Omega$	$E_8 : DDD$		
. . .	. . .	$\widetilde{E}_8 : DDE$; 0 or $2CEE$; 1 or 3 $CD\Omega$		
		$\widetilde{E}_8^{\infty}$: 1 or $3DD\Omega$; 0 or $2CEE$	$\widetilde{E}_8^0 : EEE;\ CD\Omega$	
		$T_{237} : DDE;\ CFF$	$S_{237} : EEF$	$S_{237}^0 : FFF$
		$T_{238} : DDE;\ CGG$; 0 or $2CE\Omega$	$S_{238} : EEG;\ DE\Omega$	
		$T_{239} : DDE;\ CHH$	$S_{238}^0 : EE\Omega$	$S_{239} : FFG$; 0 or $2DGG$ $S_{239}^0 : GGG$

$\widetilde{E}_7 : CCCC$; 0 or 2 or $4CCC\Omega$			
$T_{245} : CCDD$; or $2CCC\Omega$	$S_{245} : CDDD;\ CCC\Omega$		
$T_{246} : CCEE$; 0 or $2CCD\Omega$; 0 or $2CCC\Omega$	$S_{246} : CDD\Omega;\ CCC\Omega$	$T_{255} : 2CCDD$	$S_{255} : DDDD$
$T_{247} : CCFF$; 0 or $2CCC\Omega$	$S_{247} : CEEE;\ CCC\Omega$	$T_{256} : CCDD;\ CCEE$; 0 or $2CCC\Omega$	$S_{256} : DDD\Omega$

table is non-zero, but for which the entries up to $\underline{C}a^2 + Bb$ in the second row are zero and also *every* entry in the first row is zero.

Table 8.2 shows the highest codes that are attainable for a range of singularity types, where the notations for the various possibilities which then occur are those of Arnol'd (1972) or, for some of the later ones, of Brieskorn (1979). Their classifications are up to *contact equivalence* about which we have something to say in Chapter 9. It is a fact that within the range shown the probes are diagnostic of the various types.

For example A_1 germs are characterised by $A = 0$ and the non-existence of a non-zero vector a such that $Ba = 0$. In particular $B \neq 0$, so the code for any probe that exists must be a two letter one, with first letter A. There are in fact two cases according to whether the twice linear form B is elliptic or hyperbolic. It cannot be parabolic, for in that case there would be a non-zero vector a such that $Ba = 0$, which we have assumed not to be the case.

In the elliptic case there are no root directions of B. So no non-zero vector a exists such that $Ba^2 = 0$ and the best probe has the code AA. In the hyperbolic case there are two root directions, and for a in either of these $Ba^2 = 0$. Since $Ba \neq 0$ the second derivative b of φ at 0 can then be chosen such that $Ca^3 + 3Bab + Ac = Ca^3 + 3Bab = 0$; then c can be chosen such that $Da^4 + 6Ca^2b + 3Bb^2 + 4Bac = 0$, and so on for ever along the first row of the table. Thus in this case there are two probes each with code $A\Omega$. The vectors $a, b, c, \ldots$ that arise are not unique. Different choices just correspond to different parametrisations of the probe φ.

The A_2 case is the parabolic case that we have just excluded, with $A = 0$, $B \neq 0$, but $Ba = 0$ for some $a \neq 0$, with the additional condition that there does not exist a vector b such that $Ca^2 + Bb = 0$. An equivalent condition is that $Ca^3 \neq 0$, for, when $Ba = 0$, $Ca^3 = 0$ is implied by $Ca^2 + Bb = 0$, and the converse is easily established also (see Proposition 7.4 and Exercise 7.17). Thus the best probe in this case has code BB.

A_3 germs are characterised by $A = 0$, $B \neq 0$, but $Ba = 0$, for some $a \neq 0$, and $Ca^3 = 0$, with one further condition, namely that the equations $Ca^2 + Bb = 0$ and $Da^4 + 6Ca^2b + 3Bb^2 = 0$ are not simultaneously solvable for b. Now in considering these equations there is no loss of generality in supposing that b, if it exists, is orthogonal to a, since if b is a solution of either then so is $b + \lambda a$ for any λ. The second of these is then simply a quadratic equation for b. The condition that the equations do not have a common solution b is then equivalent to the assertion that the quadratic equation for b is not parabolic. In the elliptic case, one that we shall later style the *sterile* case, where the linear equation is solvable but the quadratic one is not, the best probe has code CC. In the hyperbolic case, that we shall later style the *fertile* case, such a probe also exists but additionally there are two probes with code $B\Omega$.

The A_4 case is just the parabolic case that we have just excluded, with the additional condition that $Ea^5 + \ldots = Ea^5 + 10Da^3b + 15Cab^2 \neq 0$. However, a vector c can be found such that $Da^3 + 3Cab + Bc = 0$. So the best probe in this case has code DD.

The higher A_ks are handled analogously.

Now suppose that $A = 0$ and $B = 0$. The equation $Ca^3 = 0$ is always solvable for $a \neq 0$, but there are several cases: the *elliptic* case, when the Hessian of the cubic form C is elliptic and the cubic equation determines three distinct real root directions, the *hyperbolic* case, when the Hessian of C is hyperbolic, but the cubic determines only one real root direction, the *parabolic* case, when the Hessian is parabolic and two root directions of the cubic equation coincide, the *perfect* case, when the Hessian vanishes and all three root directions of the cubic coincide, this being the case that the cubic form is a perfect cube, and finally the case that C is itself zero. The perfect case is characterised by the existence of $a \neq 0$ such that $Ca = 0$, when a is then the repeated root, while the parabolic case is characterised by the existence of $a \neq 0$ such that $Ca^2 = 0$, but $Ca \neq 0$, when a again is the repeated root.

The D_4 germs are characterised by $A = 0$ and $B = 0$ and the non-existence of any $a \neq 0$ such that $Ca^2 = 0$. Thus there are just two cases, the elliptic and the hyperbolic. In the elliptic case there are three probes with code $BB\Omega$ and in the hyperbolic case just the one.

A D_5 germ, by contrast, is completely characterised by the existence of $a \neq 0$ such that $A = 0$, $B = 0$, $Ca^2 = 0$, but $Ca \neq 0$, and $Da^4 \neq 0$, C in this case being parabolic, but not perfect. The best probe in this case thus has code BCC.

If $Da^4 = 0$ we are faced with the equations $Da^3 + 3Cab = 0$ and $Ea^5 + 10Da^3b + 15Cab^2 = 0$ for b and the situation is analogous to that encountered in the A_3 case. If there is no common root there is always a probe with code BDD. Such a probe exists in the hyperbolic case also but there are then two additional probes with code $BC\Omega$. These are the D_6 singularity types. The D_7 case is just the parabolic case where the above equations do have a common root, with the further condition that $Fa^6 + 15Ea^4b + 45Da^2b^2 + 15Cb^3 \neq 0$. The best code is then BEE. The higher D_k singularities are handled analogously.

The types E_6, E_7 and E_8 are briefly dealt with. For E_6 there exists $a \neq 0$ such that $A = 0$, $B = 0$ and $Ca = 0$, but $C \neq 0$ and $Da^4 \neq 0$, this clearly corresponding to the code CCC. As we have already remarked, the equation $Ca = 0$ is equivalent to the assertion that the cubic equation $Ca^3 = 0$ has three coincident roots; that is that the cubic form is a perfect cube. For E_7 there exists $a \neq 0$ such that $A = 0$, $B = 0$ and $Ca = 0$, but $C \neq 0$, and $Da^4 = 0$, but $Da^3 \neq 0$. But then $Ea^5 + 10Da^3b = 0$ is solvable for b, and so on for ever along the top row of the table. So the best code in this case is $CC\Omega$. Finally, for E_8 there exists $a \neq 0$ such that $A = 0$, $B = 0$, $Ca = 0$, but $C \neq 0$, and $Da^3 = 0$, but $Ea^5 \neq 0$. Then there exists b such that $Da^2 + Cb = 0$. So the best code is DDD.

In every case so far considered one can specify a *canonical form*, an example of the type considered to which any other of the same type is contact equivalent. For completeness we append here a list of canonical forms, each of map-germs at the origin in $\mathbb{R}^2$:

A_k: $x^{k+1} \pm y^2$, for example $A_3 : x^4 \pm y^2$ and $A_4 : x^5 + y^2$
D_k: $x^2y \pm y^{k-1}$, for example $D_4 : x^2y \pm y^3$ and $D_5 : x^2y + y^4$
E_6: $x^3 + y^4$
E_7: $x^3 + xy^3$
E_8: $x^3 + y^5$.

The significance of this list is that this is a complete list of the *simple singularities*, those whose canonical forms are clear cut and do not involve *moduli* in any way. A *modulus* is a real coefficient, any two distinct values of which give rise to inequivalent map-germs. Such moduli inevitably arise if one carries the classification further, as we will do in the next section. Probe analysis is not fine enough to detect such moduli in general though it may single out certain values of a modulus as allowing a different probe structure. In our treatment we have consciously played down such canonical forms, as the diagnostic probes introduced in this chapter will prove to be much more relevant to elucidating the nature of geometric features of surfaces in $\mathbb{R}^3$ such as ridges and umbilics, the main aim of the present work.

8.3 Optional reading

For the purposes of the rest of this book this is as far is one needs to probe, together with the fact from singularity theory that in each of the cases so far considered the probe structure of a map-germ characterises it up to contact equivalence, for which see Chapter 9. However, partly as an opportunity to correct several misprints in Table II of Porteous (1983a) we proceed to discuss the remaining entries of Table 8.2, prepared in collaboration with Stelios Markatis (1980). It is hoped that the details may intrigue the reader to read further elsewhere.

The next stage then is when $A = 0$, $B = 0$, $C \neq 0$, but there exists $a \neq 0$ such that $Ca = 0$, $Da^3 = 0$ and $Ea^5 = 0$. Our attention is then drawn at the same time to *three* equations for $b \in \mathbb{R}^2$, namely

$$Da^2 + Cb = 0 \in L_2(\mathbb{R}^2, \mathbb{R}),$$

$$Ea^4 + 6Da^2b + 3Cb^2 = 0 \in L(\mathbb{R}^2, \mathbb{R}),$$

$$Fa^6 + 15Ea^4b + 45Da^2b^2 + 15Cb^3 = 0 \in \mathbb{R}.$$

Note that the left-hand sides of the second and first equations are up to numerical factors the first and second 'derivatives' of the left-hand side of the third equation 'with respect to b', while since $Ca = 0$, $Da^3 = 0$ and $Ea^5 = 0$, the first equation is essentially a single linear equation for b rather than three and the second a single quadratic equation rather than two.

To get some feel for what such a map-germ might look like suppose that $a = (0, 1)$. Then $A = 0$ means that there are no linear terms. $B = 0$ means that there are no quadratic terms, $Ca = 0$ means that the only cubic term is the term involving x^3, $Da^3 = 0$ means that the only quartic terms are those involving x^4, x^3y and x^2y^2, while $Ea^5 = 0$ means that the quintic terms are those involving x^5, x^4y, x^3y^2, x^2y^3 and xy^4. An example of such a function is $x^3 + 3\lambda x^2y^2 + 3\mu xy^4 + \nu y^6$, a cubic form in x and y^2. It turns out that the map-germ is indeed reducible to one of this form, and after further changes of coordinates either λ or μ may be chosen to be 0 and ν to be equal to 1, -1 or 0. The three equations for $b = (\beta, 0)$, after cancellation of numerical factors, then become

$$8\nu + 3\mu(4\beta) + 3\lambda(2\beta^2 + \beta^3) = 0,$$

$$4\mu + 2\lambda(2\beta) + \beta^2 = 0 \text{ and } 2\lambda + \beta = 0.$$

The general case, styled $\widetilde{E}_8$, is when no two of these equations have a common root, in which case the cubic is either elliptic or hyperbolic, as also is the quadratic. There will certainly be a probe with code DDE, with $\beta = 2\lambda$. There will also be two supplementary probes of type CEE when the quadratic is hyperbolic, which is certainly the case when the cubic is elliptic and possibly so when the cubic is hyperbolic. Finally there is a further probe of type $CD\Omega$ in the case that the cubic is hyperbolic and three further such probes when the cubic is elliptic. The canonical form for such a map-germ is $x^3 + 3\mu xy^4 + y^6$, with $\mu \neq 0$ and $\mu^3 \neq -\frac{1}{4}$, the quadratic equation being elliptic if $\mu > 0$ and hyperbolic if $\mu < 0$ and the cubic equation being elliptic if $\mu^3 < -\frac{1}{4}$ and hyperbolic if $\mu^3 > -\frac{1}{4}$.

The case that the quadratic equation is parabolic is when the linear and quadratic equations have a common solution, that is when there exists b such that $Da^2 + Cb = 0$ and $Ea^4 + 3Da^2b$ $(= Ea^4 + 6Da^2b + 3Cb^2) = 0$. There is then a probe of type EEE, for since $Ca = 0$ and $Da^2 + Cb = 0$, c can be found such that $Ea^3 + 3Dab + Cc = 0$. In this case the 5-jet is completable to a perfect cube. In coordinate terms the 5-jet is completable to a perfect cube in x and y^2, by the addition of a suitable multiple of y^6, and so by an obvious coordinate change it is reducible to x^3. That is coordinates may be so chosen around 0 in $\mathbb{R}^2$ that $D = 0$ and $E = 0$ for the map-germ f. Then $b = 0$ and $c = 0$ for the probe. Provided that the common root b does not satisfy the cubic equation there is no probe of type EEF, and the canonical form reduces to $x^3 + y^6$. We denote this type by $\widetilde{E}_8^0$, the modulus μ being 0. There is a supplementary probe of type $CD\Omega$ in this case, the cubic for b being hyperbolic.

The next case to consider, styled $\widetilde{E}_8^\infty$, is where the linear and cubic equations have a common solution that is not a solution of the quadratic equation. This case turns out to be that in which the 6-jet is reducible to $x^3 \pm 3xy^4$, the sign being positive or negative according to whether the cubic is hyperbolic or elliptic or, equivalently, whether the quadratic is elliptic or hyperbolic. There are accordingly either one or three probes with code $DD\Omega$, with two supplementary probes of type CEE in the former case.

The types $\widetilde{E}_8^0$ and $\widetilde{E}_8^\infty$ are known as the *equianharmonic* and *harmonic* $\widetilde{E}_8$s respectively, preferably regarded as distinct singularity types.

The third possibility is when the quadratic and cubic equations have a common solution. Then the cubic is parabolic, the infinite run along the top row of Table 8.1 does not in general occur, but $c \in \mathbb{R}^2$ can be found such that $Fa^5 + 10Ea^3b + 15Dab^2 + 10(Da^2 + Cb)c = 0$. The 6-jet of f at 0 can then be taken to be $x^3 + \lambda x^2y^2$, with $\lambda \neq 0$, provided that the solution of the linear equation remains distinct from the others. In the case that $Ga^7 + \ldots \neq 0$ the canonical form is $x^3 + \lambda x^2y^2 + y^7$, the code CFF supplements the code DDE and the map-germ is said to be of type T_{237}. The further study of the T_{23r} map-germs is analogous to our earlier study of the D_k map-germs.

There remains the possibility that all three equations share a common root. It turns out that in this case we may choose coordinates such that $D = 0$ and $E = 0$, when the common root b is 0. Then the first three rows of Table 8.1 from and to the right of the 'E' column reduce to

0	0	Ga^7	$Ha^8 + 56Fa^5c$	$Ia^9 + \ldots$
0	Fa^5	$Ga^6 + 10Cc^2$	$Ha^7 + 35Fa^4c + 35Ccd$	$Ia^8 + \ldots$
Cc	$Fa^4 + Cd$	$Ga^5 + Ce$	$Ha^6 + 20Fa^3c + Cf$	$Ia^7 + \ldots.$

Suppose first that $Fa^5 \neq 0$ and that $Ga^7 \neq 0$. This is the first '*triangular*' map-germ, denoted by Brieskorn (1979) by S_{237}, with canonical form $x^3 + \alpha xy^5 + y^7$, with $\alpha \neq 0$. (The name arises from a context in the theory of map-germs $\mathbb{R}^3 \rightarrowtail \mathbb{R}$ involving certain 'hyperbolic triangles'. It is not possible to explain the terminology further here.) The best probe is of type EEF. If $\alpha = 0$ then $Fa^5 = 0$ and d can be found such that $Fa^4 + Cd = 0$. The best probe is now of type FFF. We denote this distinct type by S^0_{237}.

Next suppose that $Fa^5 \neq 0$, but that $Ga^7 = 0$ and that $Cc = 0$ and $Ha^8 + 56Fa^5c = 0$ are not simultaneously solvable for c, that is that $Ha^8 \neq 0$. Then the best probe is of type EEG, with a supplementary probe of type $DE\Omega$. This is the triangular map-germ of type S_{238}, with canonical form $x^3 + xy^5 + \alpha y^8$, with $\alpha \neq 0$. When $\alpha = 0$ the two equations have the common solution $c = 0$, $Ha^8 = 0$ and there is a probe of type $EE\Omega$. We denote this distinct type by S^0_{238}.

Finally suppose that $Fa^5 = 0$ and $Ga^7 = 0$, but that $Ha^8 \neq 0$. Then in general $Cc = 0$ and $Ga^6 + 10Cc^2 = 0$ do not have a common solution c. This is the triangular map-germ of type S_{239}, with canonical form $x^3 + \alpha xy^6 + y^8$, with $\alpha \neq 0$, the quadratic being elliptic or hyperbolic according as α is positive or negative. The best probe is then of type FFG, with two supplementary probes of type DGG in the case that the quadratic form is hyperbolic. In the parabolic case $c = 0$, $Ga^6 = 0$ and $e \in \mathbb{R}^2$ can be found such that $Ga^5 + Ce = 0$. The canonical form is then $x^3 + y^8$ and the best probe is then of type GGG. We denote this distinct type by S^0_{239}.

Until now we have assumed that $A = 0$ and $B = 0$ but that $C \neq 0$.

We look very briefly at the first few options when $C = 0$ also, when the first four rows of the D, E and F columns of Table 8.1 reduce to:

$$
\begin{array}{lll}
Da^4 & Ea^5 + 10Da^3b & Fa^6 + 15Ea^4b + 45Da^2b^2 + 20Da^3c \\
Da^3 & Ea^4 + 6Da^2b & Fa^5 + 10Ea^3b + 15Dab^2 + 10Da^2c \\
Da^2 & Ea^3 + 3Dab & Fa^4 + 6Ea^2b + 3Db^2 + 4Dac \\
Da & Ea^2 + Db & Fa^3 + 3Eab + Dc.
\end{array}
$$

We proceed just far enough to note the first cases of failure of probe analysis in its simple form, but nevertheless to determine characteristic codes for the remaining five triangular singularity types which can occur for map-germs $\mathbb{R}^2, 0 \rightarrowtail \mathbb{R}, 0$.

The singularity type $\widetilde{E}_7$ is characterised by the non-existence of $a \neq 0$, real or complex, such that $Da^3 = 0$. By mentioning the complex case here we are able to exclude the case that the quartic form D has repeated complex conjugate roots. Any probe satisfies $CCCC$ trivially, but there may be zero, two or four real probes of code $CCC\Omega$, according to the special nature of the quartic form D.

The case that $Da^3 = 0$ has one non-zero solution a but $Da^2 = 0$ has none leads to the T_{24r} series of singularity types, where $r > 4$, while if $Da^3 = 0$ has two distinct solutions, either both real or a complex conjugate pair, that is if the 4-jet is reducible either to x^2y^2 or to $(x^2 + y^2)^2$, then we have the T_{2qr} singularity types, with q and r both greater than 4. In the complex case they will be of type T_{2qq}. The characteristic probe codes are easily determined in these cases.

Next we suppose that $Da^2 = 0$ has a non-zero solution a but that $Da = 0$ has none. If $Ea^5 \neq 0$ we have the triangular singularity type S_{245}, with canonical form $x^3y + \alpha xy^4 + y^5$ and code $CDDD$. If $Ea^5 = 0$ but $Ea^4 \neq 0$ we have the triangular type S_{246} with canonical form $x^3y + \alpha x^2y^3 + xy^4$ and code $CDD\Omega$. If $Ea^4 = 0$ but $Fa^6 \neq 0$ we have the type S_{247} with canonical form $x^3y + \alpha xy^5 + y^6$ and code $CEEE$, for b can then be found such that $Ea^3 + 3Dab = 0$.

Finally we suppose that $Da = 0$ has a non-zero solution a but that $D \neq 0$. If $Ea^5 \neq 0$ we have the triangular type S_{255} with canonical form $x^4 + y^5 + \alpha x^2y^3$ and code $DDDD$, while if $Ea^5 = 0$ but $Ea^4 \neq 0$ we have the type S_{256} with canonical form $x^4 + xy^4 + \alpha y^6$ and code $DDD\Omega$.

In none of these five cases does any probe pick out any particular value of the modulus α as being significantly different from the others.

In conclusion we suppose still that $Da = 0$ has a non-zero solution a, with $D \neq 0$ but $Ea^4 = 0$. Here our probe analysis fails to analyse the quadratic forms in x^2 and y^3 that then turn up.

Despite this apparent failure of probe analysis, curve probes may be used to define all the *Thom–Boardman* singularities of map-germs $\mathbb{R}^2,0 \rightarrowtail \mathbb{R},0$. The only probes used in their definition are those whose code word, of finite length, consists in repetitions of a single letter. The best probes for the first few Thom–Boardman symbols are set out in Table 8.3.

One of the earliest papers in singularity theory is that of Whitney (1955) that proves that the only *generic* singularities of a smooth map of the plane $\mathbb{R}^2$ to the plane are right–left equivalent either to the *fold* or the *(Whitney) cusp* or *pleat*, the Thom–Boardman symbol being 1,0 for the fold and 1,1,0 for the pleat. Explicitly a smooth map $f: \mathbb{R}^2 \rightarrowtail \mathbb{R}^2$ has Thom–Boardman symbol 1,0 at (x, y) if at that point there is a non-zero vector $\mathbf{a}_1 \in \mathbb{R}^2$ such that $f_1\mathbf{a}_1 = 0$ but no $\mathbf{a}_2 \in \mathbb{R}^2$ such that $f_2\mathbf{a}_1{}^2 + f_1\mathbf{a}_2 = 0$. There is an alternative version of this latter condition. At a point $(x, y) \in \mathbb{R}^2$ where $f_1(x, y): \mathbb{R}^2 \to L(\mathbb{R}^2, \mathbb{R})$ has a non-zero kernel vector $\mathbf{a}_1$ there will be a non-zero $\boldsymbol{\omega} \in L(\mathbb{R}^2, \mathbb{R})$ such that $\boldsymbol{\omega} f_1(x, y) = 0$. It is then easy to prove that at (x, y) there exists a vector $\mathbf{a}_2 \in \mathbb{R}^2$ such that $f_2\mathbf{a}_1{}^2 + f_1\mathbf{a}_2 = 0$ if and only if $\boldsymbol{\omega} f_2\mathbf{a}_1{}^2 = 0$. The expression $\boldsymbol{\omega} f_2\mathbf{a}_1{}^2$ has been called the *second intrinsic derivative* of f (Porteous, 1962). Likewise a smooth map $f: \mathbb{R}^2 \rightarrowtail \mathbb{R}^2$ has Thom––Boardman symbol 1,1,0 at (x, y) if at that point there is a non-zero vector $\mathbf{a}_1 \in \mathbb{R}$ such that $f_1\mathbf{a}_1 = 0$ and a vector $\mathbf{a}_2 \in \mathbb{R}^2$ such that $f_2\mathbf{a}_1{}^2 + f_1\mathbf{a}_2 = 0$, but no vector $\mathbf{a}_3 \in \mathbb{R}^2$ such that $f_3\mathbf{a}_1{}^3 + 3f_2\mathbf{a}_1\mathbf{a}_2 + f_1\mathbf{a}_3 = 0$. An alternative version of this last equation is $\boldsymbol{\omega} f_3\mathbf{a}_1{}^3 - 3\boldsymbol{\omega} f_2\mathbf{a}_1 f_1{}^{-1} f_2\mathbf{a}_1{}^2 = 0$, where the expression $f_1{}^{-1} f_2\mathbf{a}_1{}^2$ represents any of the solutions $\mathbf{a}_2$ of the equation $f_2\mathbf{a}_1{}^2 + f_1\mathbf{a}_2 = 0$, no matter which, since any two solutions differ by a real multiple of $\mathbf{a}_1$ and we already have $\boldsymbol{\omega} f_2\mathbf{a}_1\mathbf{a}_1 = 0$. The expression $\boldsymbol{\omega} f_3\mathbf{a}_1{}^3 - 3\boldsymbol{\omega} f_2\mathbf{a}_1 f_1{}^{-1} f_2\mathbf{a}_1{}^2$ is then the *third intrinsic derivative* of f.

Full proofs of all the results of this chapter require the sophisticated theorems of singularity theory that provide effective criteria for determining when an obviously irrelevant 'tayl' of the Taylor series of a map-germ can be removed by local diffeomorphisms of the source and target spaces of the map. The trade term for such equivalence of mapgerms is *$\mathcal{A}$-equivalence*. In certain contexts, which include most of those relevant in this book, an alternative equivalence of map-germs is more appropriate. This is *contact equivalence* or *$\mathcal{K}$-equivalence*, which we discuss in the next chapter.

Table 8.3

Thom–Boardman symbol	Best probe	Singularity types included
2,0	AA	A_1
2,1,0	BB	A_2
2,1,1,0	CC	A_3,
2,1,1,1,0	DD	A_4, and so on
2,2,0	BBB	$D_k,\ k \geqslant 4$
2,2,1,0	CCC	$E_6,\ E_7$
2,2,1,1,0	DDD	$E_8,\ \widetilde{E}_8,\ \widetilde{E}_8^{\infty},\ T_{23r},\ r \geqslant 7$
2,2,1,1,1,0	EEE	$\widetilde{E}_8^0,\ S_{237},\ S_{238},\ S_{238}^0$
2,2,1,1,1,1,0	FFF	$S_{237}^0,\ S_{239}$
2,2,1,1,1,1,1,0	GGG	S_{239}^0
2,2,2,0	$CCCC$	$\widetilde{E}_7,\ T_{2qr},\ q \geqslant 4,\ r \geqslant 5,$ $S_{245},\ S_{246},\ S_{247}$
2,2,2,1,0	$DDDD$	$S_{255},\ S_{256}$

Exercises

8.1 Determine the subset $\Sigma^1 f$, $\Sigma^{1,1} f$ and $\Sigma^{1,1,1} f$ of $\mathbb{R}^3$ in the case that f is the map

$$\mathbb{R}^3 \to \mathbb{R}^3;\ (x, y, z) \mapsto (x, y, \tfrac{1}{4}z^4 - xz^2 - yz),$$

showing that each is a smooth submanifold of $\mathbb{R}^3$.

8.2 Let $h : \mathbb{R}^2 \rightarrowtail \mathbb{R}$ be a smooth function. Determine the subsets $\Sigma^1 f$ and $\Sigma^2 f$ of $\mathbb{R}^2 \times \mathbb{R}^2$ in the case that f is the map

$$\mathbb{R}^2 \times \mathbb{R}^2 \rightarrowtail \mathbb{R} \times \mathbb{R} \times \mathbb{R}^2;\ (w, x) \mapsto ((w - x)\cdot(w - x),\ h(w),\ x).$$

Verify that $(w, x) \in \Sigma^1 f$ if x lies on the normal to the level curve of h through w, or, in the special case that $h_1(w) = 0$, if w is a critical point of h and x is any point of $\mathbb{R}^2$, and that $(w, x) \in \Sigma^2 f$ if w is a critical point of h and $x = w$.

What can you say about $\Sigma^{1,1} f$ and $\Sigma^{1,1,1} f$?

(We shall discuss the locus of vertices of the level curves of a height function in a slightly different way in Chapter 15.)

9
Contact

9.0 Introduction

The study of the curvature of a plane curve may be presented as the study of the contact of the curve with circles, while the study of the curvature of a space curve may be presented as the study of the contact of the curve with spheres. In the chapters that follow we shall be studying the curvature of a surface in $\mathbb{R}^3$ and this may be presented in terms of the contact of the surface not only with spheres but also with circles. This is therefore an appropriate point to clarify what is meant by the term *contact*.

9.1 Contact equivalence

We start with two smooth submanifolds M and M' of $\mathbb{R}^n$ with a point a in common. The submanifolds need not be of the same dimension. For example if M is a surface in $\mathbb{R}^3$ then M' might be a sphere or plane of $\mathbb{R}^3$, or a circle lying in a plane through a in $\mathbb{R}^3$, or simply a line through a in $\mathbb{R}^3$. The *contact* at a point b of a second pair of smooth submanifolds W and W' of $\mathbb{R}^n$ is said to be *equivalent* to that of M and M' at a if there is a local diffeomorphism $\mathbb{R}^n, a \rightarrowtail \mathbb{R}^n, b$ that sends W to M and W' to M'.

We have lots of choice as to how to represent M and M' near a. For example M may be represented either *explicitly*, that is *parametrically*, as the image of an immersive map $\mathbf{r} : \mathbb{R}^{m'}, 0 \rightarrowtail \mathbb{R}^n, a$ or *implicitly* as the set of zeros of a submersive map $F' : \mathbb{R}^n, a \rightarrowtail \mathbb{R}^{n-m'}, 0$.

Probe structures provide useful invariants of contact. For example:

Theorem 9.1 With the above notations, the probe structures at 0 of the smooth map-germs $F'\mathbf{r} : \mathbb{R}^m,0 \rightarrowtail \mathbb{R}^{n-m'},0$ *and* $F\mathbf{r} : \mathbb{R}^{m'},0 \rightarrowtail \mathbb{R}^{n-m},0$ *are equivalent.* □

In fact it is possible to prove more, namely:

Theorem 9.2 Let $\mathbf{r} : \mathbb{R}^m,0 \rightarrowtail \mathbb{R}^n,a$, *and* $\mathbf{r}' : \mathbb{R}^{m'},0 \rightarrowtail \mathbb{R}^n,a$ *be immersive maps and* $F : \mathbb{R}^n,a \rightarrowtail \mathbb{R}^{n-m},0$ *and* $F' : \mathbb{R}^n,a \rightarrowtail \mathbb{R}^{n-m'},0$ *submersive maps. Then the probe structures of the map-germs*

$$F'\mathbf{r} : \mathbb{R}^m,0 \rightarrowtail \mathbb{R}^{n-m'},0,$$

$$F\mathbf{r}' : \mathbb{R}^{m'},0 \rightarrowtail \mathbb{R}^{n-m},0,$$

$$(F, F') : \mathbb{R}^n,a \rightarrowtail \mathbb{R}^{n-m} \times \mathbb{R}^{n-m'},(0, 0),$$

and

$$(\mathbf{r} \;-\mathbf{r}') : \mathbb{R}^m \times \mathbb{R}^{m'},(0, 0) \rightarrowtail \mathbb{R}^n,0; (w, w'),0 \mapsto \mathbf{r}(w) - \mathbf{r}'(w'),0$$

are equivalent.

Proof We prove the equivalence of the first and third, the equivalence of the second and third then being similar. The proof of the fourth with the others is left as an exercise.

Consider first a k-probe at 0 of $F'\mathbf{r}$, that is an immersive map-germ $\sigma : \mathbb{R}^k,0 \rightarrowtail \mathbb{R}^m,0$ such that $(F'\mathbf{r}\sigma)_1 0 = 0$. Then $\mathbf{r}\sigma : \mathbb{R}^k,0 \rightarrowtail \mathbb{R}^n,0$ is a k-probe at a of (F, F'), since $((F, F')\mathbf{r}\sigma)_1 0 = (0, F'\mathbf{r}\sigma)_1 0 = 0$.

Conversely, suppose that $\varphi : \mathbb{R}^k,0 \rightarrowtail \mathbb{R}^n,a$ is a k-probe at a of (F, F'). The trick is to choose a diffeomorphism-germ $h : \mathbb{R}^n,a \rightarrowtail \mathbb{R}^n,0$ such that $Fh^{-1} : \mathbb{R}^n,0 \rightarrowtail \mathbb{R}^{n-m},0$ is a restriction of a *linear* map. This we can do since F is submersive, by the Inverse Function Theorem. The map-germ $h\varphi : \mathbb{R}^k,0 \rightarrowtail \mathbb{R}^n,0$ then is a k-probe at 0 of $(Fh^{-1}, F'h^{-1})$.

Thus in searching for a k-probe at 0 of $F'\mathbf{r}$ there is no loss of generality in supposing that F is linear and that the point $\mathbf{a}$ is the origin. This we do. The image of the immersive map $\mathbf{r} : \mathbb{R}^m,0 \rightarrowtail \mathbb{R}^n,0$ then coincides near 0 with $\ker F$. That is, $\mathbf{r}$ near 0 is the composition of a diffeomorphism from $\mathbb{R}^m$ to $\ker F$ followed by the inclusion of $\ker F$ in $\mathbb{R}^n$.

Now, since F is linear, $(F\varphi)_1(0) = F\varphi_1(0)$, implying that the image of $\varphi_1(0)$ lies in the kernel of F. But this kernel coincides near the origin with the submanifold M and so with the image of the immersive map $\mathbf{r}$. But, by what we have just proved about $\mathbf{r}$, $\varphi_1(0)$ factors through $\mathbb{R}^m$ as $\mathbf{r}\sigma$, say, where $\sigma : \mathbb{R}^k,0 \rightarrowtail \mathbb{R}^m,0$ is immersive. Now $\varphi_1(0)$ is not only an alternative k-probe of F but also an alternative k-probe of F'. From this it follows at once that σ is a k-probe of $F'\mathbf{r}$.

Higher order probes are handled in the same way. In the case of a (k, j)-probe (φ, χ), for example, with F supposed linear as before, all the relevant first and second derivatives of φ have their images lying in the kernel of F, the image of $\mathbf{r}$. Replacing φ by its 2-jet in the case that χ is the identity and by the relevant part of its 2-jet in the general case, this then factoring through $\mathbb{R}^m$ as before as $\mathbf{r}\sigma$, we then have at once that (σ, χ) is a (k, j)-probe at 0 for $F'\mathbf{r}$. In the case of a (k, j, i)-probe (φ, χ, ψ) it will be the relevant 3-jet of φ that replaces φ, the (k, j, i)-probe at 0 of $F'\mathbf{r}$ being (σ, χ, ψ), and so on. □

In conclusion it hardly needs to be spelled out that we get equivalent probe structures no matter which representatives we choose for the sub-manifolds M and M' at 0. For example suppose that M near a is defined explicitly by $\mathbf{s}$, rather than by $\mathbf{r}$ and M' is defined implicitly by G' rather than by F'. Then the probe structures of $F'\mathbf{R}$ and $G'\mathbf{s}$ at 0 are equivalent, each being equivalent, by the Theorem, to the probe structure of $F\mathbf{r}'$ at 0.

Example 9.3 Consider the case of a regular plane curve and a line in the plane $\mathbb{R}^2$. Instead of representing the curve parametrically and the line as the zeros of an equation we can do the opposite. So let $F : \mathbb{R}^2,a \rightarrowtail \mathbb{R},0$ be a smooth submersive map, and $\mathbb{R} \to \mathbb{R}^2;\ t \mapsto a + tb$ a line through a, the vector b being non-zero. Then the contact of the curve $F^{-1}\{0\}$ and line at a is expressed in terms of the nullity of the map $\mathbb{R} \rightarrowtail \mathbb{R};\ t \mapsto F(a + tb)$ at 0. In particular it is A_1 at least if and only if $F_1(a)b = 0$, A_2 at least if and only if also $F_2(a)b^2 = 0$, and so on. A necessary and sufficient condition for the curve to be linear at a is then found by eliminating the vector b from these two equations. For the answer see Exercise 9.2. □

9.2 $\mathcal{K}$-equivalence

Any smooth map-germ $f : \mathbb{R}^n,a \rightarrowtail \mathbb{R}^p,b$ admits a factorisation as the composite of the immersive map-germ

$$\mathbb{R}^n,a \rightarrowtail \mathbb{R}^n \times \mathbb{R}^p,\ (a, b);\ x,a \mapsto (x, f(x)),\ (a, b)$$

followed by the submersive map-germ

$$\mathbb{R}^n \times \mathbb{R}^p, (a, b) \rightarrowtail \mathbb{R}^p, b; (x, y), (a, b) \mapsto y, b.$$

The image of the first of these is the graph of f while the fibre over b of the second is the *horizontal* affine subspace $\mathbb{R}^n \times \{b\}$. Any smooth map-germ $f' : \mathbb{R}^n, a \rightarrowtail \mathbb{R}^p, b$ that induces an equivalent submanifold pair at (a, b) is then said to be *$\mathcal{K}$-equivalent* to f at a (Mather, 1970).

Towards the end of the last section we saw that the probe structures of certain map-germs $F'\mathbf{r}$ and $G'\mathbf{s}$ at 0 that describe the contact of two submanifolds M and M' of $\mathbb{R}^n$, whether of the same dimension or not, are equivalent. These are both maps from $\mathbb{R}^m$ to $\mathbb{R}^{n-m'}$ that send 0 to 0. James Montaldi (1986a) has shown that these maps are $\mathcal{K}$-equivalent at 0. It is possible to generalise the definition of contact equivalence in such a way that all four maps of Theorem 9.2 are contact equivalent.

Map-germs f and $f' : \mathbb{R}^n, a \rightarrowtail \mathbb{R}^p, b$ that are *right–left equivalent* (*$\mathcal{A}$-equivalent* in the received jargon), in the sense that there are *diffeomorphism-germs*, h of $\mathbb{R}^n$, a and h' of $\mathbb{R}^p$, b such that $h'f' = fh$, are $\mathcal{K}$-equivalent but not conversely.

Montaldi gives the following geometrical example: Each of the maps f and $f' : \mathbb{R}^3 \rightarrow \mathbb{R}^2$, given by $(x, y, z) \mapsto (y, z)$ and $(y, z - xy)$ respectively, cuts out the x-axis as the zero set. Consider their contact at the origin with the curve $\mathbb{R} \rightarrow \mathbb{R}^3$; $t \mapsto (t, t^2, t^3)$. The contact is given by the contact classes at the origin of the maps $t \mapsto (t^2, t^3)$ and $t \mapsto (t^2, 0)$ respectively. These are certainly $\mathcal{K}$-equivalent, but not $\mathcal{A}$-equivalent. See Exercise 9.5.

We refer the reader to Montaldi's paper for further details on contact equivalence. The papers of John Mather are foundational for singularity theory but are very technical. An essential role is played in his work and in most subsequent work by the *local algebra* of a map-germ. We give further references later. It would take us too far out of our way to stop here to discuss this very important concept.

9.3 Applications

For our somewhat limited purposes in this book all the technicalities of contact equivalence are almost entirely bypassed by Theorem 9.2. In fact in almost all the cases of interest to us probes are adequate diagnostic tools, as well as being of great practical value, as we shall see.

For example, let $\mathbf{s} : \mathbb{R}^2 \rightarrowtail \mathbb{R}^3$ be a regular surface parametrically presented, and let $w \in \operatorname{dom} \mathbf{s}$. Then its contact with a sphere that passes through $\mathbf{s}(w)$, with centre $\mathbf{c}$ and radius ρ, say, is determined by the contact class of the map $\mathbb{R}^2 \rightarrowtail \mathbb{R}$; $w \mapsto (\mathbf{s}(w) - \mathbf{c}) \cdot (\mathbf{s}(w) - \mathbf{c}) - \rho^2$. Various types of contact that can occur are then determined by probe analysis of this map. It is a fact, a theorem of singularity theory that we do not prove, that for low dimensions this probe analysis does give a complete classification of the types of contact with spheres that occur generically, up to contact equivalence. The complete classification, in terms of local algebras, is to be found for example in Arnol'd, Gusein-Zade and Varchenko (1985).

As a second example we have the various types of contact that such a surface can have with a plane, this also reducing to the study of maps $\mathbb{R}^2 \rightarrowtail \mathbb{R}$ up to contact equivalence.

Also of interest are the contacts that a surface can have with circles. Now a circle in $\mathbb{R}^3$ can be presented as the zero set of a submersive map $\mathbb{R}^3 \to \mathbb{R}^2$, as well as the image of an immersive map $\mathbb{R} \to \mathbb{R}^3$. So according as the surface is presented explicitly or implicitly the contact problem reduces to the study either of maps $\mathbb{R}^2 \rightarrowtail \mathbb{R}^2$ or of maps $\mathbb{R} \rightarrowtail \mathbb{R}$. By Theorem 9.1 these will be contact equivalent.

Further study of the contact of a regular smooth surface with spheres or with circles is the subject of the chapters that follow.

Exercises

9.1 Complete the proof of Theorem 9.2.

9.2 (Example 9.3 reformulated in more traditional notations.)

Let $f(x, y) = 0$, where $f : \mathbb{R}^2 \rightarrowtail \mathbb{R}$ is smooth, be the equation of a non-singular smooth curve in $\mathbb{R}^2$. Prove that the point $(x, y) = 0$ is a linear point of a curve if and only if at (x, y) we have, for some $a, b \in \mathbb{R}$, not both zero,

$$f = 0,\ af_x + bf_y = 0 \text{ and } a^2 f_{xx} + 2ab f_{xy} + b^2 f_{yy} = 0,$$

where the subscripts indicate partial derivatives in the traditional way, or equivalently if and only if $f(x, y) = 0$ and at (x, y)

$$\begin{vmatrix} f_{xx} & f_{xy} & f_x \\ f_{xy} & f_{yy} & f_y \\ f_x & f_y & 0 \end{vmatrix} = 0.$$

(Consider the contact of the curve with the lines $\mathbb{R} \to \mathbb{R}^2$: $t \mapsto (x + at, y + bt)$ that pass through (x, y). In the second part of the exercise be sure to prove both implications.)

9.3 Let $F(x, y, z)$ be a homogeneous polynomial of degree n and let $f(x, y) = F(x, y, 1)$. By use of Euler's Theorem on F and its derivatives prove that

$$\begin{vmatrix} f_{xx} & f_{xy} & f_x \\ f_{xy} & f_{yy} & f_y \\ f_x & f_y & f \end{vmatrix} = \begin{vmatrix} F_{xx} & F_{xy} & F_{xz} \\ F_{xy} & F_{yy} & F_{yz} \\ F_{xz} & F_{yz} & F_{zz} \end{vmatrix}.$$

9.4 Exercise 9.3 has obvious application to the determination of linear points of projective curves. Show in particular that by Bézout's Theorem a generic cubic curve on $\mathbb{C}P^2$ has nine inflections. (In fact at most three of these can be real.)

9.5 The maps $f : \mathbb{R}^3 \to \mathbb{R}^2$; $(x, y, z) \mapsto (y, z)$ and $f' : \mathbb{R}^3 \to \mathbb{R}^2$ $(y, z - xy)$ each cut out the x axis as the zero set. Consider their contact with the curve $\mathbb{R} \to \mathbb{R}^3$; $t \mapsto (t, t^2, t^3)$. The contact is given by the contact classes of the maps $t \mapsto (t^2, t^3)$ and $t \mapsto (t^2, 0)$ respectively. Show that these are $\mathcal{K}$-equivalent, but not $\mathcal{A}$-equivalent.

9.6 Determine to what extent the classification of rhamphoid cusps by contact equivalence relates to their classification by the degree of nullity of the curvature, or equivalently the radius of curvature, at the cusp.

10
Surfaces in $\mathbb{R}^3$

10.0 Introduction

When one thinks of the curvature of a generic regular curve in the plane one thinks at once of its focal set or evolute, the curve of its centres of curvature, whose tangents are the normals to the curve, and which has an ordinary cusp at each ordinary vertex or critical point of curvature of the curve. Each point of a normal to the curve, as a point of that normal, carries the label A_1, except for its centre of curvature, its point of tangency with the evolute, which carries the label A_2, except for the ordinary cusps of the evolute, which carry the label A_3. The higher labels $A_4, A_5, \ldots$ appear only in exceptional circumstances or else, generically, for particular curves in families of curves. The normals may be thought of as projective lines, each with one point at infinity, generally labelled A_1 but labelled A_2 at an ordinary inflection of the curve and labelled A_3 at an ordinary undulation, though generically one only encounters the latter in at least one-parameter families of curves. These labels classify the types of contact that circles or lines locally have with the curve.

When one thinks similarly of the curvature of a generic regular curve in three-dimensional space one thinks of the focal line in each normal plane, the focal lines being the tangents to the space evolute, the curve of centres of spherical curvature, which has ordinary cusps at occasional points which we have called the vertices of the space curve. Moreover, the osculating planes of the space evolute are the normal planes to the original curve. Each point of a normal plane, as a point of that normal plane, carries the label A_1, except for the focal line, whose points carry the label A_2, except for the centre of spherical curvature, the point of tangency of the focal line with the space evolute, which carries the label A_3, except for ordinary cusps of the space evolute, which carry the label A_4.

The normal planes may be thought of as projective planes whose points at infinity carry the label A_1, except for the point at infinity on the focal line which carries the label A_2, unless it is a planar point (or point of 'zero torsion') of the curve, when it carries the label A_3. The case where the entire focal line is at infinity, which is the case at a linear point (or point of 'zero curvature') is non-generic, as is the case where the entire focal line carries the label A_3, with the exception of one point which is a non-standard A_4. However, each of these occurs generically in one-parameter families of space curves. All these labels classify the types of contact that spheres or planes locally have with the curve.

For a generic regular surface in $\mathbb{R}^3$ a similar picture emerges, with a variety of labels that classify the types of contact that spheres or planes locally have with the surface. The crucial fact is that there are now on each normal line to the surface generally *two* focal points, the *principal centres of curvature*, these points coinciding only for isolated points of the surface, known as the *umbilics* of the surface. The focal points constitute the *focal surface* or *evolute* of the original surface, each normal line being tangent to it at each of its focal points. A focal point is the centre of a sphere that has particularly strong contact with the surface at the base of the normal to the surface on which it lies. Away from umbilical centres the evolute consists of two separate sheets, which it is helpful to imagine, locally at least, as coloured with different colours – say red and blue – it does not matter which is which.

The labelling of the normals and of the evolute to the surface proceeds as follows. Points of each normal, as points of that normal, carry the label A_1, except for the focal points, which generally carry the label A_2. Figures 10.1 and 10.2 indicate two ways in which a surface patch may be related to evolute patches, one red and one blue. In Figure 10.1 both focal points on each normal lie on the same side of the surface while in Figure 10.2 the focal points on each normal lie on opposite sides of the surface. In either case each surface normal is tangential to each sheet of the evolute.

The picture is not always so simple. Just as the evolute of plane curve may have cusps, so the evolute of a surface may have cuspidal edges, and these may have cusps. For brevity we call the cuspidal edges of the evolute the *ribs* of the original surface. Their points in general carry the label A_3, except for ordinary cusps on the ribs, which

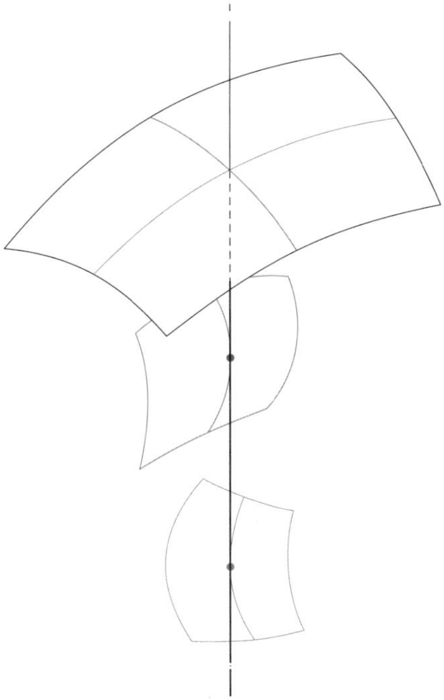

Figure 10.1

carry the label A_4. The ribs lie over curves on the surface that we call the *ridges* of the surface. These are the analogues of the vertices of a plane curve (Figure 10.3).

As in the case of a plane curve the normals to a surface may be thought of as projective lines, their points at infinity carrying the label A_1 except at points, carrying the label A_2, where one of the two focal points is at infinity, these points lying over curves on the original surface known traditionally as the *parabolic line*, though a less confusing term might be the *inflectional curves*, of the surface. The point at infinity of a normal carries the label A_3 at a point of intersection of the parabolic line with a ridge of the same colour. A point of the parabolic line is one where the surface has unusually strong contact with a plane.

All this supposes that the focal points on each normal are distinct. However, at *umbilics* they coincide. The double focal point then carries the label D_4 in general, though special types of umbilic carry other labels, such as D_5 and E_6, as we shall see later.

Each of the A_1, A_3 and D_4 points has associated with it a quadratic form, enabling us to separate points of these types further into *elliptic*

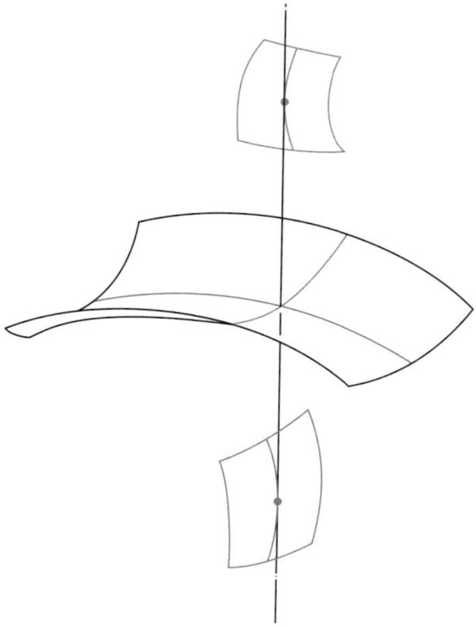

Figure 10.2

and *hyperbolic* types, the corresponding transitional *parabolic* types being those of types A_2, A_4 and D_5, respectively. Points of normals close to the surface are elliptic A_1s, those lying between the two A_2 points (on the *projective* normal line) being hyperbolic A_1s. The elliptic and hyperbolic umbilics are characterised in several different ways. Most dramatically there are three ribs through each elliptic umbilical centre (Figure 10.4), but only one through each hyperbolic umbilical centre (Figure 10.5, or Figures 10.6 and 10.7), the ribs in either case passing smoothly from one sheet of the focal surface to the other as they do so. Credits for these figures will be disclosed later! Characterisations of elliptic and hyperbolic ribs will also be given later. We only remark here that ribs consist of hyperbolic A_3s in the neighbourhood of umbilical centres and change type on passing through a generic A_4 point.

This brief catalogue of the principal features of the evolute of a generic surface in $\mathbb{R}^3$ has failed to mention many important things. For example, to each of the two focal points on a normal there is associated a tangent line to the surface at the base of the normal, and the two principal tangent lines thus defined at a non-umbilical point are mutually orthogonal. They are everywhere tangent to *lines of curvature* of the surface, these comprising two families of regular

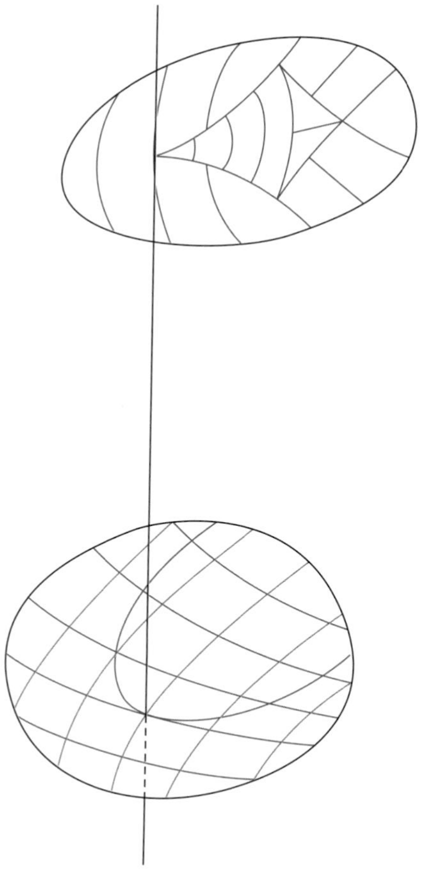

Figure 10.3

smooth curves that *foliate* the surface, one red and one blue. Away from umbilics the red lines of curvature everywhere intersect the blue lines of curvature orthogonally. Around the umbilics the lines form distinctive configurations, distinct from, though related to, the configurations formed by the ridges of the surface, the lines above which lie the ribs of the surface. In general ridges are not lines of curvature, though any line of symmetry is both a line of curvature and a ridge, the line as a ridge being blue where the line as a line of curvature is red, and vice versa.

Away from umbilics a ridge intersects the line of curvature of its own colour transversally except below an A_4 or ordinary cusp point of the corresponding rib, where it is simply tangent to the line of curvature of

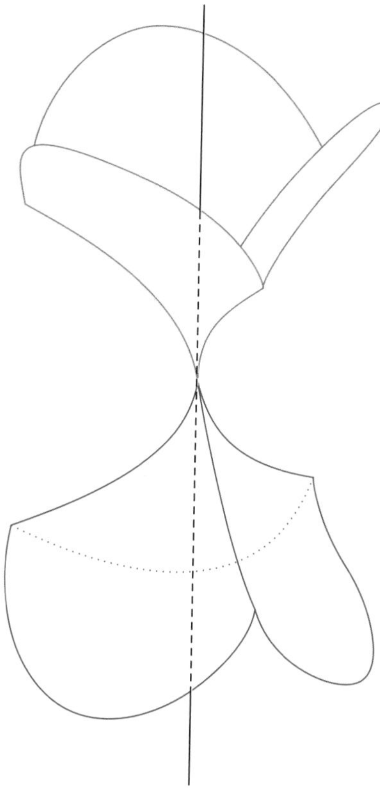

Figure 10.4

its own colour (Figure 10.3). We call such a place an *ordinary turning point* of the ridge. Along a line of curvature the relevant principal curvature has an ordinary local maximum or minimum where the line crosses a ridge of the same colour transversally, and has an ordinary stationary inflection where it is simply tangential to a ridge at an ordinary turning point of the ridge. Over a line of curvature of the surface there lies a curve in the evolute formed from the relevant principal centres of curvature. We call such a curve a *focal curve (of curvature)* of the surface. Each sheet of the focal surface is *foliated* by its focal curves.

The tangent developable of a space curve, which we encouraged the reader to make a model of at the beginning of Chapter 5 (Figure 5.1), is a surface which is not everywhere regular but has a cuspidal edge, namely the original space curve. It is easy to see that at each regular

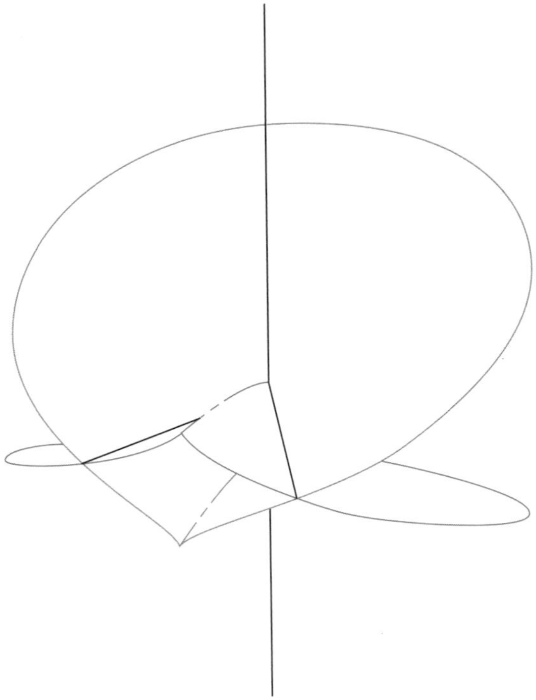

Figure 10.5

point of such a developable one of the two focal points is at infinity, one of the two principal curvatures, and therefore their product, being zero. The normal lines to a regular surface along a line of curvature form such a developable, the tangent developable of the corresponding focal curve. This curve need not be everywhere regular, having an ordinary cusp where it crosses a rib of the surface (and a cusp of higher order at a rib-cusp). The sheets of the evolute itself are not in general developable surfaces. They have in general inflectional curves (parabolic lines). However, each focal curve has nowhere zero curvature. Such a curve is a *geodesic* on the evolute, that is at each of its points the principal normal line to the curve coincides with the normal line to the evolute. Thus there is a preferred foliation of each sheet of the focal surface by geodesics.

The emphasis in most of what has been said so far has been on the contact that a regular surface has with spheres and planes. Also of interest are the types of contact that a surface can have with circles or lines, and we later prove some apparently new results on such contact.

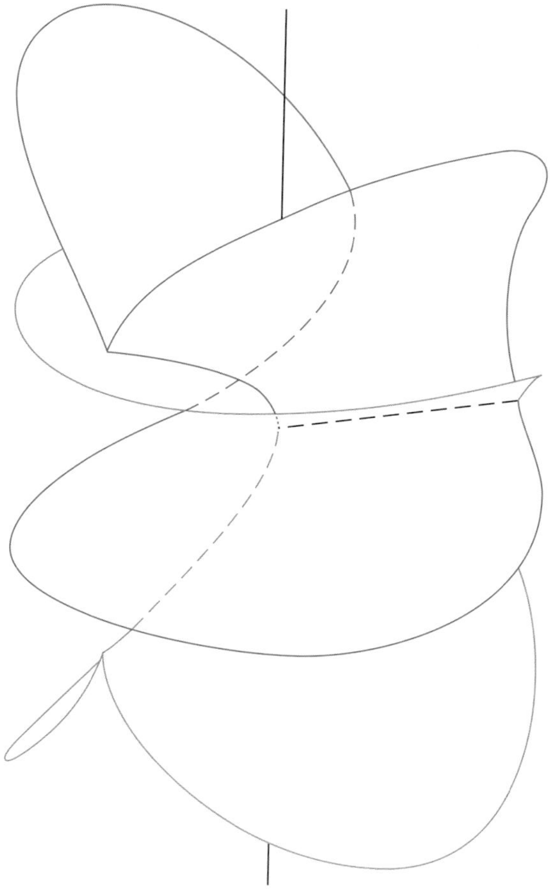

Figure 10.6

So far in this brief glance forwards we have almost exclusively considered generic phenomena. Other things can occur for particular surfaces and a number of such occurrences deserve detailed study, especially those that are generic for one-parameter families of surfaces. Such include, for example, parabolic, or D_5 umbilics, non-generic D_4 umbilics, for example birth or death points for umbilics, and non-generic A_4 points, including birth or death points for ridges or same-colour transversal intersections of ridges, the *acnodes* and *crunodes* of ridges. Also of great interest are the forms that the various types take in the neighbourhood of a plane of symmetry of the surface or of a point around which a higher symmetry group holds sway.

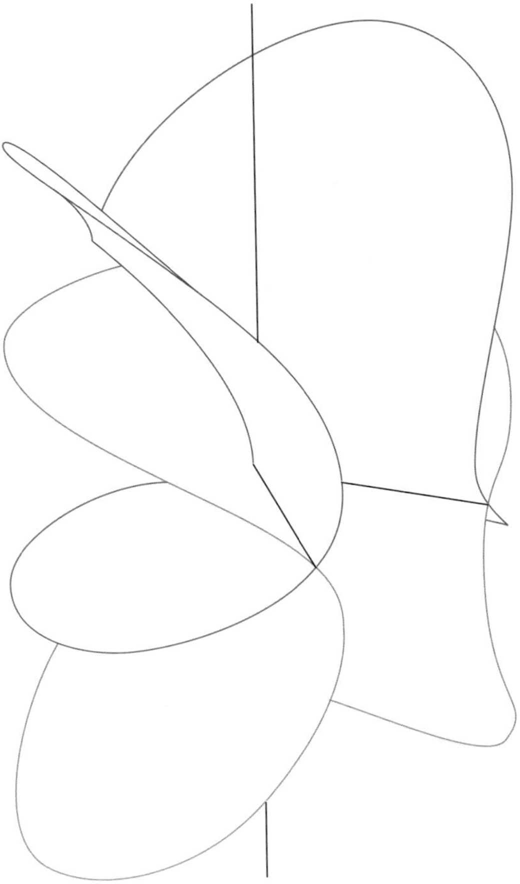

Figure 10.7

Another topic that has been very important in the development of the subject concerns those features of a surface that are *intrinsic* to the surface, depending only on the metric *first fundamental form* of the surface and remaining invariant under isometric bending of the surface. Such features include the *Gaussian curvature* of the surface, equal to the product of its two principal curvatures, and the *geodesics* of the surface.

Most of the material goes back to Euler in the eighteenth century and to Monge and Gauss and their contemporaries in the early nineteenth century but much of the detail on umbilics and ridges is of surprisingly recent date.

In the present chapter the scene is set by the construction of the focal surface of a regular surface, other aspects of the theory being developed in the chapters that follow.

10.1 Euler's formula

We begin our detailed study of a regular surface in $\mathbb{R}^3$ with an informal look at such a surface in a neighbourhood of one of its points.

We choose coordinates in $\mathbb{R}^3$ so that the point of interest is at the origin and so that the (x, y)-plane is the tangent plane to the surface there. The surface is then expressible, in *Monge form*, as the graph of a smooth function f from the (x, y) plane to the z-axis, with equation

$$z = f(x, y) = \tfrac{1}{2}(ax^2 + 2bxy + cy^2) + O(x, y)^3.$$

Moreover, by an orthogonal rotation of the (x, y)-plane we may reduce the quadratic form to a sum of squares, $\kappa x^2 + \lambda y^2$; that is we suppose that the surface near the origin consists of those points (x, y, z) for which

$$z = \tfrac{1}{2}(\kappa x^2 + \lambda y^2) + O(x, y)^3,$$

the contours, for z small, being small deformations of small ellipses or pieces of hyperbolas with the x- and y-axes as principal axes, at least if neither κ nor λ is zero. Indeed for a point on an ellipsoid these contours are ellipses.

Now the plane $y = 0$, normal to the tangent plane $z = 0$, intersects the surface near the origin in the curve

$$z = \tfrac{1}{2}\kappa x^2 + O(x)^3,$$

which has curvature κ at $x = 0$. Likewise the plane $x = 0$, also normal to the tangent plane $z = 0$, intersects the surface in the curve

$$z = \tfrac{1}{2}\lambda y^2 + O(y)^3,$$

which has curvature λ at $y = 0$. More generally, the normal plane making an angle θ with the plane $y = 0$, intersects the surface in the curve

$$z = \tfrac{1}{2}(\kappa \cos^2\theta + \lambda \sin^2\theta)r^2 + O(r)^3,$$

where $x = r\cos\theta$, $y = r\sin\theta$. As a curve in the (r, z)-plane this has curvature

$$\kappa_\theta = \kappa\cos^2\theta + \lambda\sin^2\theta$$

at the origin. This is the classical *formula of Euler*. Clearly κ and λ are the extreme values of κ_θ, occurring where $\theta \equiv 0$ or $\frac{1}{2}\pi$, mod π.

The numbers κ and λ are the *principal curvatures* of the surface at the origin and the x- and y-axes are the (mutually orthogonal) *principal tangent lines* there. At an *umbilic*, where $\kappa = \lambda$, the surface locally has the form

$$z = \tfrac{1}{2}\kappa(x^2 + y^2) + O(x, y)^3.$$

For small z the contours near the origin of this surface are small deformations of small circles. Indeed for an ellipsoid they actually are small circles. We study these small nearly circular contours in greater detail later, in Chapter 15.

At the heart of the above proof of Euler's formula there is the solution of an eigenvalue problem involving two quadratic forms, namely $x^2 + y^2$, the *first fundamental form* on the tangent plane to the surface at the origin, and $ax^2 + 2bxy + cy^2$, the *second fundamental form* on the tangent plane.

The principal curvatures κ and λ are the necessarily real eigenvalues of the real symmetric matrix

$$\begin{bmatrix} a & b \\ b & c \end{bmatrix}$$

of the second fundamental form. The corresponding necessarily mutually orthogonal eigenspaces (when the eigenvalues are distinct) are the principal tangent lines at the origin.

What we shall do in our more sophisticated treatment below, is to perform the above analysis simultaneously at all points of the surface. Indeed this is essential if we wish to consider the ways in which κ and λ vary as we move over the surface. The trick is to consider the various forms of contact that can occur between spheres and the surface, not just its contact with normal circles. We return in Chapter 15 to give consideration to the contact of a surface with circles that are not necessarily normal at the point of contact, a study that generalises the classical formula of Euler, already introduced, and that of Meusnier.

We proceed therefore to study the possible types of contact of a regular surface in $\mathbb{R}^3$ with spheres or with planes in $\mathbb{R}^3$.

10.2 The sophisticated approach

As with curves in the plane or in space we begin by concentrating almost entirely on surfaces presented parametrically.

A *smooth parametric surface* in $\mathbb{R}^3$ is a smooth map

$$\mathbf{s} : \mathbb{R}^2 \rightarrowtail \mathbb{R}^3;\ w = (u, v) \mapsto \mathbf{s}(w),$$

with domain a generally connected open subset of $\mathbb{R}^2$. Such a surface is *regular* (or *immersive*) at w if its first derivative $\mathbf{s}_1(w)$ is injective. Here $\mathbf{s}_1(w)$ is the linear map $\mathbb{R}^2 \to \mathbb{R}^3$ which (up to an additive constant) best approximates $\mathbf{s}$ at w, its matrix being the 3×2 Jacobian matrix of the partial derivatives of the components of $\mathbf{s}$ at w. Since $\mathbf{s}_1(w)$ is injective, its image is a two-dimensional linear subspace of $\mathbb{R}^3$, the *tangent vector plane* to $\mathbf{s}$ at w. Its translate by $\mathbf{s}(w)$ is then the *tangent plane* to $\mathbf{s}$ at w (or at $\mathbf{s}(w)$). The orthogonal complement in $\mathbb{R}^3$ of the tangent vector plane is the *normal vector line* of $\mathbf{s}$ at w. Its translate by $\mathbf{s}(w)$ is then the *normal line* to $\mathbf{s}$ at w (or at $\mathbf{s}(w)$). The *standard unit normal vector* $\mathbf{n}(w)$ to $\mathbf{s}$ at w is then the cross product of the images by $\mathbf{s}_1(w)$ in $\mathbb{R}^3$ of the basis vectors $(1, 0)$ and $(0, 1)$ of $\mathbb{R}^2$, normalised to be of unit length. The spherical map $\mathbf{n} : \mathbb{R}^2 \rightarrowtail \mathbb{R}^3$ is called the *Gauss map* of $\mathbf{s}$.

Let $\mathbf{s} : \mathbb{R}^2 \rightarrowtail \mathbb{R}^3;\ w = (u, v) \mapsto \mathbf{s}(w)$ be a regular surface in $\mathbb{R}^3$. By analogy with what we have done with curves we ask to what extent this surface can be approximated at w by a sphere. To do this we look at all possible spheres that pass through w and for each we measure the type of contact that it has with the surface. In so doing we proceed naively, though the reader who has mastered the chapters on probe analysis and contact may be able to anticipate ways in which the argument will develop.

As usual when measuring contact between two submanifolds one of them is taken in explicit parametric form and the other in implicit form as the set of zeros of some equation. Here it is the sphere with centre $\mathbf{c}$ and radius ρ that it is natural to take in implicit form as the set of points $\mathbf{r}$ satisfying the equation

$$(\mathbf{c} - \mathbf{r}) \cdot (\mathbf{c} - \mathbf{r}) = \rho^2,$$

where $\mathbf{r} \in \mathbb{R}^3$, or equivalently the equation

$$\mathbf{c} \cdot \mathbf{r} - \tfrac{1}{2}\mathbf{r} \cdot \mathbf{r} - \tfrac{1}{2}(\mathbf{c} \cdot \mathbf{c} - \rho^2) = 0.$$

For a parametrisation $\mathbf{s} : \mathbb{R}^2 \rightarrowtail \mathbb{R}^3$; $w \mapsto \mathbf{r} = \mathbf{s}(w)$ of this sphere therefore, or of part of it, it follows that

$$(\mathbf{c} - \mathbf{s}) \cdot \mathbf{s}_1 = 0,$$

$$(\mathbf{c} - \mathbf{s}) \cdot \mathbf{s}_2 - \mathbf{s}_1 \cdot \mathbf{s}_1 = 0,$$

$$(\mathbf{c} - \mathbf{s}) \cdot \mathbf{s}_3 - 3\mathbf{s}_1 \cdot \mathbf{s}_2 = 0,$$

and so on, provided that the derivatives $\mathbf{s}_i$ are suitably defined.

For a generic regular surface $\mathbf{s} : \mathbb{R}^2 \rightarrowtail \mathbb{R}^3$ there is no problem in finding points $\mathbf{c}$ that satisfy the first of these equations at a given point $w \in \mathbb{R}^2$. Such points are just the points of the normal to $\mathbf{s}$ at w. In general, however, by contrast to the case of plane curves, none of these points need satisfy the second equation. Suppose that we set $\mathbf{c} - \mathbf{s} = \zeta\mathbf{n}$. Then the left hand side of the second equation becomes $\zeta\mathbf{n} \cdot \mathbf{s}_2 - \mathbf{s}_1 \cdot \mathbf{s}_1$, a quadratic form on $\mathbb{R}^2$ that depends on the value of ζ. What we can expect is that for certain values of $\zeta = \rho$ this form becomes parabolic, the condition for this being that there exists a non-zero vector $\mathbf{a}_1 \in \mathbb{R}^2$, unique up to real multiples, such that $\rho\mathbf{n} \cdot \mathbf{s}_2\mathbf{a}_1 = \mathbf{s}_1\mathbf{a}_1 \cdot \mathbf{s}_1$, or equivalently

$$\mathbf{n} \cdot \mathbf{s}_2\mathbf{a}_1 = \kappa\mathbf{s}_1\mathbf{a}_1 \cdot \mathbf{s}_1,$$

where $\kappa = \rho^{-1}$. (A reason for the notation $\mathbf{a}_1$ will appear presently.)

This equation can usefully be thought of as involving two quadratic forms, or rather two symmetric twice linear forms, $\mathrm{I}_2 = \mathbf{s}_1 \cdot \mathbf{s}_1$, the *first fundamental form* on $\mathbb{R}^2$, and $\mathrm{II}_2 = \mathrm{n} \cdot \mathrm{s}_2$, the *second fundamental form* on $\mathbb{R}^2$.

The first fundamental form is *positive-definite*, that is for any $w \in \mathbb{R}^2$ and any $\mathbf{a}_1 \in \mathbb{R}^2$,

$$\mathrm{I}_2(w)\mathbf{a}_1{}^2 = \mathbf{s}_1(w)\mathbf{a}_1 \cdot \mathbf{s}_1(w)\mathbf{a}_1 \geqslant 0,$$

with $\mathrm{I}_2\mathbf{a}_1{}^2 = 0$ if and only if $\mathbf{a}_1 = 0$.

Since $\mathbf{n} \cdot \mathbf{s}_1 = 0$ everywhere it follows that $\mathbf{n} \cdot \mathbf{s}_2 + \mathbf{n}_1 \cdot \mathbf{s}_1 = 0$, implying that $\mathrm{II}_2 = -\mathbf{n}_1 \cdot \mathbf{s}_1$, an expression that sometimes is easier to compute than $\mathbf{n} \cdot \mathbf{s}_2$. See for example Exercise 10.7. It follows from the equality of $\mathbf{n}_1 \cdot \mathbf{s}_1$ with $-\mathbf{n} \cdot \mathbf{s}_2$ that the twice linear form $\mathbf{n}_1 \cdot \mathbf{s}_1$ is symmetric, a fact perhaps not otherwise obvious.

Both fundamental forms lift unambiguously to forms on the tangent spaces of the surface $\mathbf{s}$, this following in the case of the second one from the fact that the expression $-\mathbf{n}_1 \cdot \mathbf{s}_1$ involves *first* derivatives only. Generally speaking expressions involving higher derivatives do not lift to the tangent spaces. Those that do are known as *tensors*.

In the particular case that the surface **s** is in the Monge form

$$\mathbf{s}(x, y) = (x, y, \tfrac{1}{2}(ax^2 + 2bxy + cy^2) + O(x, y)^3)$$

we have

$$\mathbf{s}_1(0,0) = \begin{bmatrix} 1 & 0 \\ 0 & 1 \\ 0 & 0 \end{bmatrix}, \ \mathbf{n}(0,0) = \begin{bmatrix} 0 \\ 0 \\ 1 \end{bmatrix} \text{ and } \mathbf{s}_2(0,0) = \begin{bmatrix} 0 & 0 & 0 \\ 0 & 0 & 0 \\ \mathrm{a} & \mathrm{b} & \mathrm{c} \end{bmatrix}_2.$$

So $\mathrm{I}_2(0)(x, y)^2 = x^2 + y^2$, while $\mathrm{II}_2(0)(x, y)^2 = ax^2 + 2bxy + cy^2$, in agreement with our earlier account.

Proposition 10.1 With notations as above, the condition for $\mathrm{II}_2(w) - \kappa\mathrm{I}_2(w)$ *to have a non-zero kernel is a quadratic equation for* κ *with two real roots, distinct if* $\mathrm{II}_2(w)$ *is not a multiple of* $\mathrm{I}_2(w)$. *Let these roots be* $\kappa(w)$ *and* $\lambda(w)$ *and let* $\mathbf{a}_1$ *and* $\mathbf{b}_1$ *be corresponding non-zero kernel vectors. These are then root vectors of the Jacobian quadratic form of the forms* $\mathrm{II}_2(w)$ *and* $\mathrm{I}_2(w)$. *Moreover,*

$$\mathrm{I}_2(w)\mathbf{a}_1\mathbf{b}_1 = \mathrm{II}_2(w)\mathbf{a}_1\mathbf{b}_1 = 0.$$

In particular, when $\kappa(w) \neq \lambda(w)$ *the tangent vectors* $\mathbf{s}_1(w)\mathbf{a}_1$ *and* $s_1(\mathbf{w})b_1$ *are mutually orthogonal.*

Proof There are various ways to prove this. One direct method is as follows. Since $\mathrm{I}_2(w)$ is positive definite its eigenvalues, α, β say, are both positive real numbers, and a basis for $\mathbb{R}^2$ may be chosen so that the matrix of $\mathrm{I}_2(w)$ is the diagonal matrix

$$\begin{bmatrix} \alpha & 0 \\ 0 & \beta \end{bmatrix}.$$

Let the matrix of $\mathrm{II}_2(w)$ then be

$$\begin{bmatrix} L & M \\ M & N \end{bmatrix}.$$

Then the condition for the roots of the equation

$$\begin{bmatrix} L - \kappa\alpha & M \\ M & N - \kappa\beta \end{bmatrix} = 0$$

to be real is that $(\alpha N + \beta L)^2 - 4\alpha\beta(LN - M^2) = (\alpha N - \beta L)^2 + 4\alpha\beta M^2 \geqslant 0$, which is so, since both α and β are > 0. Moreover, this is a quadratic equation for κ, so there are two real roots, $\kappa(w)$ and $\lambda(w)$ say, coinciding if and only if the matrix of II_2 is a multiple of the matrix of I_2.

If $\mathbf{a}_1$ and $\mathbf{b}_1$ are corresponding non-zero kernel vectors of the parabolic forms $\mathrm{II}_2(w) - \kappa(w)\mathrm{I}_2(w)$ and $\mathrm{II}_2(w) - \lambda(w)\mathrm{I}_2(w)$ then $\mathrm{II}_2(w)\mathbf{a}_1 = \kappa(w)I_2(w)\mathbf{a}_1$ and $\mathrm{II}_2(w)\mathbf{b}_1 = \lambda(w)\mathrm{I}_2(w)\mathbf{b}_1$, implying that $\mathbf{a}_1$ and $\mathbf{b}_1$ are Jacobian vectors of the forms $\mathrm{II}_2(w)$ and $\mathrm{I}_2(w)$ by the remarks following Proposition 7.5. Moreover

$$\mathrm{II}_2(w)\mathbf{a}_1\mathbf{b}_1 = \kappa(w)\mathrm{I}(w)_2\mathbf{a}_1\mathbf{b}_1 = \lambda(w)\mathrm{I}(w)_2\mathbf{a}_1\mathbf{b}_1,$$

so that $\mathrm{II}_2(w)\mathbf{a}_1\mathbf{b}_1 = 0 = \mathrm{I}(w)_2\mathbf{a}_1\mathbf{b}_1$, if $\kappa(w) \neq \lambda(w)$. □

What this proposition does is to recover what we proved by more elementary means earlier in the chapter when our surface was taken to be in Monge form. The numbers $\kappa(w)$ and $\lambda(w)$ are the *principal curvatures* of the surface $\mathbf{s}$ at w, while the vectors $\mathbf{a}_1$ and $\mathbf{b}_1$ represent mutually orthogonal *principal tangent vectors* at w, $\mathbf{s}_1(w)\mathbf{a}_1$ and $\mathbf{s}_1(w)\mathbf{b}_1$, generating the *principal tangent lines* at w. It is of course possible that either of the numbers $\kappa(w)$ and $\lambda(w)$ may be zero. When this is not so their reciprocals $\rho(w)$ and $\sigma(w)$ are the *principal radii of curvature* at w, and the points $\rho(w)\mathbf{n}(w)$ and $\sigma(w)\mathbf{n}(w)$ the *principal centres of curvature* or *focal points* at w.

10.3 Lines of curvature

The principal tangent lines on $\mathbf{s}$ defined at each point other than umbilics integrate to *lines of curvature*. To see this we employ the classical notations for the first and second fundamental forms, namely

$$\mathrm{I}_2(w)\mathrm{d}w^2 = E\mathrm{d}u^2 + 2F\mathrm{d}u\mathrm{d}v + G\mathrm{d}v^2$$

and

$$\mathrm{II}_2(w)\mathrm{d}w^2 = L\mathrm{d}u^2 + 2M\mathrm{d}u\mathrm{d}v + N\mathrm{d}v^2,$$

where E, F, G, L, M, N are functions of $w = (u, v)$, and $\mathrm{d}w = (\mathrm{d}u, \mathrm{d}v)$ denotes the vector in $\mathbb{R}^2$, denoted by $\mathbf{a}_1$ above. Then the principal tangent lines are determined as the root lines of the Jacobian form of these two forms, namely (dropping irrelevant 2s as usual)

$$(EM - FL)\mathrm{d}u^2 + (EN - GL)\mathrm{d}u\mathrm{d}v + (FN - GM)\mathrm{d}v^2.$$

This provides a quadratic equation for dv/du or for du/dv. The existence of two families of integral curves away from umbilics, where the Jacobian vanishes, then follows directly from the fundamental existence theorem for ordinary differential equations of the first order. Each family of lines of curvature foliates the surface. We return later to the problem of what happens in the neighbourhood of umbilical points.

10.4 Focal curves of curvature

As we indicated in the introduction it is convenient to think of one principal curvature and its associated family of lines of curvature as coloured blue and the other as coloured red. It may be that either curvature is equal to zero, such points forming the *parabolic line*, points where both are zero being *flat* umbilics. For the moment we ignore this possibility. Then the principal radii of curvature $\rho = \kappa^{-1}$ and $\sigma = \lambda^{-1}$ are everywhere defined.

Let w be a point of the domain of a regular surface $\mathbf{s}$ in $\mathbb{R}^3$ where neither of the principal curvatures is zero, and suppose that the blue and red lines of curvature through the point $\mathbf{s}(w)$ are represented by curves $\mathbf{a} : \mathbb{R} \rightarrowtail \mathbb{R}^2$, with $\mathbf{a}(0) = w$ and $\mathbf{b} : \mathbb{R} \rightarrowtail \mathbb{R}^2$, with $\mathbf{b}(0) = w$, respectively. Then $\mathbf{a}_1 = \mathbf{a}_1(0)$ and $\mathbf{b}_1 = \mathbf{b}_1(0)$ are principal tangent vectors at w, explaining at last our earlier choice of notations.

Let $\mathbf{e} : \mathbb{R}^2 \rightarrowtail \mathbb{R}^3$ and $\mathbf{f} : \mathbb{R}^2 \rightarrowtail \mathbb{R}^3$ represent the blue and red sheets of the *focal surface* of $\mathbf{s}$, the loci of the *principal centres of curvature* of $\mathbf{s}$, $\rho\mathbf{n}$ and $\sigma\mathbf{n}$. In what follows we focus attention on one sheet of the focal surface, the blue one $\mathbf{e}$ say. The maps $\mathbf{s}$, $\mathbf{e}$ and $\mathbf{a}$ form a diagram

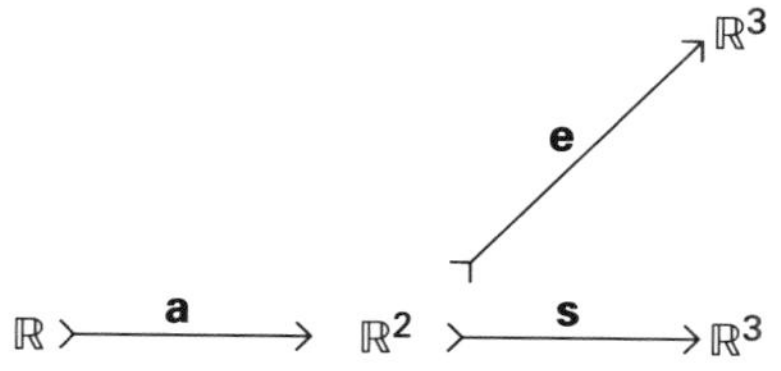

the curve $\mathbf{sa}$ being the blue line of curvature on $\mathbf{s}$ and the curve $\mathbf{ea}$ the corresponding *focal curve (of curvature)* on the blue sheet $\mathbf{e}$ of the focal surface. Moreover

$$\mathbf{ea} - \mathbf{sa} = (\rho\mathbf{a})\mathbf{na}.$$

The next theorem and its corollary provide analogues for regular surfaces of the fundamental formulas $\kappa\mathbf{r}_1 + \mathbf{n}_1 = 0$ and $\mathbf{e}_1 = \rho_1\mathbf{n}$ for a regular plane curve $\mathbf{r}$ and its evolute $\mathbf{e}$.

Theorem 10.2 (Rodrigues' Theorem) Let $\mathbf{s} : \mathbb{R}^2 \rightarrowtail \mathbb{R}^3$ *be a regular smooth surface and let* $\mathbf{a} : \mathbb{R} \rightarrowtail \mathbb{R}^2$ *represent a line of curvature on* $\mathbf{s}$ *through a point* $w = \mathbf{a}(t)$ *and let* κ *be the corresponding principal curvature function on* $\mathbf{s}$. *Then, at* t,

$$(\kappa\mathbf{a})(\mathbf{s}_1\mathbf{a})\mathbf{a}_1 + (\mathbf{n}_1\mathbf{a})\mathbf{a}_1 = 0,$$

usually abbreviated to

$$\kappa\mathbf{s}_1\mathbf{a}_1 + \mathbf{n}_1\mathbf{a}_1 = 0.$$

Proof We have, at w,

$$\mathbf{n} \cdot \mathbf{s}_2\mathbf{a}_1 = \kappa\mathbf{s}_1\mathbf{a}_1 \cdot \mathbf{s}_1.$$

Now, since $\mathbf{n} \cdot \mathbf{s}_1 = 0$, $\mathbf{n} \cdot \mathbf{s}_2 + \mathbf{n}_1 \cdot \mathbf{s}_1 = 0$ everywhere. Moreover, as we have already noted, the twice linear form $\mathbf{n}_1 \cdot \mathbf{s}_1$ is symmetric since $\mathbf{n} \cdot \mathbf{s}_2$ is. Accordingly, putting $\mathbf{a}_1$ in the appropriate slot, we find that

$$(\kappa\mathbf{s}_1\mathbf{a}_1 + \mathbf{n}_1\mathbf{a}_1) \cdot \mathbf{s}_1 = 0.$$

Since also $(\kappa\mathbf{s}_1\mathbf{a}_1 + \mathbf{n}_1\mathbf{a}_1) \cdot \mathbf{n} = 0$, the assertion follows. □

Corollary 10.3 With the above notations $(\mathbf{ea})_1 = (\rho\mathbf{a})_1\mathbf{n}$.

Proof This follows from Theorem 10.2 in the form $(\mathbf{s}_1 + \rho\mathbf{n}_1)\mathbf{a}_1 = 0$, on differentiating the equation $\mathbf{ea} - \mathbf{sa} = (\rho\mathbf{a})\mathbf{na}$. □

Further relations between the derivatives of $\mathbf{e}$, $\mathbf{s}$ and ρ follow by differentiating the above formulas and the equations

$$(\mathbf{e} - \mathbf{s}) \cdot \mathbf{s}_1 = 0$$

and

$$(\mathbf{e} - \mathbf{s}) \cdot \mathbf{s}_2\mathbf{a}_1 - \mathbf{s}_1\mathbf{a}_1 \cdot \mathbf{s}_1 = 0,$$

the first an equation on $L(\mathbb{R}^2, \mathbb{R})$ but the second an equation on $\mathbb{R}$. The most important is the following:

Proposition 10.4 Let $\mathbf{s}$ *be a regular surface without umbilics or parabolic line, let* $\mathbf{a}$ *represent the line of curvature through a point* $w = \mathbf{a}(t)$ *and let* $\mathbf{e}$ *be the relevant sheet of the focal surface. Then*

$$\mathbf{e}_1(w)\mathbf{a}_1(t) \cdot \mathbf{s}_1(w) = 0 \text{ and } \mathbf{e}_1(w) \cdot \mathbf{s}_1(w)\mathbf{a}_1(t) = 0.$$

Proof Differentiating the first of the above equations we get

$$\mathbf{e}_1 \cdot \mathbf{s}_1 + (\mathbf{e} - \mathbf{s}) \cdot \mathbf{s}_2 - \mathbf{s}_1 \cdot \mathbf{s}_1 = 0,$$

from which it follows that the twice linear form $\mathbf{e}_1 \cdot \mathbf{s}_1$ is symmetric. Putting $\mathbf{a}_1(t)$ in either slot at $w = \mathbf{a}(t)$ we have that

$$\mathbf{e}_1(w)\mathbf{a}_1(t) \cdot \mathbf{s}_1(w) = 0 \text{ and } \mathbf{e}_1(w) \cdot \mathbf{s}_1(w)\mathbf{a}_1(t) = 0. \qquad \square$$

The equations of Proposition 10.4 are generally abbreviated to

$$\mathbf{e}_1\mathbf{a}_1 \cdot \mathbf{s}_1 = 0 \text{ and } \mathbf{e}_1 \cdot \mathbf{s}_1\mathbf{a}_1 = 0.$$

The first of these confirms what we already know, namely that at a regular point the focal curve of curvature at w, $\mathbf{ea}$, has the normal line to $\mathbf{s}$ at w as its tangent line, while the second states that the principal tangent vector $(\mathbf{sa})_1$ is normal to the focal surface $\mathbf{e}$ at w.

The next two propositions concern the regularity or otherwise of the focal curve $\mathbf{ea}$ and of the focal surface $\mathbf{e}$.

Proposition 10.5 The focal curve $\mathbf{ea}$ *is nowhere linear, has an ordinary cusp, that is* $(\mathbf{ea})_3$ *is not a multiple of* $(\mathbf{ea})_2$*, where* $(\mathbf{ea})_1 = 0$ *but* $(\mathbf{ea})_2 \neq 0$*, or equivalently where* $(\rho\mathbf{a})_1 = 0$ *but* $(\rho\mathbf{a})_2 \neq 0$*, and has an ordinary kink, that is* $(\mathbf{ea})_4$ *is not a multiple of* $(\mathbf{ea})_3$*, where* $(\mathbf{ea})_1 = 0$ *and* $(\mathbf{ea})_2 = 0$ *but* $(\mathbf{ea})_3 \neq 0$*, or equivalently where* $(\rho\mathbf{a})_1 = 0$ *and* $(\rho\mathbf{a})_2 = 0$ *but* $(\rho\mathbf{a})_3 \neq 0$.

Proof All this follows at once from the equation $(\mathbf{ea})_1 = (\rho\mathbf{a})_1\mathbf{na}$ and the equations we get by differentiating it, namely

$$(\mathbf{ea})_2 = (\rho\mathbf{a})_2\mathbf{na} + (\rho\mathbf{a})_1(\mathbf{na})_1,$$

$$(\mathbf{ea})_3 = (\rho\mathbf{a})_3\mathbf{na} + 2(\rho\mathbf{a})_2(\mathbf{na})_1 + (\rho\mathbf{a})_1(\mathbf{na})_2,$$

and

$$(\mathbf{ea})_4 = (\rho\mathbf{a})_4\mathbf{na} + 3(\rho\mathbf{a})_3(\mathbf{na})_1 + 3(\rho\mathbf{a})_2(\mathbf{na})_2 + (\rho\mathbf{a})_1(\mathbf{na})_3.$$

For if $\mathbf{ea}$ is regular then $(\rho\mathbf{a})_1 \neq 0$. Moreover, by Rodrigues' Theorem it follows that where $\mathbf{e}$ exists $(\mathbf{na})_1 = \mathbf{n}_1\mathbf{a}_1$ is non-zero and so, being orthogonal to $\mathbf{n}$, is not a multiple of $\mathbf{n}$. Accordingly $(\mathbf{ea})_2$ is nowhere a mutiple of $(\mathbf{ea})_1$. That is $\mathbf{ea}$ is nowhere linear.

Likewise if $(\mathbf{ea})_1 = 0$ but $(\mathbf{ea})_2 \neq 0$, then $(\rho\mathbf{a})_1 = 0$ but $(\rho\mathbf{a})_2 \neq 0$. By an argument similar to that just given it follows that $(\mathbf{ea})_3$ is then nowhere a multiple of $(\mathbf{ea})_2$.

Finally if $(\mathbf{ea})_1 = 0$ and $(\mathbf{ea})_2 = 0$ but $(\mathbf{ea})_3 \neq 0$, then $(\rho\mathbf{a})_1 = 0$ and $(\rho\mathbf{a})_2 = 0$ but $(\rho\mathbf{a})_3 \neq 0$, from which it follows that $(\mathbf{ea})_4$ is then nowhere a multiple of $(\mathbf{ea})_3$. □

Proposition 10.6 Let **s** *be a regular smooth surface in* $\mathbb{R}^3$ *free from parabolic line or umbilics. Then the focal surface* **e** *is regular where* **ea** *is regular, and has differential of rank* 1 *where* **ea** *is not regular.*

Proof We begin by remarking that at each point w of **s** the principal vectors $\mathbf{a}_1(w)$ and $\mathbf{b}_1(w)$ are linearly independent, and form a basis for $\mathbb{R}^2$. Working always at the point w we first prove that $\mathbf{e}_1\mathbf{b}_1 \neq 0$. This we do by inserting the principal vector $\mathbf{b}_1$ in one of the slots in the equation

$$\mathbf{e}_1 \cdot \mathbf{s}_1 + (\mathbf{e} - \mathbf{s}) \cdot \mathbf{s}_2 = \mathbf{s}_1 \cdot \mathbf{s}_1.$$

Since $(\mathbf{e} - \mathbf{s}) \cdot \mathbf{s}_2\mathbf{b}_1 \neq \mathbf{s}_1\mathbf{b}_1 \cdot \mathbf{s}_1$, since $\mathbf{b}_1$ is an eigenvector for the red curvature of the surface and not the blue one, it follows that $\mathbf{e}_1\mathbf{b}_1 \cdot \mathbf{s}_1 \neq 0$ and therefore that $\mathbf{e}_1\mathbf{b}_1 \neq 0$.

By a similar argument $\mathbf{e}_1(\alpha\mathbf{a}_1 + \beta\mathbf{b}_1) \neq 0$ whenever $\beta \neq 0$.

Accordingly **e** fails to be regular at w if and only if $\mathbf{e}_1\mathbf{a}_1 = 0$ at w, and in that event the dimension of the kernel of $\mathbf{e}_1(w)$ is 1. □

The next proposition resembles a result proved earlier for the focal curves on the focal surface of a regular smooth space curve.

Proposition 10.7 Let $\mathbf{s} : \mathbb{R}^2 \rightarrowtail \mathbb{R}^3$ *be a regular surface, and let* $\mathbf{a} : \mathbb{R} \rightarrowtail \mathbb{R}^2$ *represent one of its lines of curvature. Then at any regular point of the focal curve* **ea** *the principal normal of* **ea** *is normal to the focal surface, that is* **ea** *is a* geodesic *of the focal surface.*

Proof What has to be proved is that $(\mathbf{ea})_1$, $(\mathbf{ea})_2$ and $(\mathbf{sa})_1$ are everywhere coplanar, the latter vector being normal to the focal surface **e**. This we do by showing that each is normal to $(\mathbf{sb})_1 = \mathbf{s}_1\mathbf{b}_1$. The first and third of these cases are obvious. As to the second, differentiating $(\mathbf{ea})_1 \cdot \mathbf{s}_1\mathbf{a} = 0$ we find that $(\mathbf{ea})_2 \cdot \mathbf{s}_1\mathbf{a} + (\mathbf{ea})_1 \cdot \mathbf{s}_2\mathbf{a}_1 = 0$. Then $(\mathbf{ea})_2 \cdot \mathbf{s}_1\mathbf{b}_1 = 0$ since $(\mathbf{ea})_1 \cdot \mathbf{s}_2\mathbf{a}_1\mathbf{b}_1 = (\rho\mathbf{a})_1\mathbf{n} \cdot \mathbf{s}_2\mathbf{a}_1\mathbf{b}_1 = (\rho\mathbf{a})_1\mathbf{s}_1\mathbf{a}_1 \cdot \mathbf{s}_1\mathbf{b}_1 = 0$. □

Proposition 10.8 With **s**, **e** *and* ρ *as in Proposition 10.7 let the smooth curve* $\mathbf{r} : \mathbb{R} \rightarrowtail \mathbb{R}^2$ *represent a curve of the surface* **s** *along which the* *principal curvature ρ is constant. Then the curve* $\mathbf{e}\,\mathbf{r}$ *on the focal surface* $\mathbf{e}$ *intersects the focal curves of curvature orthogonally.*

Proof Differentiating the equation $\mathbf{e} = \mathbf{s} + \rho\mathbf{n}$ along the curve $\mathbf{r}$ we have

$$\mathbf{e}_1\mathbf{r}_1 = \mathbf{s}_1\mathbf{r}_1 + \rho\mathbf{n}_1\mathbf{r}_1.$$

Then by the symmetry of the twice linear form $\mathbf{s}_1 \cdot \mathbf{n}_1$ we have

$$\begin{aligned}\mathbf{e}_1\mathbf{a}_1\mathbf{e}_1 \cdot \mathbf{r}_1 = -\rho\mathbf{s}_1\mathbf{a}_1 \cdot \mathbf{e}_1\mathbf{r}_1 &= -\rho\mathbf{s}_1\mathbf{a}_1 \cdot (\mathbf{s}_1\mathbf{r}_1 + \rho\mathbf{n}_1\mathbf{r}_1)\\ &= -\rho\mathbf{s}_1\mathbf{r}_1 \cdot (\mathbf{s}_1\mathbf{a}_1 + \rho\mathbf{n}_1\mathbf{a}_1) = 0. \qquad \square\end{aligned}$$

For an alternative proof of Proposition 10.8 see Exercise 10.6.

Later, in Chapter 14, we look in greater detail at the geodesic foliations on the focal surface formed by the focal curves of curvature.

Away from umbilical centres the non-regular points of the focal surface $\mathbf{e}$ of a regular smooth surface $\mathbf{s}$ form *ordinary cuspidal edges*, the *ribs* of $\mathbf{s}$, with occasional *swallow-tail cusps*, the former consisting of those points where the focal curves have ordinary cusps and the latter those points where the focal curves have ordinary kinks (Figure 10.3). While this may be proved directly it is more appropriate to go back and to look at the whole focal surface of $\mathbf{s}$ from a slightly different point of view, and in particular to form a picture of what it might look like in the neighbourhood of an umbilical centre. This we do in a fresh chapter, which culminates in practical new characterisations of the ribs and of the ridges and various kinds of umbilics that occur on such a surface $\mathbf{s}$. These resemble characterisations given in previous chapters of vertices of plane curves and space curves.

10.5 Historical note

For a detailed account of the contribution of Monge to the theory of surfaces the reader is referred to a biographical study by René Taton (1951). The most accessible part of Monge's work is in the fifth edition (1850) of his École Polytechnique lectures of 1795, edited by Liouville. However, Taton draws attention to the remarkable fact that Monge's original publication on this subject was part of a paper written over twenty years earlier in 1771 (published in 1774) when he was on the staff of the military college at Mézières. This paper, entitled 'Mémoire sur la théorie des déblais et des remblais', was concerned with shifting the large quantities of earth required when a fortification was under construction, a technique termed in English 'cutting and filling'. Monge remarked that in such a case unnecessary energy is used if the paths of two particles of earth cross while in transit. It was in abstracting this idea that he was led to make a twenty page digression on the behaviour of two-parameter families of straight lines in three-dimensional space, in particular the family of normals to a surface, and how they focus.

Monge is credited by Taton with introducing the word *umbilic* for a point of a surface where the two principal curvatures coincide. However the concept already had been studied by Monge's pupil and colleague Jean-Baptiste Meusnier in 1785. For an account of Meusnier's work the reader is referred to Coolidge (1940). Monge's contribution is also discussed there, though not in such detail as in Taton.

Exercises

10.1 Let $\mathbf{s} : \mathbb{R}^2 \rightarrowtail \mathbb{R}^3$ and $\mathbf{s}' : \mathbb{R}^2 \rightarrowtail \mathbb{R}^3$ be regular surfaces such that, for each $w \in \operatorname{dom} \mathbf{s} = \operatorname{dom} \mathbf{s}'$,

$$(\mathbf{s}'(w) \mp \mathbf{s}(w)) \cdot (\mathbf{s}'(w) - \mathbf{s}(w)) = \delta^2,$$

where $\delta \in \mathbb{R}$. Prove that if $\mathbf{s}'(w)$ lies on the normal to $\mathbf{s}$ at w then $\mathbf{s}(w)$ lies on the normal to $\mathbf{s}'$ at w.

10.2 Find the principal curvatures and principal tangent lines of the surface $\mathbf{s} : (x, y) \mapsto (x, y, xy)$, at the origin.

10.3 Let $\mathbf{s} : \mathbb{R}^2 \rightarrowtail \mathbb{R}^3$ be a regular smooth surface with $0 \in \operatorname{dom} f$, let $\mathbf{n}$ be a unit normal vector to the surface at $w = \mathbf{s}(0)$ and let $h : \mathbb{R}^2 \rightarrowtail \mathbb{R}^2$ be a diffeomorphism with $h(0) = 0$. Let $\mathbf{s}' = \mathbf{s}h$. Prove that

$$\mathbf{s}_1{}'(0) \cdot \mathbf{s}_1{}'(0) = \mathbf{s}_1(0)h_1(0) \cdot \mathbf{s}_1(0)h_1(0),$$

and

$$\mathbf{n} \cdot \mathbf{s}_2{}'(0) = \mathbf{n} \cdot \mathbf{s}_2(0)(h_1(0))^2.$$

Indicate how this justifies the assertion that both the fundamental forms of $\mathbf{s}$ at w may be regarded as quadratic forms on the tangent space to $\mathbf{s}$ at w, independent of the parametrisation chosen to represent the surface.

10.4 Suppose that $\mathbf{s} : \mathbb{R}^2 \rightarrowtail \mathbb{R}^3$ is a regular smooth surface, with $0 \in \operatorname{dom} \mathbf{s}$ and $(0) = 0$, and let $k : \mathbb{R}^3 \to \mathbb{R}$ be a surjective linear map whose kernel intersects the image of $\mathbf{s}_1(0)$ in a line. Prove that the intersection of the kernel with the image of $\mathbf{s}$ is a curve that is non-singular at 0 and that the parametrisation of the *surface* can be so chosen that locally near 0 the curve of intersection is parametrised by the restriction of $\mathbf{s}$ to a line through the origin $\mathbb{R} \to \mathbb{R}^3 : t \mapsto \mathbf{s}(t\mathbf{u})$, where $\mathbf{u} \in \mathbb{R}^2$, with $\mathbf{s}_1(0)\mathbf{u} \cdot \mathbf{s}_1(0)\mathbf{u} = 1$.

10.5 For any regular surface $\mathbf{s} : \mathbb{R}^2 \rightarrowtail \mathbb{R}^3$ its *Gauss map* is the map $\mathbf{n} : \mathbb{R}^2 \rightarrowtail \mathbb{R}^3$ associating to each point $w = (u, v) \in \mathbb{R}^2$ the standard unit normal vector $\mathbf{n}(w)$ to $\mathbf{s}$ at w.

Let $\mathbf{a}_1$ be a non-zero principal vector at w to a regular surface $\mathbf{s}$, the corresponding principal curvature of $\mathbf{s}$ at w being κ, and let $\mathbf{n}$ be the Gauss map of $\mathbf{s}$. Prove that

$$\kappa = 0 \Leftrightarrow \mathbf{n} \cdot \mathbf{s}_2\mathbf{a}_1 = 0 \text{ at } w \Leftrightarrow \mathbf{n}_1\mathbf{a}_1 = 0 \text{ at } w.$$

10.6 Let $\mathbf{s} : \mathbb{R}^2 \rightarrowtail \mathbb{R}^3$ be a regular surface and let $\mathbf{c} : \mathbb{R} \rightarrowtail \mathbb{R}^2$ represent a regular curve $\mathbf{s}\,\mathbf{c}$ on $\mathbf{s}$ along which one of the principal curvatures κ is constant, not zero. Prove that the curve $\mathbf{e}\,\mathbf{c}$ on the focal surface $\mathbf{e}$ intersects each focal line of curvature orthogonally.

(Hint: Differentiate the equation $(\mathbf{ec} - \mathbf{sc}) \cdot (\mathbf{ec} - \mathbf{sc}) = \text{const}$, thus showing that $(\mathbf{ec} - \mathbf{sc}) \cdot (\mathbf{ec})_1 = 0$. Then it is one line to the answer!)

This example provides an alternative proof for Proposition 10.8.

10.7 Let $\mathbf{r}$ be a nowhere linear or planar regular smooth space curve $\mathbf{r}$ parametrised by arc length and let $\mathbf{S} : \mathbb{R}^2 \rightarrowtail \mathbb{R}^3$ be the *tube* with *core* $\mathbf{r}$ and *radius* δ, namely the surface

$$\mathbf{S} : (s, \theta) \mapsto \mathbf{r}(s) + \delta(\mathbf{n}(s)\cos\theta + \mathbf{b}(s)\sin\theta).$$

Verify that $\mathbf{S}$ is regular provided that δ is everywhere less than ρ, and that then the standard surface normal at (s, θ) is

$$\mathbf{N}(s, \theta) = -\mathbf{n}(s)\cos\theta - \mathbf{b}(s)\sin\theta.$$

With this restriction on δ verify also that

$$\mathrm{I}_2 = \mathbf{S}_1 \cdot \mathbf{S}_1 = [\tau^2\delta^2 + (1 - \kappa\delta\cos)^2, \tau\delta^2, \delta^2]_2$$

and

$$\mathrm{II}_2 = -\mathbf{N}_1 \cdot \mathbf{S}_1 = [\tau^2\delta - (1 - \kappa\delta\cos)^2\,\kappa\delta\cos, \tau\delta, \delta]_2.$$

Show that one sheet of the focal surface of **S** degenerates to the core curve while the other consists of the focal surface of **r** twice over. Verify that one family of lines of curvature consists of parallels $\mathbf{r} + \delta\mathbf{m}$ to $\mathbf{r}$ and that the corresponding focal lines of curvature of the tube are just the focal curves $\mathbf{f} = \mathbf{r} + \sigma\mathbf{m}$ where $\mathbf{r}_1 + \sigma\mathbf{m}_1 = 0$.

(Feel free just to consider the case that $\delta = 1$, with $\sigma > 1$ along each parallel. The formulas are then slightly simpler. We shall be considering tubular surfaces in detail in Chapter 16.)

10.8 Suppose that the image of a regular surface $\mathbf{s} : \mathbb{R}^2 \rightarrowtail \mathbb{R}^3$ coincides with the set of zeros of a smooth function $F : \mathbb{R}^3 \rightarrowtail \mathbb{R}$, with everywhere surjective differential. By considering the first two derivatives of the composite map $F\mathbf{s}$ show that, at a point w, the second fundamental form of $\mathbf{s}$ is a multiple of $F_2(\mathbf{s}(w))(\mathbf{s}_1(w))^2$.

10.9 Let $\mathbf{s} : \mathbb{R}^2 \rightarrowtail \mathbb{R}^3$ be a *minimal* regular surface, that is for each $w \in \operatorname{dom} \mathbf{s}$, the sum $\kappa(w) + \lambda(w)$ of the principal curvatures is zero. Prove that $\mathbf{n}_1 \cdot \mathbf{n}_1 = \kappa^2 \mathbf{s}_1 \cdot \mathbf{s}_1$.

(By Rodrigues, at any $w \in \operatorname{dom} \mathbf{s}$, $\mathbf{n}_1(\mathbf{a}_1) \cdot \mathbf{n}_1(\mathbf{a}_1) = \kappa^2 \mathbf{s}_1(\mathbf{a}_1) \cdot \mathbf{s}_1(\mathbf{a}_1)$ and $\mathbf{n}_1(\mathbf{b}_1) \cdot \mathbf{n}_1(\mathbf{b}_1) = \lambda^2 \mathbf{s}_1(\mathbf{b}_1) \cdot \mathbf{s}_1(\mathbf{b}_1) = \kappa^2 \mathbf{s}_1(\mathbf{b}_1) \cdot \mathbf{s}_1(\mathbf{b}_1)$, where $\mathbf{a}_1$ and $\mathbf{b}_1$ are principal vectors at w. Moreover, not only $\mathbf{s}_1(\mathbf{a}_1) \cdot \mathbf{s}_1(\mathbf{b}_1) = 0$ but also, again by Rodrigues, $\mathbf{n}_1(\mathbf{a}_1) \cdot \mathbf{n}_1(\mathbf{b}_1) = 0$.

Accordingly, for any $\mathbf{u} \in \mathbb{R}^2$, $\mathbf{n}_1(\mathbf{u}) \cdot \mathbf{n}_1(\mathbf{u}) = \kappa^3 \mathbf{s}_1(\mathbf{u}) \cdot \mathbf{s}_1(\mathbf{u})$ and so, for any $\mathbf{u}, \mathbf{v} \in \mathbb{R}^2$, $\mathbf{n}_1(\mathbf{u}) \cdot \mathbf{n}_1(\mathbf{v}) = \kappa^2 \mathbf{s}_1(\mathbf{u}) \cdot \mathbf{s}_1(v)$.

This says that the *Gauss map* $\mathbf{s}(w) \mapsto \mathbf{n}(w)$ of a minimal surface is *conformal*, that is the angle between any two tangent vectors to $\mathbf{s}$ at a point w is equal to the angle between the induced tangent vectors to $\mathbf{n}$ at w.)

10.10 Let $\mathbf{s}$ be a non-singular point of the surface given by the equation $F(\mathbf{s}) = 0$, where $F : \mathbb{R}^3 \rightarrowtail \mathbb{R}$ is smooth. Prove that, up to a non-zero real multiple, the second fundamental form of the surface at $\mathbf{s}$ is the restriction of the form $F_2(\mathbf{s})$ to the tangent space to the surface at $\mathbf{s}$.

10.11 Compute the second fundamental form of the ellipsoid

$$\left\{(x, y, z) \in \mathbb{R}^3 : \frac{x^2}{a^2} + \frac{y^2}{b^2} + \frac{z^2}{c^2} = 1\right\}.$$

10.12 Let $\mathbf{s}$ be a non-singular point of the surface given by the equation $F(\mathbf{s}) = 0$, where $F : \mathbb{R}^3 \rightarrowtail \mathbb{R}$ is smooth, let $\mathbf{n}$ be a unit normal vector at $\mathbf{s}$ and let $\boldsymbol{\alpha}$ be a principal tangent vector there of unit length, with $\mathbf{s} + \rho\mathbf{n}$ the corresponding principal centre of curvature at s. Prove that

$$\rho F_2(\mathbf{s})\boldsymbol{\alpha}^2 + F_1(\mathbf{s})\mathbf{n} = 0.$$

11
Ridges and ribs

11.0 Introduction

This chapter is one of the most demanding in the whole book, for our fresh look at the features of the focal surface or evolute of a regular surface $\mathbf{s}$ in $\mathbb{R}^3$ involves a number of excursions into higher-dimensional space. For example the *normal bundle* of $\mathbf{s}$ is to be thought of as a smooth submanifold in *six*-dimensional space, while to establish the basic facts about ridges and ribs in the neighbourhood of an umbilic it is desirable to work in *seven*-dimensional space. Our principal tool will be the surjective criterion corollary of the Inverse Function Theorem, Theorem 4.16.

11.1 The normal bundle of a surface

Let $\mathbf{s} : \mathbb{R}^2 \rightarrowtail \mathbb{R}^3$ be a regular surface in $\mathbb{R}^3$, and, for any point $\mathbf{c}$ of $\mathbb{R}^3$, let $V(\mathbf{c}) = \mathbf{c} \cdot \mathbf{s} - \frac{1}{2}\mathbf{s} \cdot \mathbf{s}$, with

$$V(\mathbf{c})_1 = (\mathbf{c} - \mathbf{s}) \cdot \mathbf{s}_1,$$

$$V(\mathbf{c})_2 = (\mathbf{c} - \mathbf{s}) \cdot \mathbf{s}_2 - \mathbf{s}_1 \cdot \mathbf{s}_1,$$

$$V(\mathbf{c})_3 = (\mathbf{c} - \mathbf{s}) \cdot \mathbf{s}_3 - 3\mathbf{s}_1 \cdot \mathbf{s}_2,$$

and so on.

Then the *normal bundle* of $\mathbf{s}$ consists of all its normal lines, each labelled with the point of the surface at which it is normal, namely the subset

$$\{(\mathbf{c}, \mathbf{s}(w)) \in \mathbb{R}^3 \times \mathbb{R}^3 : V(\mathbf{c})_1(w) = 0\}.$$

Theorem 11.1 The normal bundle of a regular surface $\mathbf{s}$ *in* $\mathbb{R}^3$ *is everywhere locally a smooth submanifold of* $\mathbb{R}^3 \times \mathbb{R}^3$ *of dimension* 3.

Proof It is enough to prove that $\{(\mathbf{c}, w) \in \mathbb{R}^3 \times \mathbb{R}^2 : V(\mathbf{c})_1(w) = 0\}$ is a smooth submanifold of $\mathbb{R}^3 \times \mathbb{R}^2$ of dimension 3, and this follows immediately from the surjective criterion, Theorem 4.16, for the transpose of the Jacobian matrix of the map $\mathbb{R}^3 \times \mathbb{R}^2 \rightarrowtail L(\mathbb{R}^2, \mathbb{R})$; $(\mathbf{c}, w) \mapsto V(\mathbf{c})_1(w)$ at $(\mathbf{c}, w)$ is the 5×3 matrix

$$\begin{bmatrix} \mathbf{s}_1(w) \\ V(\mathbf{c})_2(w) \end{bmatrix},$$

which is injective, since $\mathbf{s}_1(w)$ is injective. So the Jacobian matrix itself is surjective. □

Suppose that we forget the labelling of the normals. Then we have a map of the normal bundle to the ambient space.

Proposition 11.2 Let $\mathbf{s}$ *be a regular surface in* $\mathbb{R}^3$. *Then the restriction to the normal bundle of* $\mathbf{s}$ *of the projection* $\mathbb{R}^3 \times \mathbb{R}^2 \rightarrowtail \mathbb{R}^3$; $(\mathbf{c}, w) \mapsto \mathbf{c}$ *is locally a diffeomorphism unless* $V(\mathbf{c})_2\mathbf{u} = 0$, *for some non-zero vector* $\mathbf{u}$.

Proof Any tangent vector $(\mathbf{c}_1, w_1)$ of the normal bundle at $(\mathbf{c}, w)$ satisfies the equation $\mathbf{s}_1(w) \cdot \mathbf{c}_1 + V(\mathbf{c})_2(w)w_1 = 0$. So if $\mathbf{c}_1 = 0$ then also $w_1 = 0$ unless there is a non-zero vector $w_1 = \mathbf{u}$ such that $V(\mathbf{c})_2(w)\mathbf{u} = 0$. □

If such a non-zero vector $\mathbf{u}$ exists then the dimension of the kernel is equal either to 1 or to 2. We consider first the latter possibility.

11.2 Isolated umbilics

Suppose therefore that not only $V(\mathbf{c})_1(w) = 0$ but also $V(\mathbf{c})_2(w) = 0$. Then $\mathbf{s}$ has an umbilic at w that is not flat, with $\mathbf{c}$ as umbilical centre. Such an umbilic will be said to be *ordinary* if $V(\mathbf{c})_3(w)$ is non-zero and neither right-angled with respect to the first fundamental form $\mathrm{I}_2(w)$ nor perfect. We recall from Section 7.3 that a thrice linear form is *right-angled* with respect to a positive-definite quadratic form if its Hessian form is right-angled with respect to that form, while it is *perfect* if its cubic form is a perfect cube. The geometrical relevance of these conditions will shortly become clear. An umbilic of a surface is said to be *isolated* if no nearby point is an umbilic of the surface.

An umbilic of a regular surface need not be isolated. For example every point of a sphere is an umbilic, but this is clearly a very special case, as the next theorem shows.

Theorem 11.3 Any ordinary umbilic of a regular surface $\mathbf{s}$ *is isolated.*

Proof It is clearly enough to prove that

$$\{(\mathbf{c}, \mathbf{s}(w)) \in \mathbb{R}^3 \times \mathbb{R}^3 : V(\mathbf{c})_1(w) = 0,\ V(\mathbf{c})_2(w) = 0\}$$

is a zero-dimensional smooth submanifold of $\mathbb{R}^3 \times \mathbb{R}^3$ or equivalently that

$$\{(\mathbf{c}, w) \in \mathbb{R}^3 \times \mathbb{R}^2 : V(\mathbf{c})_1(w) = 0,\ V(\mathbf{c})_2(w) = 0\}$$

is a zero-dimensional smooth submanifold of $\mathbb{R}^3 \times \mathbb{R}^2$.

Now the Jacobian matrix of the map

$$\mathbb{R}^3 \times \mathbb{R}^2 \rightarrowtail L(\mathbb{R}^2, \mathbb{R}) \times L_2(\mathbb{R}^2, \mathbb{R});\ (\mathbf{c}, w) \mapsto (V(\mathbf{c})_1(w),\ V(\mathbf{c})_2(w))$$

at $(\mathbf{c}, w)$ is the square 5×5 matrix

$$\begin{bmatrix} \mathbf{s}_1(w)\cdot & 0 \\ \mathbf{s}_2(w)\cdot & V(\mathbf{c})_3(w) \end{bmatrix}.$$

This is injective and so surjective. For suppose that $(\mathbf{c}_1, w_1)$ is a kernel vector of the matrix. Then, since $\mathbf{s}_1(w)\cdot\mathbf{c}_1 = 0$, $\mathbf{c}_1 = \zeta\mathbf{n}(w)$, implying that $\zeta\mathbf{s}_2(w)\cdot\mathbf{n} + V(\mathbf{c})_3(w)w_1 = \zeta\kappa\mathbf{s}_1(w)\cdot\mathbf{s}_1(w) + V(\mathbf{c})_3(w)w_1 = 0$. Now if this were so for ζ and w_1 not both zero then either $V(\mathbf{c})_3$ is perfect, when $\zeta = 0$ but $w_1 \neq 0$, or $V(\mathbf{c})_3(w)w_1$ is a non-zero multiple of $\mathrm{I}_2(w)$, implying by Proposition 7.9 that $V(\mathbf{c})_3(w)$ is right-angled with respect to $\mathrm{I}_2(w)$. But these are exactly the cases excluded. So $\mathbf{c}_1 = \zeta\mathbf{n} = 0$ and $w_1 = 0$, implying that the kernel of the matrix is zero and therefore that the matrix is the matrix of an injective map, as had to be proved. □

11.3 The normal focal surface

Next suppose that there exists a non-zero vector $\mathbf{u}$ such that $V(\mathbf{c})_2(w)\mathbf{u} = 0$, where $V(\mathbf{c})_2(w)$ is not necessarily zero. This is the condition for $\mathbf{c}$ to be a focal point on the normal to $\mathbf{s}$ at w.

Theorem 11.4 Let $\mathbf{s}$ *be a regular surface in* $\mathbb{R}^3$ *all of whose umbilics are ordinary. Then* even where there is an ordinary umbilic *at* w *the set*

$\{(\mathbf{c}, w, \mathbf{u}) \in \mathbb{R}^3 \times \mathbb{R}^2 \times \mathbb{R}^2 : V(\mathbf{c})_1(w) = 0, V(\mathbf{c})_2(w)\mathbf{u} = 0, \mathbf{u}\cdot\mathbf{u} - 1 = 0\}$ is a smooth submanifold of $\mathbb{R}^3 \times \mathbb{R}^2 \times \mathbb{R}^2$ of dimension 2.

Proof We prove the surjectivity of the derivative of the map

$$\mathbb{R}^3 \times \mathbb{R}^2 \times \mathbb{R}^2 \rightarrowtail \mathbb{R}^2 \times \mathbb{R}^2 \times \mathbb{R};$$

$$(\mathbf{c}, w, \mathbf{u}) \mapsto (V(\mathbf{c})_1(w), V(\mathbf{c})_2(w)\mathbf{u}, \mathbf{u}\cdot\mathbf{u} - 1)$$

by proving the injectivity of its transposed matrix, namely the 7×5 matrix

$$\begin{bmatrix} \mathbf{s}_1 & \mathbf{s}_2\mathbf{u} & 0 \\ V_2 & V_3\mathbf{u} & 0 \\ 0 & V_2 & 2\mathbf{u}\,\cdot \end{bmatrix} = \begin{bmatrix} \mathbf{s}_1(w) & \mathbf{s}_2(w)\mathbf{u} & 0 \\ V(\mathbf{c})_2(w) & V(\mathbf{c})_3(w)\mathbf{u} & 0 \\ 0 & V(\mathbf{c})_2(w) & 2\mathbf{u}\,\cdot \end{bmatrix},$$

where the abbreviated form on the left may be easier to follow than the full expression.

Suppose first that we are not at an umbilic, so that $V(\mathbf{c})_2(w) \neq 0$, and suppose that $(\beta, \alpha, \mu) \in \mathbb{R}^2 \times \mathbb{R}^2 \times \mathbb{R}$ is a kernel vector of this matrix. From the third row $V(\mathbf{c})_2(w)\alpha + 2\mu\mathbf{u}\,\cdot = 0$. But since $\mathbf{u}\cdot\mathbf{u} = 1$ it follows that $2\mu = -V(\mathbf{c})_2(w)\alpha\mathbf{u} = 0$. So $V(\mathbf{c})_2(w)\alpha = 0$, implying that $\alpha = v\mathbf{u}$ for some $v \in \mathbb{R}$. Then from the first row $\mathbf{s}_1(w)\beta + v\mathbf{s}_2(w)\mathbf{u}^2 = 0$. By Exercise 11.1 it follows that $v = 0$ and that $\beta = 0$.

Suppose next that we are at an umbilic. Then the matrix reduces to

$$\begin{bmatrix} \mathbf{s}_1 & \mathbf{s}_2\mathbf{u} & 0 \\ 0 & V_3\mathbf{u} & 0 \\ 0 & 0 & 2\mathbf{u}\,\cdot \end{bmatrix} = \begin{bmatrix} \mathbf{s}_1(w) & \mathbf{s}_2(w)\mathbf{u} & 0 \\ 0 & V(\mathbf{c})_3(w)\mathbf{u} & 0 \\ 0 & 0 & 2\mathbf{u}\,\cdot \end{bmatrix}.$$

For a kernel vector (β, α, μ) we then have $\mathbf{s}_1(w)\beta + \mathbf{s}_2(w)\mathbf{u}\alpha = 0$, implying that $\mathbf{n}\cdot\mathbf{s}_2(w)\mathbf{u}\alpha = \kappa \mathrm{I}_2(w)\mathbf{u}\alpha = 0$ and that $V(\mathbf{c})_3(w)\mathbf{u}\alpha = 0$. But the umbilic is ordinary, so these only occur if $\alpha = 0$, for otherwise the cubic is either right-angled or perfect. Then by the injectivity of $\mathbf{s}_1(w)$ we also have $\beta = 0$. Finally $\mu\mathbf{u}\,\cdot = 0 \Rightarrow \mu = 0$. □

This theorem has immediate application to the *normal focal surface* of a regular surface $\mathbf{s}$. This consists of the subset of the normal bundle of $\mathbf{s}$ consisting of the focal points on each labelled normal line.

Corollary 11.5 Let $\mathbf{s}$ *be a regular surface in* $\mathbb{R}^3$ *all of whose umbilical points are ordinary. Then the normal focal surface of* $\mathbf{s}$ *is a smooth submanifold of* $\mathbb{R}^3 \times \mathbb{R}^3$ *of dimension 2, except at umbilical centres at each of which it has a* conic singularity.

Proof Let M be the submanifold of the theorem. Then the normal focal surface is the image of M by the projection $\mathbb{R}^3 \times \mathbb{R}^2 \times \mathbb{R}^2 \rightarrowtail \mathbb{R}^3 \times \mathbb{R}^2$ that forgets the last factor. Now away from an umbilic the restriction to M of this projection is everywhere regular, that is its differential everywhere is injective. For any tangent vector $(\mathbf{c}_1, w_1, \mathbf{u}_1)$ of M at $(\mathbf{c}, w, \mathbf{u})$ satisfies

$$\mathbf{c}_1 \cdot \mathbf{s}_1(w) + V_2(w)w_1 = 0,$$

$$\mathbf{c}_1 \cdot \mathbf{s}_2\mathbf{u} + V_3(\mathbf{c})(w)\mathbf{u}w_1 + V(\mathbf{c})_2(w)\mathbf{u}_1 = 0,$$

$$2\mathbf{u} \cdot \mathbf{u} = 0.$$

Clearly if $\mathbf{c}_1 = 0$ and $w_1 = 0$ then also $\mathbf{u}_1 = 0$, for from the second equation $\mathbf{u}_1 = \lambda\mathbf{u}$ and then from the third $\lambda = 0$. It follows that away from umbilics the normal focal surface is a smooth submanifold of dimension 2.

On the other hand at an umbilic an entire circle of points of M map to $(\mathbf{c}, w)$. The tangent vectors $(\mathbf{c}_1, w_1, \mathbf{u}_1)$ then satisfy the equations

$$\mathbf{c}_1 \cdot \mathbf{s}_1(w) = 0,$$

$$\mathbf{c}_1 \cdot \mathbf{s}_2(w)\mathbf{u} + V(\mathbf{c})_3(w)\mathbf{u}w_1 = 0,$$

$$2\mathbf{u} \cdot \mathbf{u}_1 = 0.$$

These project to vectors $(\mathbf{c}_1, w_1)$ satisfying the first two of these. The first implies that $\mathbf{c}_1 = \zeta\mathbf{n}(w)$ and the second then reduces to

$$(\zeta\kappa \mathrm{I}_2(w) + V(\mathbf{c})_3(w)w_1)\mathbf{u} = 0,$$

where $\mathbf{u} \neq 0$.

To see what this last equation is saying, suppose that we choose a parametrisation of the surface such that at the umbilic $\mathrm{I}_2(w)$ has the identity matrix as matrix, let $w_1 = (\xi, \eta)$ and $V(\mathbf{c})_3(w) = [p, q, r, s]_3$. Then the existence of the vector $\mathbf{u}$ implies that the determinant of the matrix of $\zeta\kappa\mathrm{I}_2(w) + V(\mathbf{c})_3(w)w_1$, namely

$$\begin{bmatrix} p\xi + q\eta + \kappa\zeta & q\xi + r\eta \\ q\xi + r\eta & r\xi + s\eta + \kappa\zeta \end{bmatrix},$$

is zero. This, by Proposition 7.11, is the equation of a non-degenerate quadric cone, since the umbilic is ordinary. It can be proved, though we do not give the proof here, that in that case there is a diffeomorphism-germ of $\mathbb{R}^3 \times \mathbb{R}^2$ at $(\mathbf{c}, w)$ mapping the normal focal surface to this cone. This is what is implied in the statement of the corollary that at an ordinary umbilic the normal focal surface has a *conic singularity*. □

As with the normal bundle so with the normal focal surface we ask what happens when we project it to $\mathbb{R}^3$.

Proposition 11.6 With $\mathbf{s}$ *as above the restriction to the normal focal surface of* $\mathbf{s}$ *of the projection* $\mathbb{R}^3 \times \mathbb{R}^2 \rightarrowtail \mathbb{R}^3 : (\mathbf{c}, w) \mapsto \mathbf{c}$ *is immersive except at umbilics of the surface or where* $V(\mathbf{c})_3\mathbf{u}^3 = 0$.

Proof Any tangent vector $(\mathbf{c}_1, w_1)$ of the normal focal surface away from an umbilic satisfies the equations

$$\mathbf{s}_1(w)\cdot\mathbf{c}_1 + V(\mathbf{c})_2(w)w_1 = 0,$$

$$\mathbf{s}_2(w)\mathbf{u}\cdot\mathbf{c}_1 + V(\mathbf{c})_3(w)\mathbf{u}w_1 + V(\mathbf{c})_2(w)\mathbf{u}_1 = 0 \text{ for some } \mathbf{u}_1,$$

$$2\mathbf{u}\cdot\mathbf{u}_1 = 0.$$

When $\mathbf{c}_1 = 0$ then from the first equation $w_1 = \lambda\mathbf{u}$, for some $\lambda \in \mathbb{R}$. On substituting this in the second equation $\lambda V(\mathbf{c})_3(w)\mathbf{u}^2 + V(\mathbf{c})_2(w)\mathbf{u}_1 = 0$, implying that $\lambda V(\mathbf{c})_3(w)\mathbf{u}^3 = 0$. Then $\lambda = 0$ and therefore $w_1 = 0$ unless $V(\mathbf{c})_3(w)\mathbf{u}^3 = 0$. □

11.4 Ridges and ribs

The regular surface $\mathbf{s}$ is said to have a *ridge point* at w and the focal surface a *rib point* at $\mathbf{c}$ if $V(\mathbf{c})_1(w) = 0$ and $V(\mathbf{c})_2(w)\mathbf{u} = 0$, for some non-zero vector $\mathbf{u}$, and $V(\mathbf{c})_3(w)\mathbf{u}^3 = 0$. According to the definition any umbilic w and its centre $\mathbf{c}$ satisfy these conditions, for $\mathbf{u}$ may then be taken to be any non-zero root vector of the thrice linear form $V(\mathbf{c})_3(w)$.

As we have anticipated in the introduction to Chapter 10 and later in Proposition 10.7, there are, away from umbilics, alternative characterisations of ridge points and rib points that involve the lines of curvature and the associated focal curves of the surface $\mathbf{s}$. These are the subject of the next theorem and its corollaries.

Theorem 11.7 Let $\mathbf{a} : \mathbb{R} \rightarrowtail \mathbb{R}^2$ *represent a line of curvature through a point* $w = \mathbf{a}(t)$ *of a regular surface* $\mathbf{s}$, *the point not being an umbilic of the surface, and let* $\mathbf{e}$ *be a parametrisation of the corresponding sheet of the focal surface. Moreover, let* V_i, *for* $i \geqslant 1$, *denote* $V(\mathbf{c})_i(w)$ *with* $\mathbf{c}$ *afterwards put equal to* $\mathbf{e}$, *so that* $V_1 = 0$ *and* $(V_1\mathbf{a})_1 = V_2\mathbf{a}_1 = 0$. *Then, at* t,

$$(\mathbf{ea})_1 = 0 \Leftrightarrow (V_1\mathbf{a})_2 = V_3\mathbf{a}_1{}^2 + V_2\mathbf{a}_2 = 0 \Leftrightarrow V_3\mathbf{a}_1{}^3 = 0,$$

and, all this being so,

$$\begin{aligned}(\mathbf{ea})_2 = 0 \Leftrightarrow (V_1\mathbf{a})_3 &= V_4\mathbf{a}_1{}^3 + 3V_3\mathbf{a}_1\mathbf{a}_2 + V_2\mathbf{a}_3 = 0,\\ &\Leftrightarrow V_4\mathbf{a}_1{}^4 + 3V_3\mathbf{a}_1{}^2\mathbf{a}_2 = 0,\end{aligned}$$

and indeed, for all $k \geqslant 1$,

$$(\mathbf{ea})_j = 0, \text{ for all } 1 \leqslant j \leqslant k \Leftrightarrow (V_1\mathbf{a})_j = 0, \text{ for all } 1 \leqslant j \leqslant k.$$

Proof The equation $V_2\mathbf{a}_1 = 0$, more strictly $(V_2\mathbf{a})\mathbf{a}_1 = 0$, is defined on $\mathbb{R}$. Differentiating it we get $\mathbf{e}_1\mathbf{a}_1 \cdot \mathbf{s}_2\mathbf{a}_1 + V_3\mathbf{a}_1{}^2 + V_2\mathbf{a}_2 = 0$, so that

$$(\mathbf{ea})_1 = 0 \Rightarrow V_3\mathbf{a}_1{}^2 + V_2\mathbf{a}_2 = 0 \Rightarrow V_3\mathbf{a}_1{}^3 = 0.$$

Conversely, putting $\mathbf{a}_1$ in the empty slot in the equation

$$\mathbf{e}_1\mathbf{a}_1 \cdot \mathbf{s}_2\mathbf{a}_1 + V_3\mathbf{a}_1{}^2 + V_2\mathbf{a}_2 = 0$$

we find that $(\mathbf{ea})_1 \cdot \mathbf{s}_2\mathbf{a}_1{}^2 + V_3\mathbf{a}_1{}^3 = 0$. Now $(\mathbf{ea})_1 = (\rho\mathbf{a})_1\mathbf{na}$, so

$$\begin{aligned}V_3\mathbf{a}_1{}^3 = 0 &\Rightarrow (\mathbf{ea})_1 \cdot \mathbf{s}_2\mathbf{a}_1{}^2 = (\rho\mathbf{a})_1\mathbf{n} \cdot \mathbf{s}_2\mathbf{a}_1{}^2 = (\rho\mathbf{a})_1\mathbf{s}_1\mathbf{a}_1 \cdot \mathbf{s}_1\mathbf{a}_1 = 0\\ &\Rightarrow (\rho\mathbf{a})_1 = 0 \Rightarrow (\mathbf{ea})_1 = 0.\end{aligned}$$

Of course if $\mathbf{a}_1$ is a vector such that $V_2\mathbf{a}_1 = 0$ and $V_3\mathbf{a}_1{}^3 = 0$ then by Proposition 7.4 there exists a vector $\mathbf{a}_2$ such that $V_3\mathbf{a}_1{}^2 + V_2\mathbf{a}_2 = 0$, such a vector not being unique since $\mathbf{a}_2 + \lambda\mathbf{a}_1$ will do equally well, for any $\lambda \in \mathbb{R}$. In the present context the geometrical meaning of such a vector $\mathbf{a}_2$ as the second derivative of the line of curvature $\mathbf{a}$ is clear. The ambiguity in its choice corresponds to the freedom one has in parametrising the lines of curvature of the surface.

The rest follows similarly after one or more further differentiations. □

Corollary 11.8 The surface $\mathbf{s}$ *has a ridge point and its focal surface a rib point where a focal curve fails to be regular.*

Proof Cf. Proposition 10.5. □

In the introduction to Chapter 10 we asserted that along a line of curvature the relevant principal curvature has an ordinary local maximum or minimum, that is an ordinary critical point, where the line crosses a ridge of the same colour transversally, and has an ordinary stationary inflection where it is simply tangential to a ridge of the same colour at an ordinary turning point of the ridge. Part of this follows at once from Proposition 10.7, as a second corollary of Theorem 11.7, but the part concerning transversality or tangency is not yet accessible, being part of Theorem 11.10 below. Both parts are set as Exercise 11.7.

11.5 A classification of focal points

It is convenient at this point to introduce notations for surfaces and their focal points analogous to the A_k notations introduced in previous chapters for plane and space curves and anticipated for surfaces at the beginning of Chapter 10. These notations have in fact already been introduced in the discussion of probes of smooth maps $\mathbb{R}^2 \rightarrowtail \mathbb{R}$ in Chapter 8, and are applied here to the smooth map $V(\mathbf{c}) : \mathbb{R}^2 \rightarrowtail \mathbb{R}$ that measures the degree of contact of the surface $\mathbf{s}$ at a point w with the sphere with centre $\mathbf{c}$ that passes through $\mathbf{s}(w)$.

Given a regular smooth surface $\mathbf{s} : \mathbb{R}^2 \rightarrowtail \mathbb{R}^3$ and a point $\mathbf{c} \in \mathbb{R}^3$ then, abbreviating $V(\mathbf{c})_k(w)$ to V_k, $\mathbf{c}$ is said to be

A_1 for $\mathbf{s}$ at w if $V_1 = 0$ and $V_2\mathbf{u} = 0$ only for $\mathbf{u} = 0$, $V_1 = 0$ implying that $\mathbf{c}$ lies on the normal line to $\mathbf{s}$ at w.

A_2 for $\mathbf{s}$ at w if $V_1 = 0$ and $V_2\mathbf{u} = 0$ for some $\mathbf{u} \neq 0$, but $V_2 \neq 0$, and $V_3\mathbf{u}^3 \neq 0$. In this case we may take $\mathbf{u}$ to be the tangent vector $\mathbf{a}_1$ of one of the lines of curvature $\mathbf{a}$ through w. Then $V_2\mathbf{u} = (V_1\mathbf{a})_1$, while $V_3\mathbf{u}^3 = (V_1\mathbf{a})_2\mathbf{a}_1$.

A_3 for $\mathbf{s}$ at w if $V_1 = 0$, but $V_2 \neq 0$, with $(V_1\mathbf{a})_1 = 0$ and $(V_1\mathbf{a})_2\mathbf{a}_1 = 0$ or equivalently $(V_1\mathbf{a})_2 = 0$, but $(V_1\mathbf{a})_3\mathbf{a}_1 \neq 0$.

A_4 for $\mathbf{s}$ at w if $V_1 = 0$, but $V_2 \neq 0$, with $(V_1\mathbf{a})_1 = 0$, $(V_1\mathbf{a})_2\mathbf{a}_1 = 0$ and $(V_1\mathbf{a})_3\mathbf{a}_1 = 0$ or equivalently $(V_1\mathbf{a})_3 = 0$, but $(V_1\mathbf{a})_4\mathbf{a}_1 \neq 0$.

Remark 11.9 When $V_3\mathbf{u}^2 + V_2\mathbf{v} = 0$ and $V_4\mathbf{u}^3 + 3V_3\mathbf{u}\mathbf{v} + V_2 w = 0$, where $\mathbf{u} = \mathbf{a}_1$, $\mathbf{v} = \mathbf{a}_2$, $\mathbf{w} = \mathbf{a}_3$, as is the case when $\mathbf{c}$ is A_4, then it follows that $V_3\mathbf{u}^2\mathbf{w} = V_4\mathbf{u}^3\mathbf{v} + 3V_3\mathbf{u}\mathbf{v}^2$. This will be relevant shortly. □

Moreover, the point $\mathbf{c}$ is said to be

$D_4{}^+$ for $\mathbf{s}$ at w if $V_1 = 0$, $V_2 = 0$, and V_3 is elliptic,
$D_4{}^-$ for $\mathbf{s}$ at w if $V_1 = 0$, $V_2 = 0$, and V_3 is hyperbolic,
D_5 for $\mathbf{s}$ at w if $V_1 = 0$, $V_2 = 0$, and V_3 is parabolic, with $V_3\mathbf{u}^2 = 0$ for some $\mathbf{u} \neq 0$ but $V_4\mathbf{u}^4 \neq 0$.
E_6 for $\mathbf{s}$ at w if $V_1 = 0$, $V_2 = 0$, and V_3 is perfect, that is $V_3\mathbf{u} = 0$ for some $\mathbf{u} \neq 0$, but $V_4\mathbf{u}^4 \neq 0$,

$\mathbf{s}$ in each of these cases having an umbilic at w with umbilical centre $\mathbf{c}$.

It has to be added that an A_1 point $\mathbf{c}$ on the normal to $\mathbf{s}$ at a point w is said to be either *elliptic*, $A_1{}^+$ or *hyperbolic*, $A_1{}^-$, according as the quadratic form $V_2 = V(\mathbf{c})_2(w)$ is elliptic or hyperbolic. Close to the surface points on the normal are elliptic, becoming hyperbolic between the two parabolic, A_2, points in the event that these are both on the same side of the surface, or beyond either of them when they are on opposite sides. If the normal has a point at infinity added to make it a projective line then in all cases the normal is hyperbolic in the segment of the line between the A_2 points that is away from the surface.

Likewise an A_3 point $\mathbf{c}$ on the normal to $\mathbf{s}$ at a point w is said to be either *elliptic*, $A_3{}^+$ or *hyperbolic*, $A_3{}^-$, according as the quadratic form for $\mathbf{a}_2$ (modulo $\mathbf{a}_1$)

$$V_4\mathbf{a}_1{}^4 + 6V_3\mathbf{a}_1{}^2\mathbf{a}_2 + 3V_2\mathbf{a}_2{}^2$$

is elliptic or hyperbolic. In this case we shall say that the point is *sterile* if it is elliptic and *fertile* if it is hyperbolic, the reason for the terminology being that, as we shall later see in greater detail in Chapter 15, umbilics can only be born on fertile ribs. We shall also discover there what the curves are in the fertile case whose first derivative is $\mathbf{a}_1$ and whose second derivatives are the roots of this quadratic. Be patient!

At an umbilical centre then for a root vector $\mathbf{a}_1$ of V_3 for which $V_2\mathbf{a}_1{}^2 \neq 0$ the quadratic for $\mathbf{a}_2$ (mod $\mathbf{a}_1$) that we noted at A_3 points reduces to the linear form $V_4\mathbf{a}_1{}^4 + 6V_3\mathbf{a}_1{}^2\mathbf{a}_2$. Once again the reader must be patient – we shall identify later the curve $\mathbf{a}$ whose first derivative is $\mathbf{a}_1$ and second derivative is the root of this form.

Though it is strictly the point $\mathbf{c}$ of the normal at a point w that acquires the label it is customary to attach the label also to the point w itself, points of ridges being $A_3{}^+$, $A_3{}^-$ or A_4 or worse, and umbilics being $D_4{}^+$, $D_4{}^-$, D_5, E_6 or worse.

11.6 More on ridges and ribs

To gain further information about the ridge points and rib points of a regular surface we return to seven-dimensional space.

Theorem 11.10 Let $\mathbf{s}$ *be a regular surface in* $\mathbb{R}^3$ *with only ordinary umbilics. Then these points* $(\mathbf{c}, w, \mathbf{u})$ *of* $\mathbb{R}^3 \times \mathbb{R}^2 \times \mathbb{R}^2$ *where*

$$V(\mathbf{c})_1(w) = 0,\ V(\mathbf{c})_2(w)\mathbf{u} = 0,\ V(\mathbf{c})_3(w)\mathbf{u}^3 = 0,\ \mathbf{u}\cdot\mathbf{u} - 1 = 0$$

from a smooth submanifold of $\mathbb{R}^3 \times \mathbb{R}^2 \times \mathbb{R}^2$ *of dimension 1, with certain exceptions that will appear in the course of the proof.*

There are then the following corollaries:

(i) at a point $w \in \mathbb{R}^2$ for which $\mathbf{c}$ is A_3 the surface $\mathbf{s}$ has a non-singular ridge and associated non-singular rib, the tangent line to the ridge being transversal to the relevant principal tangent line at w and the tangent line to the rib coinciding with the *focal line* of the relevant line of curvature, regarded as a space curve in $\mathbb{R}^3$ (not to be confused with the focal *curve (of curvature)* on the focal surface);

(ii) at a point $w \in \mathbb{R}^2$ where $\mathbf{c}$ is A_4 *and the relevant line of curvature is not planar* the ridge on $\mathbf{s}$ is non-singular and the rib has an ordinary cusp, the tangent line to the ridge at w coinciding with the relevant principal tangent line, points of the ridge or rib on one side of the w being sterile and on the other side fertile, while, if the relevant line of curvature has an ordinary planar point, the ridge has either an *acnode* or a *crunode*;

(iii) at an umbilic $w \in \mathbb{R}^2$ for which $\mathbf{c}$ is D_4^{+} the surface $\mathbf{s}$ has a single non-singular ridge and associated non-singular rib, the rib having tangent normal to $\mathbf{s}$ at w and passing there from one sheet of the focal surface to the other, with any tangent vector $\mathbf{t}$ to the ridge satisfying the equation $V_3\mathbf{tuv} = 0$ where $V_3\mathbf{u}^3 = 0$ and $I_2(w)\mathbf{uv} = 0$;

(iv) at an umbilic $w \in \mathbb{R}^2$ for which $\mathbf{c}$ is D_4^{-} the surface $\mathbf{s}$ has three non-singular ridges and associated non-singular ribs, the ribs each having tangent normal to $\mathbf{s}$ at w and passing there from one sheet of the focal surface to the other, while any tangent vector $\mathbf{t}$ to a ridge satisfies the equation $V_3\mathbf{tuv} = 0$ where $V_3\mathbf{u}^3 = 0$, and

$\mathrm{I}_2(w)\mathbf{uv} = 0$, there being three distinct root lines of the equation $V_3\mathbf{u}^3 = 0$;

(v) at an umbilic $w \in \mathbb{R}^2$ for which $\mathbf{c}$ is D_5 the surface $\mathbf{s}$ has one non-singular ridge and associated non-singular rib, the rib having tangent normal to $\mathbf{s}$ at w and passing there from one sheet of the focal surface to the other, the tangent vector $\mathbf{t}$ to the ridge satisfying the equation $V_3\mathbf{tuv} = 0$, where $\mathbf{u}$ determines the non-repeated root line of the cubic and $\mathrm{I}_2(w)\mathbf{uv} = 0$, and also has a ridge having an ordinary cusp at the umbilic with limiting tangent line spanned by $\mathbf{u}$ with $V_3\mathbf{u}^2 = 0$, the corresponding rib having a high order cusp with limiting tangent line the surface normal.

Most of the proof The Jacobian matrix of the map

$$\mathbb{R}^3 \times \mathbb{R}^2 \times \mathbb{R}^2 \rightarrowtail L(\mathbb{R}^2, \mathbb{R}) \times L(\mathbb{R}^2, \mathbb{R}) \times \mathbb{R} \times \mathbb{R}$$

$$(\mathbf{c}, w, \mathbf{u}) \mapsto (V(\mathbf{c})_1(w), V(\mathbf{c})_2(w)\mathbf{u}, V(\mathbf{c})_3(w)\mathbf{u}^3, \mathbf{u}\cdot\mathbf{u} - 1)$$

is the 6×7 matrix

$$\begin{bmatrix} \mathbf{s}_1\cdot & V_2 & 0 \\ \mathbf{s}_2\mathbf{u}\cdot & V_3\mathbf{u} & V_2 \\ \mathbf{s}_3\mathbf{u}^3\cdot & V_4\mathbf{u}^3 & 3V_3\mathbf{u}^2 \\ 0 & 0 & 2\mathbf{u}\cdot \end{bmatrix} = \begin{bmatrix} \mathbf{s}_1(w)\cdot & V(\mathbf{c})_2(w) & 0 \\ \mathbf{s}_2(w)\mathbf{u}\cdot & V(\mathbf{c})_3(w)\mathbf{u} & V(\mathbf{c})_2(w) \\ \mathbf{s}_3(w)\mathbf{u}^3\cdot & V(\mathbf{c})_4(w)\mathbf{u}^3 & 3V(\mathbf{c})_3(w)\mathbf{u}^2 \\ 0 & 0 & 2\mathbf{u}\cdot \end{bmatrix}.$$

The question we have to answer is under what conditions the kernel of this matrix at $(\mathbf{c}, w, \mathbf{u})$ has dimension 1. Now for a kernel vector $(\mathbf{c}_1, w_1, \mathbf{u}_1)$ of this matrix we have the equations

$$\mathbf{c}_1\cdot\mathbf{s}_1 + V_2w_1 = 0,$$

$$\mathbf{c}_1\cdot\mathbf{s}_2\mathbf{u} + V_3w_1\mathbf{u} + V_2\mathbf{u}_1 = 0,$$

$$\mathbf{c}_1\cdot\mathbf{s}_2\mathbf{u}^3 + V_4w_1\mathbf{u}^3 + 3V_3\mathbf{u}^2\mathbf{u}_1 = 0,$$

$$\mathbf{u}\cdot\mathbf{u}_1 = 0,$$

where for clarity we have used the abbreviated forms illustrated in the left-hand form of the matrix.

Suppose first that $V_2 \neq 0$. If we represent the line of curvature through w with $\mathbf{u}$ as tangent vector by $\mathbf{a}$, so that $\mathbf{u} = \mathbf{a}_1$ (up to a non-zero real multiple) we find, using $V_3\mathbf{a}_1{}^2 + V_2\mathbf{a}_2 = 0$ twice, that

$$\mathbf{c}_1\cdot(\mathbf{sa})_1 = \mathbf{c}_1\cdot\mathbf{s}_1\mathbf{a}_1 = 0,$$

$$\mathbf{c}_1\cdot(\mathbf{sa})_2 = \mathbf{c}_1\cdot(\mathbf{s}_2\mathbf{a}_1{}^2 + \mathbf{s}_1\mathbf{a}_2) = 0,$$

$$\mathbf{c}_1\cdot(\mathbf{sa})_3 + (V_1\mathbf{a})_3w_1 = \mathbf{c}_1\cdot(\mathbf{s}_3\mathbf{a}_1{}^3 + 3\mathbf{s}_2\mathbf{a}_1\mathbf{a}_2 + \mathbf{s}_1\mathbf{a}_3) + (V_1\mathbf{a})_3w_1 = 0.$$

Now suppose that $\mathbf{c}$ is A_3. Then w_1 is not a multiple of $\mathbf{u}$ and $\mathbf{c}_1 \neq 0$. For if w_1 is a non-zero multiple of $\mathbf{u}$ then $\mathbf{c}_1 \cdot \mathbf{s}_1 = 0$ and $\mathbf{c}_1 \cdot \mathbf{s}_2\mathbf{u}^2 = 0$ implying that $\mathbf{c}_1 = 0$. But, since $\mathbf{c}$ is A_3, $(V_1\mathbf{a})_3\mathbf{a}_1 \neq 0$ at w. The contradiction is not resolved if only $w_1 = 0$, for in that case not only $\mathbf{c}_1 = 0$ but also $\mathbf{u}_1 = 0$ and the dimension of the kernel of the linear equations would be zero, which is not the case. With $V_2 w_1 \neq 0$, $\mathbf{c}_1 \cdot \mathbf{s}_1 \neq 0$, so that $\mathbf{c}_1$ is not normal to the surface at w. However, $\mathbf{c}_1$ is orthogonal to the osculating plane to the line of curvature $\mathbf{sa}$, that is the tangent to the rib coincides with the focal line of the line of curvature, regarded as a space curve. The vector w_1 is determined by $\mathbf{c}_1 \cdot (\mathbf{sa})_3 + (V_1\mathbf{a})_3 w_1 = 0$ while $\mathbf{u}_1$ is determined by the equation $\mathbf{u} \cdot \mathbf{u}_1 = 0$ and the one preceding it. It follows that the set of linear equations has a one-dimensional kernel in this case, generated by a vector $(\mathbf{c}_1, w_1, \mathbf{u}_1)$ with both $\mathbf{c}_1$ and w_1 non-zero. Projection either to $\mathbb{R}^2$ or to $\mathbb{R}^3$ then implies that the ridge and rib are one-dimensional smooth submanifolds of $\mathbb{R}^3$ at w in this case. This completes the proof of corollary (i).

Next suppose that $\mathbf{c}$ is A_4. Then $(V_1\mathbf{a})_3 = 0$ so that $\mathbf{c}_1 \cdot (\mathbf{sa})_3 = 0$. There are then two cases.

The 'generic' case is that the space curve $\mathbf{sa}$ is not planar at w. In that case $\mathbf{c}_1$ is zero and then from $V_2 w_1 = 0$ it follows that w_1 is a multiple of $\mathbf{u}$. The linear equations again have a one-dimensional kernel, implying that the ridge is a one-dimensional submanifold of $\mathbb{R}^3$ at w in this case, but not the rib, since $\mathbf{c}_1 = 0$. Along the ridge the quadratic form at A_3 points introduced above is non-parabolic except at the A_4 point which becomes a point of transition between the hyperbolic or fertile case and the elliptic or sterile case.

Differentiating the original linear equations along the ridge at the A_4 point, taking $\mathbf{c}_1 = 0$ and $w_1 = \mathbf{u} = \mathbf{a}_1$ and $\mathbf{u}_1 = \mathbf{a}_2$ there, as we may, and using Remark 11.9, we find that

$$\begin{aligned} \mathbf{c}_2 \cdot (\mathbf{sa})_1 &= 0, \\ \mathbf{c}_2 \cdot (\mathbf{sa})_2 &= 0, \\ \mathbf{c}_2 \cdot (\mathbf{sa})_3 + (V_1\mathbf{a})_4\mathbf{a}_1 &= 0, \end{aligned}$$

the last implying that $\mathbf{c}_2 \neq 0$, since $(V_1\mathbf{a})_4\mathbf{a}_1 \neq 0$ at an A_4 point. A further differentiation leads to $\mathbf{c}_3 \cdot (\mathbf{sa})_2 \neq 0$, implying that the rib has an ordinary cusp at the A_4 point, pointing along the focal line of the line of curvature $\mathbf{sa}$.

The alternative is that **sa** is planar at w. In that case the kernel of the linear equations has dimension 2 and the argument for the ridge or rib to be one-dimensional fails. Further analysis in this case, which we omit, discloses that there is a quadratic equation for the limiting values of the tangent directions as the A_4 point is reached, and provided this equation does not vanish one is faced with a crossing point of ridges of the same colour, or *ridge crunode* or *cross node*, the hyperbolic case, or with a birth or death point of a ridge, or *ridge acnode* or *spot node*, the elliptic case, or, less probably, with a *cusp* on a ridge, the parabolic case. This completes our discussion of corollary (ii).

Next, suppose that w is an umbilic of the surface, with umbilical centre $\mathbf{c}$, where not only $V_1(\mathbf{c}) = 0$ but also $V_2(\mathbf{c}) = 0$. Then for any of the non-zero root vectors $\mathbf{u}$ of $V_3(\mathbf{c})$ the equations for possible tangent vectors reduce to

$$\mathbf{c}_1 \cdot \mathbf{s}_1 = 0$$

$$\mathbf{c}_1 \cdot \mathbf{s}_2\mathbf{u} + V_3 w_1 \mathbf{u} = 0,$$

$$\mathbf{c}_1 \cdot \mathbf{s}_2\mathbf{u}^3 + V_4 w_1 \mathbf{u}^3 + 3V_3\mathbf{u}^2\mathbf{u}_1 = 0,$$

$$\mathbf{u} \cdot \mathbf{u}_1 = 0.$$

From the first of these $\mathbf{c}_1 = \zeta\mathbf{n}$, implying that $\mathbf{c}_1 \cdot \mathbf{s}_2\mathbf{u} = \zeta\kappa\mathbf{s}_1\mathbf{u} \cdot \mathbf{s}_1$. Moreover, if $\mathbf{c}$ is elliptic or hyperbolic, so that $V_3\mathbf{u}^2 \neq 0$, then, for any $\mathbf{u}$, any non-zero kernel vector has $\mathbf{c}_1 \neq 0$ since otherwise either $\mathbf{u}$ is a Hessian vector of V_3, not the case since V_3 is not parabolic, or $w_1 = 0$. In the former case let $\mathbf{v}$ be such that $\mathbf{s}_1\mathbf{u} \cdot \mathbf{s}_1\mathbf{v} = 0$. Then $V_3 w_1 \mathbf{u}\mathbf{v} = 0$, determining w_1 up to a scalar multiple. The equation $\mathbf{c}_1 \cdot \mathbf{s}_2\mathbf{u} + V_3 w_1 \mathbf{u} = 0$ then determines w_1 fully. The other two equations determine $\mathbf{u}_1$. In the latter case it follows from the fact that $V_3\mathbf{u}^2 \neq 0$ that $\mathbf{u}_1 = 0$, which contradicts the hypothesis that the kernel vector is non-zero. That is the kernel is one-dimensional. That the ribs in either case pass from one sheet to the other follows at once from the fact that the singularity of the normal focal surface is *conic* in either case.

That deals with corollaries (iii) and (iv). In cases (v), the D_5 parabolic case, where there exists $\mathbf{u} \neq 0$ such that $V_3\mathbf{u}^2 = 0$, but $V_4\mathbf{u}^4 \neq 0$, $\mathbf{c}_1 = 0$ and $w_1 = \eta\mathbf{u}$. Since $V_4 w_1 \mathbf{u}^3 = 0$ it follows that $\eta V_4 \mathbf{u}^4 = 0$, and therefore $\eta = 0$, that is $w_1 = 0$. However, $\mathbf{u}_1$ may be any vector orthogonal to $\mathbf{u}$. So the kernel once again is one-dimensional. That the ridge actually has an ordinary cusp at the umbilic in this case with limiting tangent spanned by $\mathbf{u}$ follows from further differentiations, which we omit.

For $\mathbf{u}$ on the other root line of the cubic V_3 the argument goes as in the elliptic or hyperbolic case, there being a regular ridge through the umbilic with $V_3 w_1 \mathbf{uv} = 0$ as usual. □

In Theorem 11.10 we assumed that the surface $\mathbf{s}$ had only ordinary umbilics. The argument of corollary (iii) of that theorem also goes through without change at a right-angled umbilic where the classifying cubic V_3 is not a perfect cube, as is generally the case. Moreover, the single ridge through such an umbilic has a distinctive tangent line. By Proposition 7.9 there is a non-zero vector, w_1 say, in $\mathbb{R}^2$ such that $V_3 w_1$ is a non-zero multiple of the first fundamental form.

Proposition 11.11 Let $\mathbf{s}$ *be a regular surface and suppose that* $\mathbf{s}$ *has at a point* w *a right-angled umbilic where the cubic* V_3 *is not a perfect cube, that is where there is no vector* $\mathbf{u}$ *for which* $V_3 \mathbf{u} = 0$. *Then the unique ridge through the umbilic is non-singular and at the umbilic the vectors* w_1, *such that* $V(\mathbf{c})_3 w_1$ *is a multiple of the form* $\mathrm{I}_2(w)$, *are the tangent vectors to this ridge.* □

Some feeling for what the focal surface of a regular smooth surface looks like near an elliptic or hyperbolic umbilical centre may be gained from a series of computer pictures of Thompson and Hunt already exhibited as Figures 10.4–10.7. These originate remarkably from a parallel but quite independent study in engineering concerning the theory of the buckling of beams (Thompson and Hunt, 1975). The underlying mathematics in the two theories is the same. The symmetries of the various surfaces as shown in these figures are of course not realised in practice, but they do give a correct topological feel for what happens at ordinary umbilical centres more vividly than any other illustrations that I have come across. The two versions of the hyperbolic umbilic focal surfaces illustrated in Figure 10.5 and in Figures 10.6 and 10.7 appear intriguingly different. Whether there is an essential difference between these and whether the one is a star and the other a lemon or a monstar, as defined in the next chapter, where umbilics and their neighbourhoods are studied in detail, I leave it to the reader to determine!

Exercises

11.1 Let $\mathbf{s}$ be a regular smooth surface in $\mathbb{R}^3$, let κ be a non-zero principal curvature of $\mathbf{s}$ at a point $w \in \operatorname{dom} \mathbf{s}$, with $\mathbf{u} \in \mathbb{R}^2$

representing a non-zero principal tangent vector there, and let $\mathbf{v} \in \mathbb{R}^2$ and $v \in \mathbb{R}$ be such that

$$\mathbf{s}_1(w)\mathbf{v} + v\mathbf{s}_2(w)\mathbf{u}^2 = 0.$$

Prove that $v = 0$ and that $\mathbf{v} = 0$. (That is the vector $\mathbf{s}_2(w)\mathbf{u}^2$ does not lie in the tangent plane to $\mathbf{s}$ at w.)

11.2 Prove Corollary 11.8.

11.3 Let $\mathbf{s}$ be a regular smooth surface through the origin in $\mathbb{R}^3$ presented in Monge form, with $z = 0$ the tangent plane at the origin, that is

$$\mathbf{s}(x, y) = (x, y, \tfrac{1}{2}\kappa x^2 + \tfrac{1}{2}\lambda y^2 + \tfrac{1}{6}(ax^3 + 3bx^2y + 3cxy^2 + dy^3) + \text{higher order terms}),$$

and suppose that $\kappa \neq \lambda$, neither being zero. Verify that at the origin $\mathbf{s}_1 \cdot \mathbf{s}_2 = 0$ and that for either principal centre of curvature $\mathbf{c}$ at the origin the thrice linear form $V(\mathbf{c})_3$ is a multiple of $[a, b, c, d]$. Deduce that there is a ridge of the surface through the origin if either $a = 0$ or $d = 0$.

(Remember that $(1, 0)$ and $(0, 1)$ represent principal tangent vectors at the origin.)

11.4 Consider the smooth surface in $\mathbb{R}^3$ given by

$$(x, y) \mapsto (x, y, \tfrac{1}{2}\kappa x^2 + \tfrac{1}{2}\lambda y^2 + \tfrac{1}{2}bx^2y + \tfrac{1}{2}cxy^2),$$

where κ, λ, b and c are real numbers and $\rho = \kappa^{-1}$. Verify that

$$V(0, 0, \rho)_2(0, 0) = \rho\begin{bmatrix} 0 & 0 \\ 0 & \lambda - \kappa \end{bmatrix},$$

$$V(0, 0, \rho)_3(0, 0)\begin{bmatrix} 1 \\ 0 \end{bmatrix}^2 = \rho[0 \quad b],$$

$$V(0, 0, \rho)_4(0, 0)\begin{bmatrix} 1 \\ 0 \end{bmatrix}^4 = -3\kappa^2.$$

Verify that there are two ridges through the origin, and that the one with principal direction vector $(1, 0)$ there has a turning point (that is an A_4 point) at the origin provided that $b^2 = \kappa^3(\kappa - \lambda)$.

11.5 Find the ridges and ribs of the tube $\mathbf{S}$ of radius δ and core $\mathbf{r}$ discussed in Exercise 10.7.

11.6 Prove Proposition 11.10.

11.7 Verify that along a line of curvature of a regular surface the relevant principal curvature has an ordinary maximum or minimum where the line crosses a ridge of the same colour transversally, and has an ordinary stationary inflection where it is simply tangential to a ridge at an ordinary turning point of the ridge.

11.8 The following formulae for the ridge points of a surface, that apply when the surface is expressed implicitly as the zero set of an equation rather than parametrically, are given by Markatis (1980).

Let $\mathbf{s}$ be a non-singular point of the surface given by the equation $F(\mathbf{s}) = 0$, where $F : \mathbb{R}^3 \rightarrowtail \mathbb{R}$ is smooth, suppose that $\mathbf{s}$ is not an umbilic, and let $\boldsymbol{\alpha}$ and $\boldsymbol{\beta}$ be mutually orthogonal principal tangent vectors and $\mathbf{n}$ a normal vector to the surface at $\mathbf{s}$. Prove that if $\mathbf{s}$ lies on a ridge of the surface then either

$$(F_1(\mathbf{s})\mathbf{n})(F_3(\mathbf{s})\boldsymbol{\alpha}^3) - 3(F_2(\mathbf{s})\boldsymbol{\alpha}\mathbf{n})(F_2(\mathbf{s})\boldsymbol{\alpha}^2) = 0$$

or

$$(F_1(\mathbf{s})\mathbf{n})(F_3(\mathbf{s})\boldsymbol{\beta}^3) - 3(F_2(\mathbf{s})\boldsymbol{\beta}\mathbf{n})(F_2(\mathbf{s})\boldsymbol{\beta}^2) = 0.$$

12
Umbilics

12.0 Introduction

Umbilics of regular surfaces are points where the principal curvatures coincide. So they are the points of the surface most closely approximable by a sphere, or in the case that both curvatures are zero, by a plane. We learned quite a lot about umbilics in Chapter 11. In particular we learned in Theorem 11.3 that ordinary umbilics are isolated, an umbilic being regarded as *ordinary* if it is not flat and if the cubic $V(\mathbf{c})_3$ is not right-angled or perfect or zero. We also determined in Theorem 11.10 the possible configurations of ridges and ribs at various types of umbilic. After some preliminary remarks we shall continue the exploration of the rich geometry of a regular surface around an umbilic.

So let $\mathbf{s}: \mathbb{R}^2 \rightarrowtail \mathbb{R}^3$ be a regular surface, with an umbilic, not flat, at $w \in \mathbb{R}^2$. Then we have not only $V(\mathbf{c})_1(w) = 0$ but also $V(\mathbf{c})_2(w) = 0$, for some point $\mathbf{c} \in \mathbb{R}^3$, the *umbilical centre*. These two equations are equivalent to the equation

$$\mathbf{n} \cdot \mathbf{s}_2 = \kappa \mathbf{s}_1 \cdot \mathbf{s}_1, \text{ at } w,$$

which holds also when $\kappa = 0$ and $\mathbf{c}$ is 'at infinity', that is when the umbilic is *flat*. The following proposition gives an alternative criterion for w to be an umbilic.

Proposition 12.1 With $\mathbf{s}$ *and* w *as above the point* w *is an umbilic, with curvature* κ, *if and only if* $\mathbf{n}_1 + \kappa \mathbf{s}_1 = 0$ *at* w.

Proof Exercise. □

Just about the only thing proved about umbilics in most elementary books on differential geometry is the following intuitively obvious theorem, due to Jean-Baptiste Meusnier (1785) – see Coolidge (1940) pp. 328–9.

Theorem 12.2 Let $\mathbf{s}$ *be a connected regular surface in* $\mathbb{R}^3$ *such that every point is an umbilic. Then the image of* $\mathbf{s}$ *is an open subset either of a plane or of a sphere.*

Proof We have everywhere $\mathbf{n}_1 + \kappa \mathbf{s}_1 = 0$. Accordingly

$$\mathbf{n}_2 + \kappa \mathbf{s}_2 + \kappa_1 \mathbf{s}_1 = 0.$$

Since both $\mathbf{n}_2$ and $\mathbf{s}_2$ are twice linear it follows that $\kappa_1 \mathbf{s}_1$ also is twice linear. Now let $\mathbf{u} = (1, 0)$ and $\mathbf{v} = (0, 1)$, the vectors $\mathbf{u}$ and $\mathbf{v}$ then being a basis for $\mathbb{R}^2$. From the twice linearity of $\kappa_1 \mathbf{s}_1$ we have

$$\kappa_1(\mathbf{u})\mathbf{s}_1(\mathbf{v}) = \kappa_1(\mathbf{v})\mathbf{s}_1(\mathbf{u}).$$

But $\mathbf{s}_1$ is everywhere injective, so that everywhere $\mathbf{s}_1(\mathbf{u})$ and $\mathbf{s}_1(\mathbf{v})$ are linearly independent. It follows that $\kappa_1(\mathbf{u}) = 0 = \kappa_1(\mathbf{v})$, everywhere, implying that κ is constant, since the domain of $\mathbf{s}$ is connected.

There are then two cases. If $\kappa = 0$ then $\mathbf{n}_1 = 0$ and $\mathbf{n}$ is constant, implying that the image of $\mathbf{s}$ is an open subset of a plane, with normal $\mathbf{n}$. The alternative is that $\kappa \neq 0$, when $\mathbf{s}_1 + \rho \mathbf{n}_1 = 0$ with $\rho = \kappa^{-1}$. Then for $\mathbf{c} = \mathbf{s} + \rho \mathbf{n}$ we have $\mathbf{c}_1 = \rho_1 \mathbf{n} = 0$, since $\rho_1 = 0$, ρ being constant. So $\mathbf{c}$ is constant and the image of $\mathbf{s}$ is an open subset of the sphere, centre $\mathbf{c}$ and radius ρ. □

This is, of course, a very special case. It is also possible that a surface has umbilics all along some curve, as we shall see later from examples (Exercises 16.5 and 16.7). However, the generic situation is that umbilics are isolated. This is to be expected since the condition for the twice linear form $V(\mathbf{c})_2$ to be zero is three conditions on $\mathbf{c}$, the vector space of symmetric 2×2 real matrices being of dimension three.

12.1 Curves through umbilics

Much of the geometry of a regular surface around an umbilic concerns the patterns made by curves of various types that pass through it.

Let $\mathbf{s} : \mathbb{R}^2 \rightarrowtail \mathbb{R}^3$ be a regular smooth surface with an ordinary umbilic at the origin and suppose that $\mathbf{r} : \mathbb{R} \rightarrowtail \mathbb{R}^2$ represents a regular smooth curve on the surface that passes through the umbilic. Then there are two regular smooth curves on the focal surface each lying over the curve $\mathbf{sr}$ and crossing at the umbilic. We use one of these in the proof of the following proposition.

Proposition 12.3 Let $\mathbf{s}$ and $\mathbf{r}$ *be as above, and suppose that as one passes along the curve* $\mathbf{r}$ *through the umbilic* $w = \mathbf{r}(t)$ *the limiting principal directions are spanned by vectors* $\mathbf{u}$ *and* $\mathbf{v}$*, where necessarily*

$$I_2(w)\mathbf{uv} = \mathbf{s}_1(w)\mathbf{u} \cdot \mathbf{s}_1(w)\mathbf{v} = 0,$$

and suppose that the tangent line to $\mathbf{r}$ *at the umbilic is spanned by the vector* $\mathbf{t}$*. Then* $V_3(c)\mathbf{tuv} = 0$*, where* $\mathbf{c}$ *is the umbilical centre. This determines* $\mathbf{t}$ *up to a real multiple when* $\mathbf{u}$ *and* $\mathbf{v}$ *are known.*

Proof Let $\mathbf{c} : \mathbb{R} \rightarrowtail \mathbb{R}^3$ be either of the smooth curves on the focal surface lying over the curve $\mathbf{sr}$ and let $\mathbf{u} : \mathbb{R}^2 \rightarrowtail \mathbb{R}^2$ denote a unit principal vector chosen smoothly along the curve. Along the curve $V_1(\mathbf{c}) = (\mathbf{c} - \mathbf{sr}) \cdot \mathbf{s}_1 = 0$ and $V_2(\mathbf{c})\mathbf{u} = (\mathbf{c} - \mathbf{sr}) \cdot \mathbf{s}_2\mathbf{u} - \mathbf{s}_1\mathbf{u} \cdot \mathbf{s}_1 = 0$.
Differentiating along the curve we then have

$$\mathbf{c}_1 \cdot \mathbf{s}_1 + V_2(\mathbf{c})\mathbf{r}_1 = 0 \text{ and } \mathbf{c}_1 \cdot \mathbf{s}_2\mathbf{u} + V_3(\mathbf{c})\mathbf{r}_1\mathbf{u} + V_2(\mathbf{c})\mathbf{u}_1 = 0,$$

reducing at the umbilic to $\mathbf{c}_1 \cdot \mathbf{s}_1 = 0$ and $\mathbf{c}_1 \cdot \mathbf{s}_2\mathbf{u} + V_3(\mathbf{c})\mathbf{r}_1\mathbf{u} = 0$, from the first of which it follows that at the umbilic $\mathbf{c}_1 = \zeta\mathbf{n}$ for some $\zeta \in \mathbb{R}$. The second then becomes

$$\zeta\mathbf{n} \cdot \mathbf{s}_2\mathbf{u} + V_3(\mathbf{c})\mathbf{r}_1\mathbf{u} = \zeta\mathbf{s}_1\mathbf{u} \cdot \mathbf{s}_1 + V_3(\mathbf{c})\mathbf{r}_1\mathbf{u} = 0,$$

implying that at the umbilic $V_3(\mathbf{c})\mathbf{r}_1\mathbf{uv} = 0$ where $\mathbf{s}_1\mathbf{u} \cdot \mathbf{s}_1\mathbf{v} = 0$. □

Remarks: (i) The result extends to the case of a curve passing through an isolated flat umbilic, with $V_3(\mathbf{c})$ replaced by $\mathbf{n} \cdot \mathbf{s}_3$. The precise formulation and the reworking of the proof are left as exercises.
(ii) This result generalises part of Theorem 11.9 which gave the above formula for the tangent vectors to a regular *ridge* passing through an ordinary umbilic.

The tangent directions to the ridges through an umbilic are thus computable directly in terms of the form $V(\mathbf{c})_3$. It should be noted that ridge tangents may coincide even when the root lines of $V(\mathbf{c})_3$ are distinct. In particular two of the ridges through an elliptic umbilic pass through the umbilic in the same direction if and only if two of the root lines of the cubic $V(\mathbf{c})_3$ are mutually orthogonal with respect to the first fundamental form. For cubics represented in the β plane this occurs on the circle $|\beta| = \frac{1}{3}$.

We have already in Proposition 11.10 identified the distinctive tangent line to the single ridge through a right-angled umbilic.

As a corollary of Proposition 12.3 we have all possible directions for lines of curvature that pass through an isolated umbilic.

Proposition 12.4 Let $\mathbf{s}$ *be a regular smooth surface in* $\mathbb{R}^3$ *with an isolated umbilic at* $w \in \mathbb{R}^2$. *Then the possible tangents to lines of curvature through* w *are the root vectors of the cubic form that is the Jacobian of the principal cubic form* $V(\mathbf{c})_3$ *and the first fundamental form* $I_2(w)$, *namely vectors* $\mathbf{u}$ *such that* $V(\mathbf{c})_3\mathbf{u}^2\mathbf{v} = 0$, *where* $I_2(w)\mathbf{uv} = 0$.

Proof Exercise. □

12.2 Classifications of umbilics

There are several distinct classifications of umbilics, each depending on the thrice linear form $\mathbf{n} \cdot \mathbf{s}_3 - 3\kappa\mathbf{s}_2 \cdot \mathbf{s}_1$ at the umbilic, equal to $\kappa V(\mathbf{c})_3$, where $\mathbf{c}$ is the umbilical centre, if $\kappa \neq 0$. When the surface is in Monge form

$$z = \tfrac{1}{2}\kappa(\mathbf{x}^2 + \mathbf{y}^2) + \tfrac{1}{6}C_3(x, y)^3 + O(x, y)^4$$

with an umbilic at the origin, $\mathbf{s}_2 \cdot \mathbf{s}_1 = 0$ at the origin and the classifying form there is just the form C_3.

The original classification is due to Darboux (1896) and concerns the patterns that lines of curvature make in the neighbourhood of an umbilic. Associated to this is the classification by index, generic umbilics being either of index $\frac{1}{2}$ or of index $-\frac{1}{2}$. This is rather harder to date. A reference is a paper by Donald Barnes (1967). See also Bruce and Fidal (1989). The classification into elliptic, parabolic, hyperbolic and perfect umbilics is a by-product of René Thom's work on elementary catastrophes in the 1960s (see Thom (1975)), the word umbilic being borrowed by him, by analogy from its use in differential geometry, to describe certain of the simplest types of critical point of smooth functions. It is the object of the following sections to introduce these various classifications and to show how they are interrelated. It is not all new. One finds much of the detail spelt out by Allvar Gullstrand (1904, 1911) who created the necessary higher order differential geometry in order to understand the optics of the eye lens. He considers higher order umbilics as well as the ordinary umbilic considered here. (Amusingly, Gullstrand invents a new letter after w, namely 'ɯ', writing the cubic form $C_3(x, y)^3$ in the Monge form of a surface as $ux^3 + 3vx^2y + 3wxy^2 + ɯy^3$!)

12.3 The main classification

For reasons of technical simplicity we shall assume to begin with that at points of interest neither principal curvature of the surface under study is zero anywhere. The main classification of the umbilics of a smooth surface $\mathbf{s}$ is then by the type of the *classifying form* $V(\mathbf{c})_3$ at an umbilic, whether elliptic, parabolic, hyperbolic or perfect. What we have to show is what this means in geometrical terms. The clue lies in the fact that we have encountered the form $V(\mathbf{c})_3$ already on the ridges of the surface, for if $\mathbf{a}_1$ is a principal tangent vector of the appropriate colour at a ridge point then $V(\mathbf{c})_3\mathbf{a}_1{}^3 = 0$. It follows that if a ridge passes through an umbilic of the surface then the limiting principal tangent line at the umbilic, supposing this to be well defined, must be a root-line there of the classifying form. In fact in Theorem 11.10 we proved that at an *elliptic umbilic*, where all three root lines of the form are real and distinct, there are three ridges passing through the umbilic and correspondingly three ribs of the focal surface passing through the umbilical centre, while at a *hyperbolic umbilic*, where just one of the root lines of the form is real, there is just one ridge passing through the umbilic and just one rib passing through the umbilical centre. Moreover, with one exception the ribs of the surface are representable as regular space curves that pass from one sheet of the evolute to the other as they pass through the umbilical centre, the ridges likewise being representable as regular curves on the surface, the exception being the case of a right-angled hyperbolic umbilic, where the rib belongs to one of the sheets only. Generically an umbilic will be of one of these two types, the transitional case, the *parabolic umbilic*, being non-generic, while the *perfect* (called by some the *symbolic*) *umbilic,* where the cubic form is a perfect cube, is even more special, as indeed clearly are those umbilics where the cubic form vanishes completely.

12.4 Darboux's classification

The main classification of umbilics takes almost no account of the metric on $\mathbb{R}^2$, though in one detail above exception had to be made of the right-angled case, which does for its definition depend on interplay between the thrice-linear form $V(\mathbf{c})_3$ and the first fundamental form I_2.

That this is an important special case becomes clear if we take a look at the configuration of lines of curvature round an umbilic, though all we can do here is to give a rough account of what occurs. To do this we assume that the surface is given in Monge form at an umbilic, the umbilic lying at the origin in $\mathbb{R}^3$ with the surface having the equation

$$z = \tfrac{1}{2}\kappa(x^2 + y^2) + \tfrac{1}{6}(px^3 + 3qx^2y + 3rxy^2 + sy^3) + O(x, y)^4.$$

Then to the first order in x and y in the coefficients the first and second fundamental forms, in traditional notations, are

$$\mathrm{I}_2(\mathrm{d}x, \mathrm{d}y)^2 = \mathrm{d}x^2 + \mathrm{d}y^2$$

and

$$\mathrm{II}_2(\mathrm{d}x, \mathrm{d}y)^2 = (\kappa + px + qy)\mathrm{d}x^2 + 2(qx + ry)\mathrm{d}x\mathrm{d}y \\ + (\kappa + rx + sy)\mathrm{d}y^2.$$

The principal tangent lines at (x, y) are then, by Proposition 10.1, given approximately as the root lines of the Jacobian quadratic form, namely

$$\mathrm{d}x((qx + ry)\mathrm{d}x + (\kappa + rx + sy)\mathrm{d}y) - \mathrm{d}y((\kappa + px + qy)\mathrm{d}x \\ + (qx + ry)\mathrm{d}y) \\ = (qx + ry)\mathrm{d}x^2 + (rx + sy - px - qy)\mathrm{d}x\mathrm{d}y - (qx + ry)\mathrm{d}y^2,$$

these lines being mutually orthogonal everywhere since the sum of the coefficients of $\mathrm{d}x^2$ and $\mathrm{d}y^2$ is zero. Straight line solutions through the origin of the first order differential equation obtained by setting this form equal to zero are to be found by substituting y/x for $\mathrm{d}y/\mathrm{d}x$, or equivalently y for $\mathrm{d}y$ and x for $\mathrm{d}x$ in the differential equation. The cubic in x and y that one gets is just the Jacobian of $x^2 + y^2$ and the original cubic.

The differential equation degenerates to one with constant coefficients if the coefficient of $\mathrm{d}x\mathrm{d}y$ is a constant multiple of $qx + ry$ or vice versa, this being the case if and only if

$$q(s-q) - r(r-p) = pr - q^2 + qs - r^2 = 0.$$

But this is just the condition for the Hessian of $C_3 = [p, q, r, s]_3$ to have orthogonal root lines, or equivalently for two of the root lines of the Jacobian cubic to be mutually orthogonal. As one expects, the twice linear form $C_3(r, -q)$ is then equal to $[pr - q^2, 0, r^2 - qs]_2$, a multiple of $[1, 0, 1]_2$ since $pr - q^2 + qs - r^2 = 0$.

We noted earlier that when the surface is in Monge form the classifying form is equal to C_3. So the degenerate case just noted occurs exactly where that form is right-angled.

Clearly there are two main cases according as the Jacobian cubic is hyperbolic or elliptic, but it follows from the above remarks that there is a further fundamental distinction that must be made in the elliptic case according as the three root lines of the Jacobian cubic are contained, or are not contained, within a right angle, since the case where two of these lines are at right angles is clearly special. Sketch pictures of the patterns of lines of curvature can be conjectured in the various cases, bearing in mind that blue and red solution curves in $\mathbb{R}^2$ must intersect each other approximately at right angles. Figures 12.1–12.3 provide examples of the three main types, known as the

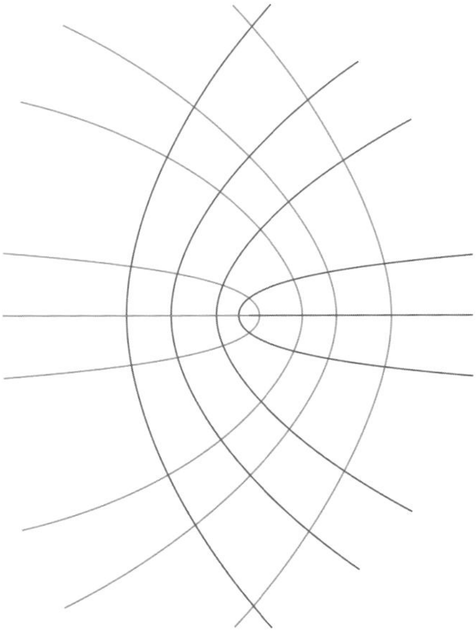

Figure 12.1

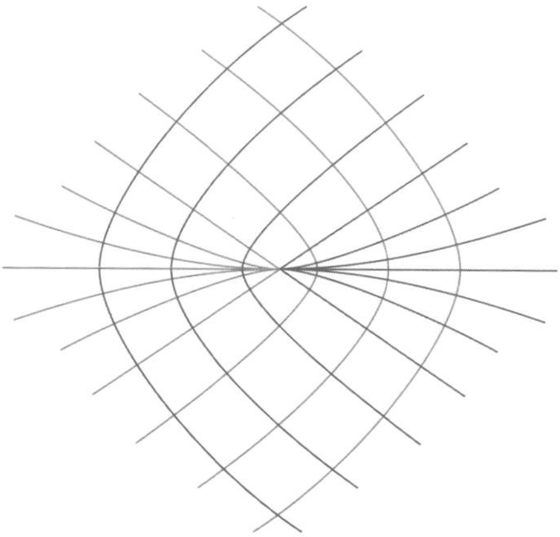

Figure 12.2

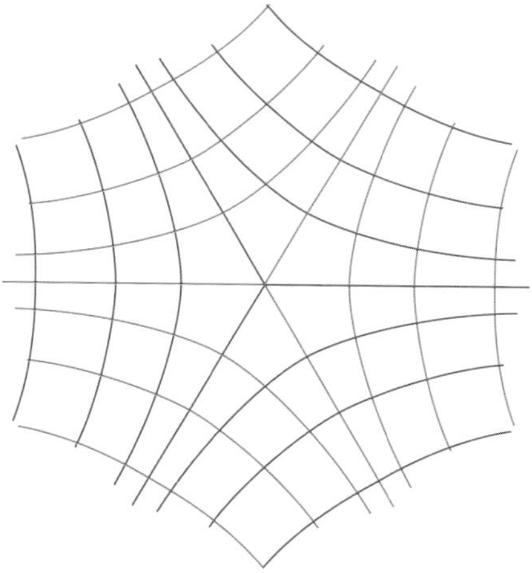

Figure 12.3

lemon, where the Jacobian cubic is hyperbolic, the *lemonstar*, or more briefly and conveniently since it has a different initial letter, the *monstar*, where the Jacobian cubic is elliptic but the three root lines are contained within a right angle, and the *star*, where the Jacobian cubic is elliptic but the three root lines are not contained within a right angle. The names are due to Hannay (Berry and Hannay, 1977), though the pictures go back to Darboux (1896) and Gullstrand (1904). Note in the monstar case that though there are three 'directions' for lines of curvature through the umbilic there are more than three such lines.

There is an interesting somewhat earlier paper by Frost (1870) who, after noting that there is only one line of curvature through an umbilic of an ellipsoid, purports to produce an example of a surface with an umbilic through which there pass three such lines. It is his misfortune that the cubic form of his chosen example is right-angled! A note (1870) in the same volume by Cayley argues that any umbilic should be regarded as a triple point of the line(s) of curvature. I was previously of the opinion that what Cayley was pointing out was that the theory provides for three directions of lines of curvature, rather than three such lines – the point made above, and illustrated dramatically by the monstar – but on rereading the note I am convinced that this is not the case.

The classifications of the classifying cubic form and of the Jacobian cubic form do not coincide, though the classifications are not unrelated. The most convenient way to expose their relationship is by means of the β-plane representation of the classifying form. See Figure 12.4, in which L denotes 'lemon', M denotes 'monstar', HS denotes 'hyperbolic star' and E denotes 'elliptic (star)', the umbilic in each of

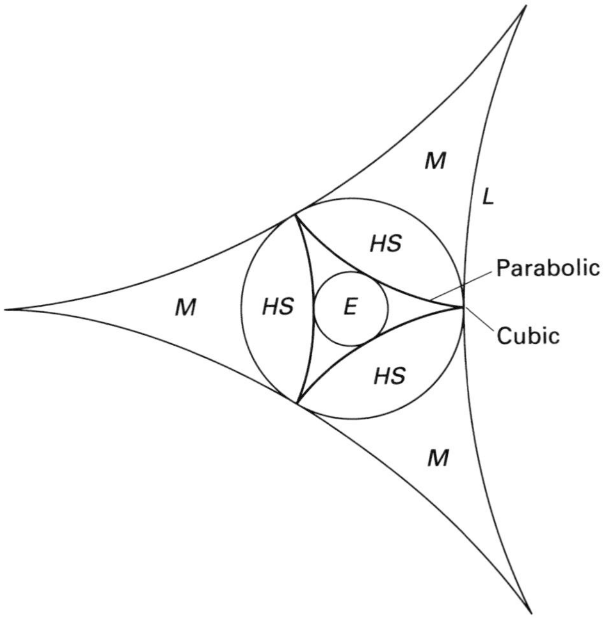

Figure 12.4

these cases being of type D_4. Parabolic umbilics are generally of type D_5 and perfect umbilics generally of type E_6.

One case in which the lines of curvature differential equation can be integrated explicitly is the case I call the *pure lemon* case in which the original cubic is a multiple of the quadratic form $x^2 + y^2$. In complex form (see Section 7.4) this is equivalent to the case where $\alpha = 0$, so that the cubic has to be thought of as lying at infinity in the 'β-plane'. The harmonic part of such a cubic is equal to zero.

Example 12.5 Consider the case where $C_3(x, y)^3 = 3x^3 + 3xy^2$; that is the case where $p = 3$, $q = 0$, $r = 1$ and $s = 0$. Then the quadratic differential equation describing the lines of curvature approximately round the umbilic is

$$y\mathrm{d}x^2 - 2x\mathrm{d}x\mathrm{d}y - y\mathrm{d}y^2 = 0.$$

When one replaces $\mathrm{d}x$ by x and $\mathrm{d}y$ by y this reduces to

$$-x^2y - y^3 = 0,$$

giving the single straight line solution $y = 0$. A guess is that the solutions might be parabolas with this line as axis. The equation of such a parabola is $y^2 = 4ax + c$. Then $2y\,\mathrm{d}y/\mathrm{d}x = 4a$. Substituting this in the differential equation we find that for this to be a solution one must have

$$y^2 - 4ax - 4a^2 = c - 4a^2 = 0.$$

So for each a the parabola with equation $y^2 = 4ax + 4a^2$ is a solution of the differential equation, as is readily checked, this being a quadratic equation for a, and therefore for $\mathrm{d}y/\mathrm{d}x = 2a/y$, at each point (x, y) off the x-axis, the two gradients one gets for parabolas through the point necessarily being mutually orthogonal. For $y = 0$ one gets $a = 0$ or $-x$, yielding $\mathrm{d}y/\mathrm{d}x = 0$ or ∞ except at the origin. So the approximate lines of curvature in this case are represented by two families of mutually orthogonal parabolas in $\mathbb{R}^2$. □

The whole of the above section has been admittedly sketchy and open to the criticism implied earlier of the work of Frost! For detailed studies of the configurations of lines of curvature around an isolated umbilic the reader is referred to papers and a recent book by Sotomayor Teilo and Gutierrez, for example (Sotomayor Teilo and Gutierrez 1982; Gutierrez and Sotomayor Teilo, 1991) and also to a paper by Bruce and Fidal (1989).

12.5 Index

There is a striking difference between the patterns of lines of curvature in the star case and those in the monstar and lemon cases. As one makes a circuit of the umbilic in the star case the orientation of either family of lines of curvature rotates in the opposite or negative direction through one half of a full rotation. One says that the umbilic has *index* $-\frac{1}{2}$. In either of the other two cases the orientation of lines of curvature rotates in the same or positive direction through one half of a full rotation. One says that the umbilic has *index* $\frac{1}{2}$. In the transitional right-angled case where the pattern of lines is approximated by a plain rectangular grid the index is zero. Umbilics that are represented by cubics equivalent to cubics in the β-plane with $|\beta| < 1$ have index $-\frac{1}{2}$ while those represented by cubics equivalent to cubics in the β-plane with $|\beta| > 1$ have index $\frac{1}{2}$. Under smooth deformation of a surface the total index must remain constant so long as the number of umbilics remains finite. The total index for a compact surface with isolated umbilics has been shown by Donald Barnes (1967) to be equal to the Euler–Poincaré index of the surface, which is equal to 2 for an ellipsoid. The generic ellipsoid has four umbilics all of index $\frac{1}{2}$, as we show later. In fact, as we shall see in Section 16.2, they are all of pure lemon type. The rugby ball, or ellipsoid of revolution, with two semiaxes equal, has just two umbilics, one at each pole. Each of these umbilics has index 1, the classifying cubic in this case vanishing.

12.6 Straining a surface

Suppose that a regular surface $\mathbf{s}$ in $\mathbb{R}^3$ is *strained* by being moved normal to itself a distance δ. Provided that δ is nowhere as great as the smallest principal radius of curvature of the surface the offset strained surface $\mathbf{s} + \delta\mathbf{n}$ will also be regular. As the next proposition shows, the straining determines the principal tangent lines on the surface $\mathbf{s}$ and also its ribs and umbilics.

Proposition 12.6 Let $\mathbf{s}$ *be a regular surface in* $\mathbb{R}^3$ *and let* $\delta \in \mathbb{R}$, *where* δ *is everywhere not equal to either of the principal radii of curvature of* $\mathbf{s}$. *Then the eigenspaces at a point* w *of the first fundamental form of the strained surface* $\mathbf{s} + \delta\mathbf{n}$, *namely* $(\mathbf{s}_1 + \delta\mathbf{n}_1) \cdot (\mathbf{s}_1 + \delta\mathbf{n}_1)$, *with respect to the first fundamental form of the surface* **s**, *namely* $\mathbf{s}_1 \cdot \mathbf{s}_1$, *are the principal tangent lines of* **s** *at* w.

Proof This is an immediate consequence of Rodrigues' Theorem 10.2. For let $\mathbf{a}_1$ be a principal vector at w with principal curvature κ. Then

$$\begin{aligned}(\mathbf{s}_1 + \delta\mathbf{n}_1)\mathbf{a}_1 \cdot (\mathbf{s}_1 + \delta\mathbf{n}_1) &= (1 - \delta\kappa)\mathbf{s}_1\mathbf{a}_1 \cdot (\mathbf{s}_1 + \delta\mathbf{n}_1)\\ &= (1 - \delta\kappa)\mathbf{s}_1 \cdot (\mathbf{s}_1 + \delta\mathbf{n}_1)\mathbf{a}_1\\ &= (1 - \delta\kappa)^2\mathbf{s}_1\mathbf{a}_1 \cdot \mathbf{s}_1.\end{aligned}$$

Likewise $(\mathbf{s}_1 + \delta\mathbf{n}_1)\mathbf{b}_1 \cdot (\mathbf{s}_1 + \delta\mathbf{n}_1) = (1 - \delta\lambda)^2\mathbf{s}_1\mathbf{b}_1 \cdot \mathbf{s}_1$, where $\mathbf{b}_1$ is a principal vector at w with principal curvature λ. □

The straining therefore determines the lines of curvature of the surface **s**. It also determines the ridges, since

$$(1 - \delta\kappa)_1{}^2 = 2(1 - \delta\kappa)(-\delta\kappa_1) = 0$$

if and only if $\kappa_1 = 0$, since $1 - \delta\kappa \neq 0$.

A general strain of a surface superficially is similar. However, the analogue of the classifying form is no longer necessarily symmetric in all three slots, but only in two, so that more varied phenomena can occur at the analogues of the umbilic points of a surface in $\mathbb{R}^3$, namely the isotropic points of a two-dimensional field of stress or strain. It is, however, still the case that around such a point the principal lines of stress or strain take generically one of the three forms, star, lemon or monstar. The analogy is closest in the special case of statical equilibrium without body forces, an example of such being the strain induced in a smooth surface when offset parallel to itself, when, as we have seen, the strain lines coincide with the lines of curvature of the surface. A full classification of what can occur generically has been made by Thorndike, Cooley and Nye (1978), with interesting applications to glaciology in Nye (1983, 1986). For a good discussion see Hutchison, Nye and Salmon (1983). The latter paper gives the following illustration of the practical significance of isotropic points:

> Suppose a plate is to be subjected to plane stress or plane strain and one wishes to bore small bolt holes in it that will remain circular after deformation. A small circular hole made at a general point will become elliptical. To solve the problem, one should place the holes at the isotropic points, for they will then remain circular (if small enough) under the prescribed load, even though they will be expanded or contracted.

12.7 The birth of umbilics

Under smooth deformation of a surface umbilics may be born or die. Examples show that in a family of surfaces it is possible to have a path in which two umbilics are brought together to annihilate each other or, in the reverse direction, it is possible for umbilics to be born in pairs. At such a moment of birth or death the classifying form must be right-angled, and therefore not parabolic, as is evident from the map of cubic forms, and the subsequent umbilics must be of opposite index, one of index $\frac{1}{2}$ and the other of index $-\frac{1}{2}$, for the total index cannot change under smooth deformations of the surface. By Proposition 11.10 at the moment of birth or death the vectors w_1 that feature in that proposition are tangent to the ridge on which the birth or death takes place.

Suppose that we are given a piece of regular surface without any redeeming features such as ridges, parabolic lines or umbilics, and suppose we decide smoothly to deform part of it where the Gaussian curvature is positive so as to create an umbilic there. Intuitively one imagines a ball-bearing being pressed into the surface in an attempt to make the surface more spherical at some chosen point. By what we have proved the first event to occur must be the birth of a ridge, let us suppose a blue one. After the moment of birth it will be a simple closed loop with two A_4 points on it separating a fertile hyperbolic arc from a sterile elliptic one. Presently twin umbilics may be born on the fertile arc, one of them after the moment of birth being a monstar of index $\frac{1}{2}$ and the other a hyperbolic star of index $-\frac{1}{2}$, the ridge between them now being red. The monstar is then free to mature into a lemon. Before the other hyperbolic one can become elliptic another ridge has to be born, perhaps this time a red one, with a fertile and a sterile arc as before. The fertile arc of this ridge then deforms towards the umbilic on the 'blue' side, acquiring a cusp there momentarily when the umbilic becomes parabolic. Further deformation extends the new ridge by a blue loop on the far side of the umbilic. Later two of the ridges near the umbilic of opposite colour may change places so that finally the ridges through the elliptic umbilic alternate in colour. By this time the classifying form at the umbilic is inside the 'inner circle' on the map of cubic forms. See Figures 12.5–12.12 (the annihilation of two A_4 points at a transitional A_5 point is an optional extra in this sequence). One possible *perestroika* of the grid of lines of curvature at the birth of a pair of umbilic twins is illustrated in Figures 12.13–12.16.

Figure 12.5

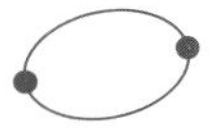

Figure 12.6

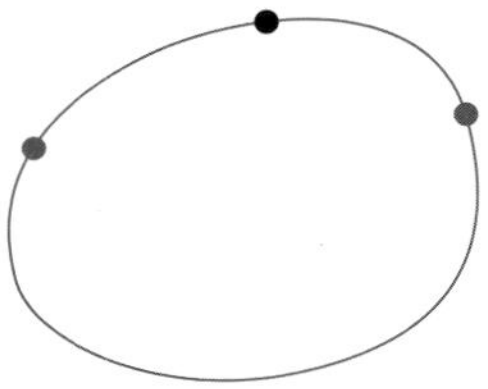

Figure 12.7

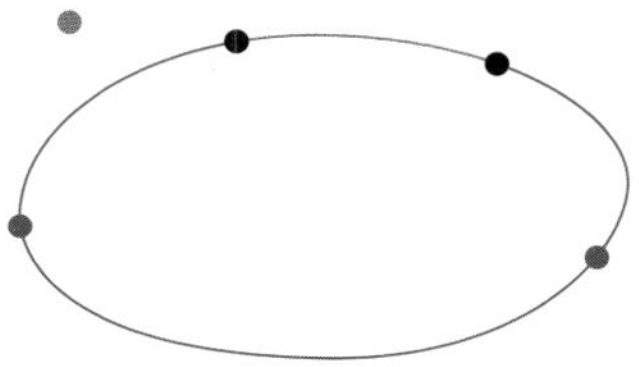

Figure 12.8

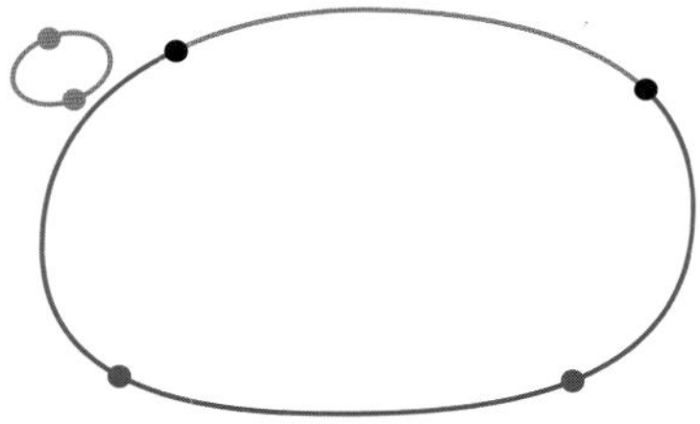

Figure 12.9

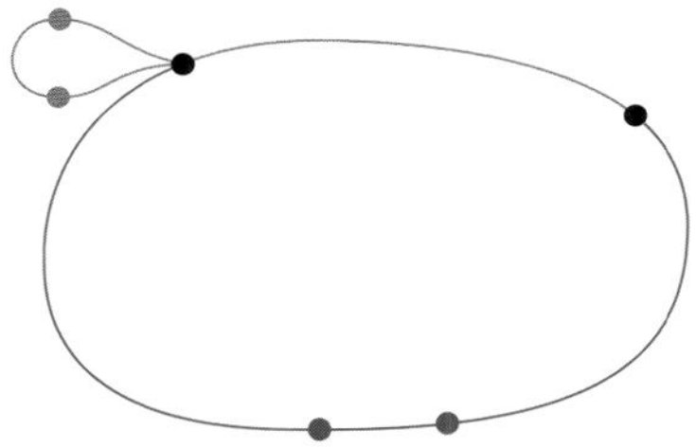

Figure 12.10

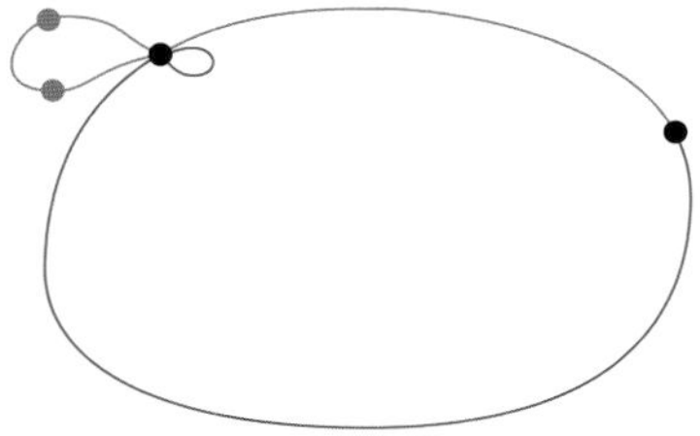

Figure 12.11

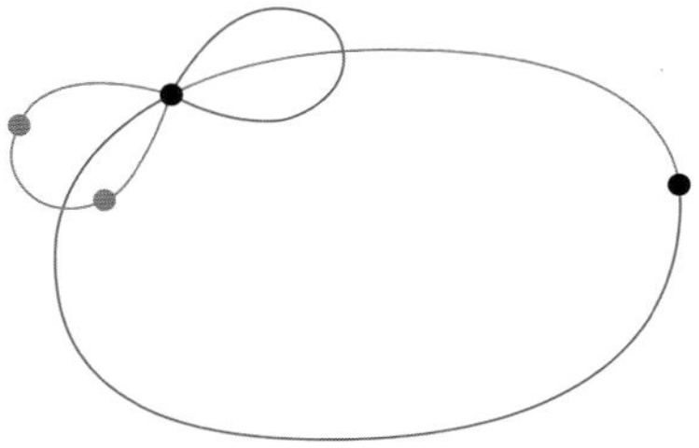

Figure 12.12

Examples of deformations of surfaces, that involve the birth or death of umbilics, are provided by the 'bumpy spheres' of Stelios Markatis which we describe in Chapter 16.

Our study of the geometry of the surface **s** around an umbilic is not yet complete. We shall in the next chapter be looking at the *subparabolic lines* of a surface and how they behave in the neighbourhood of an isolated umbilic.

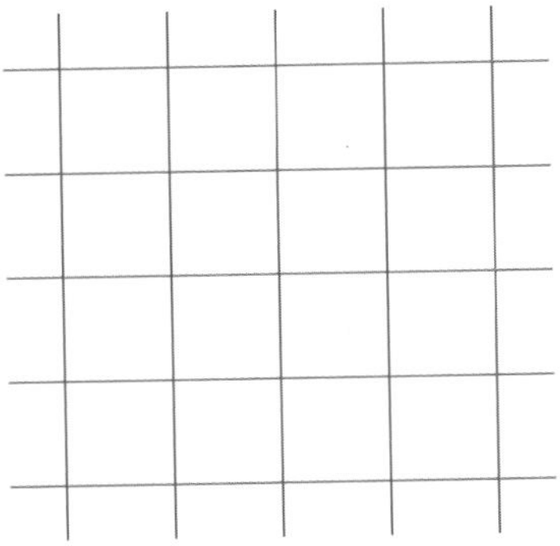

Figure 12.13

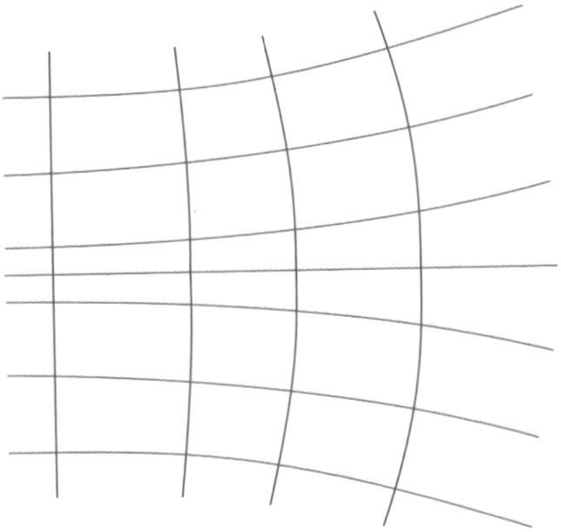

Figure 12.14

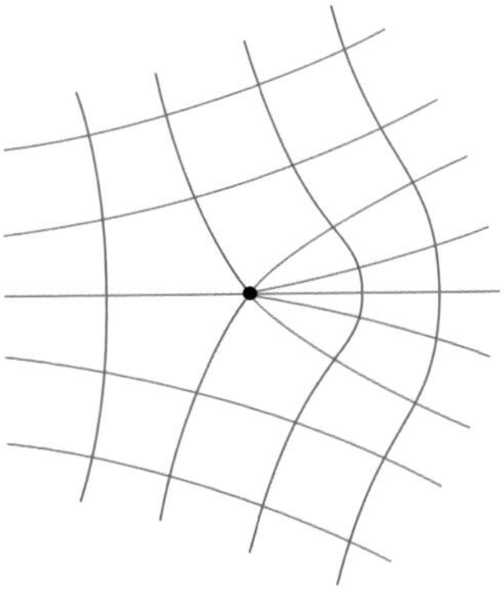

Figure 12.15

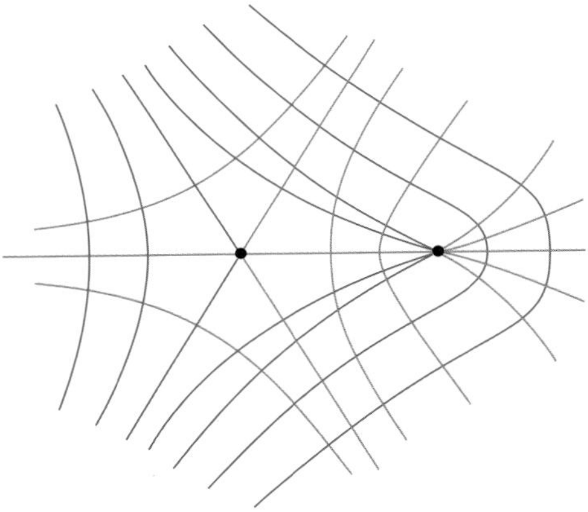

Figure 12.16

Exercises

12.1 Prove Proposition 12.1.

12.2 Extend Proposition 12.3 to the case of a flat umbilic.

12.3 Prove Proposition 12.4.

12.4 Verify that certain parabolas of the form $y^2 = 4ax + c$ are solutions of the differential equation

$$y\mathrm{d}x^2 - 2x\mathrm{d}x\mathrm{d}y - y\mathrm{d}y^2 = 0,$$

and that away from the origin there are exactly two such curves through each point, intersecting each other orthogonally.

12.5 Consider the ellipsoid

$$\frac{x^2}{a^2} + \frac{y^2}{b^2} + \frac{z^2}{c^2} = 1, \quad \text{where } a > b > c > 0.$$

Prove that the map

$$\mathbf{s} : \mathbb{R}^2 \rightarrowtail \mathbb{R}^3; (\theta, \varphi) \mapsto (a \cos \varphi \sin \theta, b \cos \theta, c \sin \varphi \sin \theta)$$

where $0 < \theta < \pi$, is a regular parametrisation of the ellipsoid, apart from two points.

Hence or otherwise verify that the ellipsoid has umbilics at the four points where $\theta = \frac{1}{2}\pi$ and $b^2 = a^2 \sin^2 \varphi + c^2 \cos^2 \varphi$.

What happens if $a = b$ or if $b = c$?

12.6 Find all the umbilics of the ellipsoid $\frac{1}{4}x^2 + y^2 + z^2 = 1$.

12.7 Find all umbilics

(i) of the hyperboloid of one sheet

$$\frac{x^2}{a^2} + \frac{y^2}{b^2} - \frac{z^2}{c^2} = 1, \; a \neq b,$$

(ii) of the hyperboloid of two sheets

$$\frac{x^2}{a^2} - \frac{y^2}{b^2} - \frac{z^2}{c^2} = 1, \; b \neq c.$$

12.8 For which values of the real number c is the cubic $x^3 + 3cxy^2$ of star type, of monstar type, of lemon type?

What are the corresponding values of β (refer to Section 7.4)?

12.9 Let $\mathbf{s}$ be an umbilic of the surface given by the equation $F(\mathbf{s}) = 0$, where $F : \mathbb{R}^3 \rightarrowtail \mathbb{R}$ is smooth. Prove that, up to a real multiple, the classifying cubic form at $\mathbf{s}$ is the restriction to the tangent space there of the form

$$(F_1(\mathbf{s})\mathbf{n})(F_3(\mathbf{s})) - 3(F_2(\mathbf{s})\mathbf{n})(F_2(\mathbf{s})).$$

Deduce that any umbilic of a quadratic for which this form is not identically zero is a pure lemon.

13
The parabolic line

13.0 Introduction

Up till now we have for the most part ignored the possibility that at some point or points either of the principal curvatures of our surface **s**, and therefore their product, the *Gaussian curvature* of the surface, is zero. This omission we now rectify. The points of zero Gaussian curvature in general form a curve on the surface known as the *parabolic line* of the surface, though a better name might be the *inflectional line* or *inflectional curve* of the surface. A remarkable pair of recent theorems due to Jan Koenderink shows how the Gaussian curvature is relevant to how we see a surface either from far away or close up. We also touch in this chapter on recent work on the curvature of the focal surface of a regular surface, studying in particular the *subparabolic lines* of a surface, the traces on the surface of the parabolic lines on the sheets of the evolute, and in particular the configurations of subparabolic lines that arise at umbilics.

13.1 Gaussian curvature

Before embarking on the study of points of zero Gaussian curvature it is appropriate for us at this point to prove Gauss's *excellent* theorem, the *Theorema egregium* (1828), for Gauss wrote in Latin, concerning what ever since has been called the *Gaussian curvature* of a surface, the product of its two principal curvatures. This states that this product depends on the first fundamental form and its derivatives only and not on the second fundamental form, despite the fact that both the fundamental forms are required to define the individual factors κ and λ. A corollary is that under an isometric deformation of the surface in $\mathbb{R}^3$ the Gaussian curvature remains invariant.

Among the derivatives of the first fundamental form I_2 of a regular surface $\mathbf{s}$ in $\mathbb{R}^3$ a special role is played by the thrice linear form $\mathbf{s}_2 \cdot \mathbf{s}_1$, known as the *Christoffel form* of the surface. Despite its name it depends essentially on the parametrisation, as is generally the case with higher derivatives, and unlike the second fundamental form cannot be considered as a form on the tangent spaces of $\mathbf{s}$†). The Christoffel form is not fully symmetric. It is symmetric in the first two slots, clearly, but not in either of these with the third. Accordingly it is ambiguous to write the first derivative of I_2 as $2\mathbf{s}_2 \cdot \mathbf{s}_1$. Explicitly, at w, with $\mathbf{a} \in \mathbb{R}^2$,

$$(I_2)_1(w)\mathbf{a} = \mathbf{s}_2(w)\mathbf{a} \cdot \mathbf{s}_1(w) + \mathbf{s}_1(w) \cdot \mathbf{s}_2(w)\mathbf{a},$$

this being symmetric in the remaining two slots. That is for $\mathbf{b}, \mathbf{c} \in \mathbb{R}^2$,

$$(I_2)_1(w)\mathbf{abc} = \mathbf{s}_2(w)\mathbf{ab} \cdot \mathbf{s}_1(w)\mathbf{c} + \mathbf{s}_1(w)\mathbf{b} \cdot \mathbf{s}_2(w)\mathbf{ac}.$$

On the other hand the cubic form $\mathbf{a} \mapsto \mathbf{s}_2\mathbf{a}^2 \cdot \mathbf{s}_1\mathbf{a}$ is indeed induced by a fully symmetric thrice linear form C_3, namely that defined by

$$C_3\mathbf{abc} = \tfrac{1}{3}(\mathbf{s}_2\mathbf{bc} \cdot \mathbf{s}_1\mathbf{a} + \mathbf{s}_2\mathbf{ca} \cdot \mathbf{s}_1\mathbf{b} + \mathbf{s}_2\mathbf{ab} \cdot \mathbf{s}_1\mathbf{c}).$$

Although $\mathbf{s}_2 \cdot \mathbf{s}_1$ and $(\mathbf{s}_1 \cdot \mathbf{s}_1)_1$ are distinct, each determines the other, via the common cubic form which each induces.

The derivative of the Christoffel form $\mathbf{s}_2 \cdot \mathbf{s}_1$ is the form $\mathbf{s}_3 \cdot \mathbf{s}_1 + \mathbf{s}_2 \cdot \mathbf{s}_2$, where the slots get filled as follows:

$$\mathbf{s}_3\mathbf{abd} \cdot \mathbf{s}_1\mathbf{c} + \mathbf{s}_2\mathbf{ab} \cdot \mathbf{s}_2\mathbf{cd}.$$

Theorem 13.1 (Gauss' Theorem) *Let κ, λ be the principal curvatures of a regular smooth surface* $\mathbf{s}$. *Then their product $K = \kappa\lambda$ depends only on the first fundamental form $\mathbf{I}_2 = \mathbf{s}_1 \cdot \mathbf{s}_1$ and its derivatives.*

Proof For any $w \in \mathbb{R}^2$, choose a basis $\mathbf{e}$, $\mathbf{f}$ for $\mathbb{R}^2$, orthonormal with respect to $I_2(w)$, that is such that the three vectors $\mathbf{s}_1(w)\mathbf{e}$, $\mathbf{s}_1(w)\mathbf{f}$, $\mathbf{n}(w)$ form an orthonormal basis for $\mathbb{R}^3$. In the above equation for the derivative of $\mathbf{s}_2 \cdot \mathbf{s}_1$ (for brevity omitting the explicit reference to the point w) set $\mathbf{a}$, $\mathbf{b}$, $\mathbf{c}$ and $\mathbf{d}$ first of all equal to $\mathbf{e}$, $\mathbf{e}$, $\mathbf{f}$ and $\mathbf{f}$ and then equal to $\mathbf{e}$, $\mathbf{f}$, $\mathbf{f}$ and $\mathbf{e}$. Then subtract to get

$$\mathbf{s}_2\mathbf{e}^2 \cdot \mathbf{s}_2\mathbf{f}^2 - \mathbf{s}_2\mathbf{ef} \cdot \mathbf{s}_2\mathbf{ef}.$$

† In the traditional jargon of differential geometry it is not a *tensor*.

Now

$$\mathbf{s}_2\mathbf{e}^2 = \mathbf{n}(\mathbf{n}\cdot\mathbf{s}_2\mathbf{e}^2) + \mathbf{s}_1\mathbf{e}(\mathbf{s}_1\mathbf{e}\cdot\mathbf{s}_2\mathbf{e}^2) + \mathbf{s}_1\mathbf{f}(\mathbf{s}_2\mathbf{f}\cdot\mathbf{s}_2\mathbf{e}^2)$$
$$\mathbf{s}_2\mathbf{f}^2 = \mathbf{n}(\mathbf{n}\cdot\mathbf{s}_2\mathbf{f}^2) + \mathbf{s}_1\mathbf{e}(\mathbf{s}_1\mathbf{e}\cdot\mathbf{s}_2\mathbf{f}^2) + \mathbf{s}_1\mathbf{f}(\mathbf{s}_2\mathbf{f}\cdot\mathbf{s}_2\mathbf{f}^2)$$
$$\mathbf{s}_2\mathbf{ef} = \mathbf{n}(\mathbf{n}\cdot\mathbf{s}_2\mathbf{ef}) + \mathbf{s}_1\mathbf{e}(\mathbf{s}_1\mathbf{e}\cdot\mathbf{s}_2\mathbf{ef}) + \mathbf{s}_1\mathbf{f}(\mathbf{s}_2\mathbf{f}\cdot\mathbf{s}_2\mathbf{ef}).$$

So

$$\mathbf{s}_2\mathbf{e}^2\cdot\mathbf{s}_2\mathbf{f}^2 - \mathbf{s}_2\mathbf{ef}\cdot\mathbf{s}_2\mathbf{ef} = \mathbf{n}\cdot\mathbf{s}_2\mathrm{e}^2\mathbf{n}\cdot\mathbf{s}_2\mathbf{f}^2 - (\mathbf{n}\cdot\mathbf{s}_2\mathbf{ef})^2 + \text{an expression}$$

depending on the Christoffel form and **e** and **f** only.

Now let **a** and **b** be unit principal vectors at w. Then

$$\mathbf{e} = \mathbf{a}\cos\theta + \mathbf{b}\sin\theta \text{ and } \mathbf{f} = -\mathbf{a}\sin\theta + \mathbf{b}\cos\theta, \text{ for some angle } \theta,$$

and

$$\begin{aligned}\mathbf{n}\cdot\mathbf{s}_2\mathbf{e}^2\,\mathbf{n}\cdot\mathbf{s}_2\mathbf{f}^2 - (\mathbf{n}\cdot\mathbf{s}_2\mathbf{ef})^2 &= (\kappa\cos^2\theta + \lambda\sin^2\theta)(\kappa\cos^2\theta + \lambda\sin^2\theta)\\ &\quad - ((-\kappa+\lambda)\cos\theta\sin\theta)^2\\ &= \kappa\lambda\,(\cos^4\theta + \sin^4\theta + 2\cos^2\theta\sin^2\theta)\\ &= \kappa\lambda.\end{aligned}$$

(We couldn't originally choose **a**, **b** as basis for $\mathbb{R}^2$, since their definition involves the second fundamental form II_2 as well as I_2.)

So, *mirabile dictu*, $\kappa\lambda$ depends on I_2 and its derivatives only. □

The product $K = \kappa\lambda$ is known as the *Gaussian curvature* of the surface. It is zero wherever either κ or λ is zero.

The Gaussian curvature of a regular smooth surface $\mathbf{s} : \mathbb{R}^2 \rightarrowtail \mathbb{R}^3$ is also definable in terms of the *Gauss* map, namely the map

$$\mathbf{n} : \mathbb{R}^2 \rightarrowtail S^2 \subset \mathbb{R}^3;\ w \mapsto \mathbf{n}(w),$$

that associates to each point $w \in \mathbb{R}^2$ one of the unit normal vectors to the surface **s** there. More properly the Gauss map is the induced map from the actual surface im **s** to the unit sphere, and the Gaussian curvature is then equal to its local area change, the determinant of its differential. This is easily verifiable from the fact that if one identifies the tangent spaces at source and target by parallel transport along the common normal then, by the theorem of Rodrigues, Theorem 10.2, the eigendirections are the principal directions, the determinant of the differential being the product of the eigenvalues, namely $\kappa\lambda = K$.

The Gaussian curvature of a regular surface is a *bending invariant*, remaining unchanged by isometric deformations of the surface. A practical manifestation of that is our paper model of the tangent developable of a space curve (see Figure 6.1), this surface everywhere having Gaussian curvature zero, one of the principal curvatures being zero.

Much of the development of differential geometry since the time of Gauss is concerned with following the roads opened up by Gauss' Theorem, roads that we choose not to follow here. For a reprint of Gauss' paper of 1828, written in Latin, a translation of it into English and an assessment of it by Peter Dombrowski 150 year on, see Volume 62 of the journal *Astérisque*, published by the Société mathématique de France.

13.2 The parabolic line

There are several distinct intuitions concerning the parabolic line of a regular smooth surface $\mathbf{s} : \mathbb{R}^2 \rightarrowtail \mathbb{R}^3$. We start with the Gauss map

$$\mathbf{n} : \mathbb{R}^2 \rightarrowtail S^2 \subset \mathbb{R}^3;\ w \mapsto \mathbf{n}(w),$$

that, as we have just remarked, associates to each point w one of the unit normal vectors to the surface $\mathbf{s}$ there. Since $\mathbf{n} \cdot \mathbf{n} = 1$ it follows that

$$\mathbf{n} \cdot \mathbf{n}_1 = 0,\ \mathbf{n} \cdot \mathbf{n}_2 + \mathbf{n}_1 \cdot \mathbf{n}_1 = 0, \text{ etc.}$$

Moreover, since $\mathbf{n} \cdot \mathbf{s}_1 = 0$ it follows that

$$\mathbf{n} \cdot \mathbf{s}_2 + \mathbf{n}_1 \cdot \mathbf{s}_1 = 0,\ \mathbf{n} \cdot \mathbf{s}_3 + 2\mathbf{n}_1 \cdot \mathbf{s}_2 + \mathbf{n}_2 \cdot \mathbf{s}_1 = 0, \text{ etc.},$$

from the first of which it follows that $\mathbf{n}_1 \cdot \mathbf{s}_1$ is symmetric.

In particular for any $\mathbf{u} \in \mathbb{R}^2$ it follows that $\mathbf{n}_1\mathbf{u} \cdot \mathbf{s}_1 = \mathbf{n}_1 \cdot \mathbf{s}_1\mathbf{u}$.

We next recall Rodrigues' Theorem, proved earlier as Theorem 10.2:

Theorem 13.2 (Rodrigues' Theorem) *Let* $\mathbf{a}_1$ *be a principal vector at* w *to the regular surface* $\mathbf{s}$, *the corresponding principal curvature of* $\mathbf{s}$ *being* κ. *Then*

$$\kappa\mathbf{s}_1(w)\mathbf{a}_1 + \mathbf{n}_1(w)\mathbf{a}_1 = 0. \qquad \square$$

Corollary 13.3 Let $\mathbf{a}_1$ *be a principal vector at* w *to the regular surface* $\mathbf{s}$, *the corresponding principal curvature of* $\mathbf{s}$ *being* κ. *Then*

$$\kappa = 0 \Leftrightarrow \mathbf{n} \cdot \mathbf{s}_2\mathbf{a}_1 = 0 \text{ at } w \Leftrightarrow \mathbf{n}_1\mathbf{a}_1 = 0 \text{ at } w. \qquad \square$$

In fact, if there is a non-zero vector $\mathbf{u}$ such that either $\mathbf{n} \cdot \mathbf{s}_2\mathbf{u} = 0$ or, equivalently, $\mathbf{n}_1\mathbf{u} = 0$ at w, then κ is a principal curvature at w with $\mathbf{u}$ a principal vector, and $\mathbf{u} = \mathbf{a}_1$ for some regular parametrisation $\mathbf{a}$ of the relevant line of curvature through w.

The set of points of the surface $\mathbf{s}$ where one of the principal curvatures is zero is called the *parabolic curve*, or more commonly the *parabolic line* on $\mathbf{s}$. By Corollary 13.3 it is also the set of points where the first derivative of the Gauss map $\mathbf{n}$ has non-zero kernel, for which reason it is often thought of as the *fold* of the Gauss map. The reason for the term *parabolic* is that the parabolic line consists of those points of the surface $\mathbf{s}$ where the second fundamental form $\mathrm{II} = \mathbf{n} \cdot \mathbf{s}_2$ is parabolic (or zero). The parabolic line divides the *elliptic* points of the surface, where II is elliptic and the *Gaussian curvature*, the product of the principal curvatures, is positive, from the *hyperbolic* points, where II is hyperbolic and the Gaussian curvature is negative. The root lines of II at a hyperbolic point are called the *asymptotic tangent lines* at the point, and a curve on the hyperbolic part of the surface each of whose tangent lines is asymptotic is called an *asymptotic line* of the surface.

A parabolic point may be thought of as one at which one of the principal centres of curvature lies at infinity, and the A_k classification extends accordingly. A parabolic point of $\mathbf{s}$ is said to be an A_2 *parabolic* point if $\mathbf{n} \cdot \mathbf{s}_2\mathbf{a}_1 = 0$ for some non-zero $\mathbf{a}_1$, or equivalently if $\mathbf{s}_1\mathbf{a}_1 \cdot \mathbf{n}_1 = 0$ for some non-zero $\mathbf{a}_1$, or equivalently if $\mathbf{n}_1\mathbf{a}_1 = 0$ for some non-zero $\mathbf{a}_1$, with $\mathbf{n} \cdot \mathbf{s}_2 \neq 0$ and $\mathbf{n} \cdot \mathbf{s}_3\mathbf{a}_1{}^3 \neq 0$, and to be an A_3 *parabolic* point if $\mathbf{n} \cdot \mathbf{s}_2\mathbf{a}_1 = 0$, or equivalently $\mathbf{n}_1\mathbf{a}_1 = 0$ for some non-zero $\mathbf{a}_1$, and $\mathbf{n} \cdot \mathbf{s}_3\mathbf{a}_1{}^3 = 0$, or equivalently $\mathbf{n} \cdot \mathbf{s}_3\mathbf{a}_1{}^2 + \mathbf{n} \cdot \mathbf{s}_2\mathbf{a}_2 = 0$, or equivalently $\mathbf{n}_2\mathbf{a}_1{}^2 + \mathbf{n}_1\mathbf{a}_2 = 0$ for some vector $\mathbf{a}_2$, with $\mathbf{n} \cdot \mathbf{s}_2 \neq 0$ and

$$\mathbf{n} \cdot \mathbf{s}_4\mathbf{a}_1{}^4 + 3\mathbf{n} \cdot \mathbf{s}_2\mathbf{a}_1{}^2\mathbf{a}_2 \neq 0.$$

A point at which $\mathbf{n}_1 = 0$, or equivalently where $\mathrm{II} = \mathbf{n} \cdot \mathbf{s}_2 = 0$ is said to be a *flat umbilic* (*not* a *parabolic* umbilic!).

Proposition 13.4 Let $\mathbf{s}$ *be a regular smooth surface. Then the set of* A_2 *parabolic points is a smooth one-dimensional submanifold of the surface. The condition for the parabolic line to be locally a smooth one-dimensional submanifold also holds at certain* A_3 *parabolic points.*

Proof There are alternative proofs according to which form of the condition for A_2 parabolicity one adopts as the starting point.

Consider first the map

$$\mathbb{R}^2 \times \mathbb{R}^2 \rightarrowtail L(\mathbb{R}^2, \mathbb{R}) \times \mathbb{R};\ (w, \mathbf{u}) \mapsto (\mathbf{n}(w) \cdot \mathbf{s}_2(w)\mathbf{u},\ \mathbf{u} \cdot \mathbf{u} - 1),$$

for which the linear equations of the differential are

$$\mathbf{n}_1 w_1 \cdot \mathbf{s}_2 \mathbf{u} + \mathbf{n} \cdot \mathbf{s}_3 w_1 \mathbf{u} + \mathbf{n} \cdot \mathbf{s}_2 \mathbf{u}_1 = 0,$$

$$\mathbf{u} \cdot \mathbf{u}_1 = 0.$$

The first of these implies that

$$\mathbf{n}_1 w_1 \cdot \mathbf{s}_2 \mathbf{u}^2 + \mathbf{n} \cdot \mathbf{s}_3 w_1 \mathbf{u}^2 = 0,$$

from which w_1 is determined up to a scalar multiple provided that $\mathbf{u}$ is so determined, that is provided that, at w, $\mathbf{n} \cdot \mathbf{s}_2 \neq 0$ and

$$\mathbf{n}_1 \cdot \mathbf{s}_2 \mathbf{u}^2 + \mathbf{n} \cdot \mathbf{s}_3 \mathbf{u}^2 \neq 0,$$

implying that the dimension of the kernel of the linear equations is one.

This is certainly the case at an A_2 parabolic point of the surface, since $\mathbf{n}_1 \cdot \mathbf{s}_2 \mathbf{u}^2 + \mathbf{n} \cdot \mathbf{s}_3 \mathbf{u}^2 = 0 \Rightarrow \mathbf{n} \cdot \mathbf{s}_3 \mathbf{u}^3 = 0$. Moreover, at such a point w_1, if non-zero, is not a multiple of $\mathbf{u}$, since if that is so then $\mathbf{n} \cdot \mathbf{s}_3 \mathbf{u}^3 = 0$, implying that the point is A_3 parabolic at least.

At an A_3 parabolic point at which $\mathbf{n}_1 \cdot \mathbf{s}_2 \mathbf{u}^2 + \mathbf{n} \cdot \mathbf{s}_3 \mathbf{u}^2 \neq 0$ the dimension of the kernel of the linear equations is still one and since $w_1 = \mathbf{u}$ is a solution it follows that at such point the tangent line to the parabolic line is spanned by the principal vector $\mathbf{u}$, and generically this is so.

Moreover, the map $\mathbb{R}^2 \rightarrowtail L_2(\mathbb{R}^2, \mathbb{R})$: $w \mapsto \mathbf{n}(w) \cdot \mathbf{s}_2(w)$ cannot have surjective differential anywhere so that generically $\mathbf{n} \cdot \mathbf{s}_2 \neq 0$; that is flat umbilics do not occur generically.

Alternatively starting from the condition $\mathbf{n}_1 \mathbf{u} = 0$ one arrives at the condition $\mathbf{s}_1 \mathbf{u} \cdot \mathbf{n}_2 \mathbf{u} \neq 0$, which, with a bit of care as to which slot is which in an expression that is symmetric in two of the slots but not all three, is seen to be equivalent to the condition $\mathbf{n}_1 \cdot \mathbf{s}_2 \mathbf{u}^2 + \mathbf{n} \cdot \mathbf{s}_3 \mathbf{u}^2 \neq 0$.

□

To summarise, a generic parabolic point is an A_2 parabolic point of the surface, where the Gauss map has a fold. An A_3 parabolic point of the surface is one where the Gauss map has a Whitney cusp, in the sense that the image of the fold on the unit sphere is cuspidal there (cf. Exercise 4.11). Such a point is commonly referred to as a *cusp of Gauss* (Banchoff, Gaffney and McCrory, 1982). A cusp of Gauss is a point where the non-singular parabolic line crosses a ridge of the same colour, the tangent to the parabolic line there being principal and therefore transverse to the ridge. Since ridge points are either fertile or sterile, so also are cusps of Gauss.

From the proof of Proposition 13.4 it follows that in a generic one-parameter family of surfaces, the birth or death point of a component of the parabolic line, or *acnode* of the parabolic line, or a level crossing of the parabolic line, or *crunode* of the parabolic line, must occur on a ridge, where $\mathbf{s}_1\mathbf{u}\cdot\mathbf{n}_2\mathbf{u}=0$ or equivalently $\mathbf{n}_1\cdot\mathbf{s}_2\mathbf{u}^2+\mathbf{n}\cdot\mathbf{s}_3\mathbf{u}^2=0$. So, for example, if one wants to create a parabolic component in a featureless part of a surface the first event in the deformation has to be the birth of a ridge. The two events do not in general occur contemporaneously. An A_4 parabolic point is in general a birth or death point of a pair of cusps of Gauss, a point where the parabolic line touches a ridge of the same colour simply, the common tangent being principal to the surface. Since an A_4 point divides points on the ridge that are fertile from points that are sterile it follows that after birth one of the pair of cusps of Gauss that have been born will be fertile and the other sterile. An acnode or crunode of the parabolic line also is associated with the birth or death of a pair of cusps of Gauss, but in that case the birth or death point is not necessarily A_4 and accordingly both the cusps of Gauss involved will in general both be fertile or both sterile.

Conditions for all these events can be given locally if the surface is taken in Monge form, with the point of interest at the origin.

Proposition 13.5 Consider the surface in $\mathbb{R}^3$ *with equation*

$$z=\tfrac{1}{2}\lambda y^2+\tfrac{1}{6}(\alpha x^3+3\beta x^2y+3\gamma xy^2+\delta y^3)+O(x,y)^4.$$

Then the origin is a parabolic point, with principal tangent vector $(1,0)$, *the tangent line to the parabolic line there having equation* $\alpha x+\beta y=0$. *The point is a cusp of Gauss if* $\alpha=0$, *but* $\beta\neq 0$, *when the tangent line to the parabolic line has equation* $y=0$ *and so is principal.*

Proof Since for the Monge form $\mathbf{s}_1\cdot\mathbf{s}_2$ at 0 and since $\mathbf{n}\cdot\mathbf{n}_1=0$ it follows that $\mathbf{n}_1\cdot\mathbf{s}_2=0$ at 0. So the equation for w_1 reduces to $\mathbf{n}\cdot\mathbf{s}_3\mathbf{u}^2w_1=0$. Now $\mathbf{n}\cdot\mathbf{s}_3\mathbf{u}^2=(\alpha\;\beta)$. All the assertions then follow easily. □

For a birth or death point (acnode) or crossing point (crunode) of the parabolic line $\beta = 0$, and there is a quadratic equation for the tangents to the parabolic line, namely

$$\tfrac{1}{2}\lambda(Ax^2 + 2Bxy + Cy^2) - \gamma^2 y^2 = 0,$$

where the terms of fourth degree in the equation of the surface are

$$\tfrac{1}{24}(Ax^4 + 4Bx^3y + 6Cx^2y^2 + 4xy^3 + Ey^4).$$

It ought to be said that the parabolic line on a surface is just one of the fibres of the principal curvature function κ on the surface and its acnodes and crunodes are amongst the critical points of κ. From this point of view it is clear that such nodes must lie on ridges of the surface. Each fibre other than the parabolic line lifts to a curve on the focal surface that is orthogonal to each of the focal lines of curvature, as is easily verified. This remark will be of relevance in Chapter 15.

13.3 Koenderink's theorems

The following theorems which involve both the Gaussian curvature and the geodesic curvature, and relate to how one actually sees a surface, are of surprisingly recent date (Koenderink, 1984, 1990). The first is concerned with distant views.

Let M be a non-singular smooth surface in three-dimensional space and let $\pi : X \to Y$ be the orthogonal projection of X on to a plane Y in X. Then the *rim* of M as *viewed* by π is the critical set of the map $\pi|M$, this being in general a curve on M, its image by π in Y being the *apparent contour* of M as viewed by π. In practice, of course, only parts of the apparent contour are visible, unless the surface is transparent.

Theorem 13.6 Suppose that M is a non-singular smooth surface in $\mathbb{R}^3$ *and let* $\pi : \mathbb{R}^3 \to Y$ *be the orthogonal projection of* $\mathbb{R}^3$ *onto a plane Y in* $\mathbb{R}^3$. *Suppose further that a is a point of the* rim *of M as viewed by* π, *let* κ_t *be the curvature at* $\pi(a)$ *of the apparent contour of M, and let* κ_r *be the curvature at a of the normal section of M at a by the plane that contains the kernel of* π. *Then* $\kappa_r\kappa_t$ *is equal to the Gaussian curvature of M at a.*

Proof There is no loss of generality in taking the point a to be the origin in $\mathbb{R}^3$, π to be the projection $\mathbb{R}^3 \to \mathbb{R}^2$, $(x, y, z) \mapsto (y, z)$, so that $\pi(0) = 0$, and to have M in Monge form as the parametric surface

$\mathbb{R}^2 \rightarrowtail \mathbb{R}^3$; $(x, y) \mapsto (x, y, z)$, with $z = \frac{1}{2}(\alpha x^2 + 2\beta xy + \gamma y^2) + O(x, y)^3$.

Then the apparent contour is the set of critical values of the map $\mathbb{R}^2 \rightarrowtail \mathbb{R}^2$; $(x, y) \mapsto (y, z)$, whose Jacobian matrix is

$$\begin{bmatrix} 0 & 1 \\ \alpha x + \beta y + O(x, y)^2 & \beta x + \gamma y + O(x, y)^2 \end{bmatrix},$$

singular where $\alpha x + \beta y + O(x, y)^2 = 0$, that is where $x = -\alpha^{-1}\beta y + O(y^2)$ (provided, of course, that $\alpha \neq 0$). But then, since

$$\alpha x^2 + 2\beta xy + \gamma y^2 = (\alpha x + \beta y)x + (\beta x + \gamma y)y,$$

the apparent contour takes the form near the origin:

$$z = \tfrac{1}{2}(\beta(-\alpha^{-1}\beta y) + \gamma y)y + O(y^3) = \tfrac{1}{2}\alpha^{-1}(\alpha y - \beta^2)y^2 + O(y^3),$$

with curvature at the origin κ_t equal to $\alpha^{-1}(\alpha\gamma - \beta^2)$. Now the curve of intersection of M with the plane normal to M at the origin and containing the kernel of π takes the form near the origin:

$$z = \tfrac{1}{2}\alpha x^2 + O(x^3),$$

with curvature at the origin κ_t equal to α. Accordingly $\kappa_t\kappa_r = \alpha\gamma - \beta^2$. But this is just the Gaussian curvature of M at the origin, being the product of the eigenvalues of the quadratic form $\alpha x^2 + 2\beta xy + \gamma y^2$. □

It follows that if one looks at M from a distance then at a point where the parabolic line of M crosses the rim the apparent contour in general has an ordinary inflection, for since in general $\kappa_r \neq 0$ it follows that at such a point $\kappa_t = 0$. Moreover, at a point where the apparent contour has a cusp the Gaussian curvature is negative, for if $\kappa_t = \infty$ then $\kappa_r = 0$, this being the case only if the kernel of projection is an asymptotic direction on M, such directions existing only where the Gaussian curvature is negative.

Our intuition tells us that the corollaries of the theorem should remain true when the surface is viewed from some finite point of space, rather than from infinity, the main point to clarify being the appropriate analogue in this context of the curvature κ_t of the apparent contour. The precise form of the theorem depends on where we put the apparent contour. Koenderink's choice is to put it on the unit sphere with centre the point of view, the appropriate curvature then being its geodesic curvature (as defined in Chapter 5).

Theorem 13.7 Suppose that M is a non-singular smooth surface in $\mathbb{R}^3$ and let $\pi : X \to Y$ be the central projection of $\mathbb{R}^3$ from a point p of $\mathbb{R}^3$ onto the unit sphere in $\mathbb{R}^3$ with centre the point p. Suppose further that a is a point of the rim *of M as viewed by π, the distance of a from p being d, let κ_t be the geodesic curvature at $\pi(a)$ of the apparent contour of M, and let κ_r be the curvature at a of the normal section of M at a by the plane that contains the kernel of π. Then $\kappa_r\kappa_t/d$ is equal to the Gaussian curvature of M at a.*

Proof Exercise. □

It may help to see how this works out in the particular case that M is a sphere of radius ρ and consequently of Gaussian curvature κ^2 where κ is the reciprocal of ρ, the point of view p being external to the sphere. Then the rim of the sphere as viewed from p is a circle of radius r say. Let θ be the semiangle subtended by this circle at p (see Figure 13.1). Then the geodesic curvature of the rim at any of its points is the component of the curvature vector tangent to the sphere centre p and radius d, namely $(1/r)\cos\theta = (1/r)(r/\rho) = 1/\rho = \kappa$. The curvature of the normal section of the sphere containing the line joining p to the point of the rim is also κ. The product of the two curvatures is then κ^2 the Gaussian curvature of the sphere. If one regards the image of π as lying on the unit sphere with centre the point of view, rather than that of radius d, then the geodesic curvature gets multiplied by d. This explains the term d in the denominator of the formula in the statement of the theorem.

In the above account we have only begun the study of the apparent contour of a regular surface, and we must refer the reader elsewhere for the full story, which begins with an article by J. J. Koenderink and A. J. van Doorn (1976). Quite independently the problem was first solved by T. Gaffney and M. Ruas (unpublished, see Gaffney (1983)). Another independent account is by E. E. Landis (1981). For a unified approach, with many references, see Bruce and Giblin (1985).

13.4 Subparabolic lines

The focal surface of a regular smooth surface in $\mathbb{R}^3$ may well have parabolic points. Indeed in the case of a tubular surface with core a regular space curve (see Chapter 15) the sheet of the focal surface coinciding with the focal surface of the core is everywhere parabolic. In

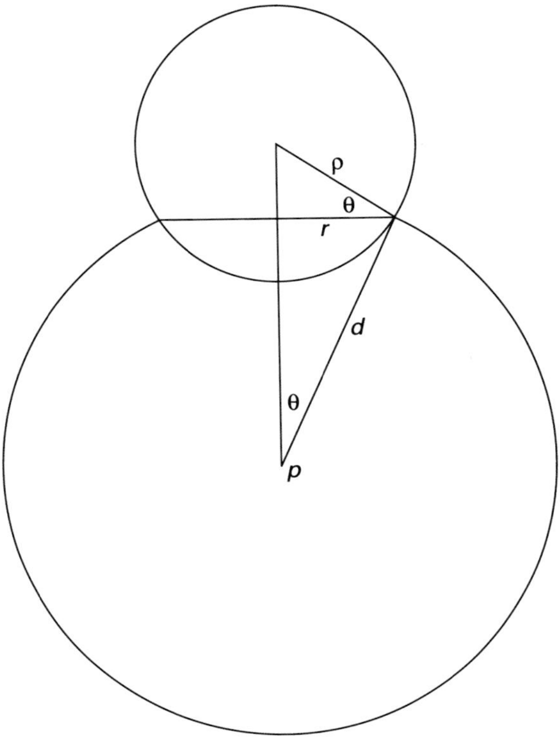

Figure 13.1

general, however, the focal surface has a parabolic line. The curve on the surface underlying this curve we shall call the *subparabolic curve* of the surface, components of this being referred to as *subparabolic lines* of the surface. Away from umbilics these are to be thought of as either blue or red according to which sheet of the focal surface they are associated with. As the next proposition shows they are the loci of points of (geodesic) inflection of the lines of curvature of the surface. What is surprising about them is their behaviour near umbilical points of the surface. What we shall prove is that in general there are either one or three branches of the subparabolic curve through such a point, the tangents to these branches coinciding with the tangent lines to the lines of curvature through the umbilic. This result is, of course, obviously true in the very special case that all the lines of curvature through the umbilic are planar lines about which the surface has reflectional symmetry, and so are geodesics of the surface. We shall have more to say about this in Chapter 16.

Away from an umbilic the characterisation of subparabolic points is straightforward. We note first the following:

Proposition 13.8 At a parabolic point of the red *sheet* **f** *of the focal surface of a regular smooth surface* **s** *the kernel of the Gauss map is the* conjugate (blue) *principal tangent line of the surface* **s** *associated to that point.*

Proof We recall that at any point of the sheet **f** of the focal surface of **s** the principal vector $\mathbf{s}_1\mathbf{b}_1$ is normal to the sheet. Suppose we denote by **v** the vector $\mathbf{b}_1$ normalised so that the vector $\mathbf{s}_1\mathbf{v}$ has length 1. Then $\mathbf{s}_1\mathbf{v} : \mathbb{R}^2 \rightarrowtail \mathbb{R}^3$ is the Gauss map of the sheet **f**. At a point of the parabolic line of **f** the derivative of the this must have kernel rank equal to 1. But

$$(\mathbf{s}_1\mathbf{v})_1 = \mathbf{s}_2\mathbf{v} + \mathbf{s}_1\mathbf{v}_1$$

and **u** will be a kernel vector of this only if $\mathbf{s}_2\mathbf{vu} + \mathbf{s}_1(\mathbf{v}_1\mathbf{u}) = 0$, and applying $\mathbf{n}\cdot$ to this it follows at once that in that case $\mathbf{s}_1\mathbf{v}\cdot\mathbf{s}_1\mathbf{u} = 0$. □

An alternative characterisation is provided by the next proposition.

Proposition 13.9 Let **s** *be a regular surface in* $\mathbb{R}^3$. *Then a point of the surface lies on a* red *subparabolic line if and only if the* blue *line of curvature passing through that point is geodesic there.*

Proof Let **u** and **v** be principal curvature vectors in the blue and red directions respectively, normalised of unit length. Then $\mathbf{s}_1\mathbf{u}\cdot\mathbf{s}_1\mathbf{v} = 0$ and $\mathbf{n}\cdot\mathbf{s}_1\mathbf{v} = 0$ everywhere. Differentiating these in the blue direction we find that $(\mathbf{s}_1\mathbf{u})_1\mathbf{u}\cdot\mathbf{s}_1\mathbf{v} + \mathbf{s}_1\mathbf{u}\cdot(\mathbf{s}_1\mathbf{v})_1\mathbf{u} = 0$ and $\mathbf{n}\cdot(\mathbf{s}_1\mathbf{v})_1\mathbf{u} + \mathbf{n}_1\mathbf{u}\cdot\mathbf{s}_1\mathbf{v} = 0$. Then $(\mathbf{s}_1\mathbf{v})_1\mathbf{u} = 0$ if and only if $(\mathbf{s}_1\mathbf{u})_1\mathbf{u}\cdot\mathbf{s}_1\mathbf{v} = 0$, and $\mathbf{n}_1\mathbf{u}\cdot\mathbf{s}_1\mathbf{v} = 0$, the last of these being always true, since $\mathbf{n}_1\cdot\mathbf{s}_1 = -\mathrm{II}_2$, while the other is the condition for the blue line of curvature to be geodesic there. □

In the section that follows it is convenient not only to let $V_k(\mathbf{e})$ denote $V(\mathbf{c})_k$ with **c** afterwards put equal to **e** but also to let $V_k(\mathbf{f})$ denote $V(\mathbf{c})_k$ with **c** afterwards put equal to **f**. For convenience also we shall continue to refer to the sheet **e** as the blue sheet and the sheet **f** as the red sheet of the focal surface of the surface **s**.

Proposition 13.10 (Morris, 1990) *Let* $\mathbf{s}$ *be a regular surface in* $\mathbb{R}^3$, *and let* $\mathbf{u}$ *and* $\mathbf{v}$ *be principal tangent vectors to* $\mathbf{s}$ *at* w. *Then the point* $\mathbf{s}(w)$ *of the surface lies on a blue subparabolic line if and only if*

$$2(\mathbf{s}_1\mathbf{u}\cdot\mathbf{s}_2\mathbf{uv})V_2(\mathbf{f})\mathbf{u}^2 = (\mathbf{s}_1\mathbf{u}\cdot\mathbf{s}_1\mathbf{u})V_3(\mathbf{f})\mathbf{u}^2\mathbf{v}.$$

Proof From the form of the proposition the surface $\mathbf{s}$ does not have an umbilic at w. Without loss of generality we may suppose that the principal vectors $\mathbf{u}$ and $\mathbf{v}$ are smoothly chosen near w, each normalised to be of unit length with respect to the first fundamental form.

Differentiating the equations

$$V_1(\mathbf{f}) = 0,$$

$$V_2(\mathbf{f})\mathbf{v} = 0$$

in the direction $\mathbf{u}$ we obtain the equations

$$\mathbf{f}_1\mathbf{u}\cdot\mathbf{s}_1 + V_2(\mathbf{f})\mathbf{u} = 0,$$

$$\mathbf{f}_1\mathbf{u}\cdot\mathbf{s}_2\mathbf{v} + V_3(\mathbf{f})\mathbf{uv} + V_2(\mathbf{f})(\mathbf{v}_1\mathbf{u}) = 0.$$

From these we obtain the equations

$$\mathbf{f}_1\mathbf{u}\cdot\mathbf{s}_1\mathbf{u} + V_2(\mathbf{f})\mathbf{u}^2 = 0,$$

$$\mathbf{f}_1\mathbf{u}\cdot\mathbf{s}_1(\mathbf{v}_1\mathbf{u}) + V_2(\mathbf{f})\mathbf{u}(\mathbf{v}_1\mathbf{u}) = 0,$$

$$\mathbf{f}_1\mathbf{u}\cdot\mathbf{s}_2\mathbf{uv} + V_3(\mathbf{f})\mathbf{u}^2\mathbf{v} + V_2(\mathbf{f})\mathbf{u}(\mathbf{v}_1\mathbf{u}) = 0.$$

From the first of these $\mathbf{f}_1\mathbf{u} = -(V_2(\mathbf{f})\mathbf{u}^2)\mathbf{s}_1(\mathbf{u}) + \beta\mathbf{n}$, for some $\beta \in \mathbb{R}$, and then using this the second and third become

$$-(V_2(\mathbf{f})\mathbf{u}^2)\mathbf{s}_1\mathbf{u}\cdot\mathbf{s}_1(\mathbf{v}_1\mathbf{u}) + V_2(\mathbf{f})\mathbf{u}(\mathbf{v}_1\mathbf{u}) = 0,$$

$$-(V_2(\mathbf{f})\mathbf{u}^2)\mathbf{s}_1\mathbf{u}\cdot\mathbf{s}_2\mathbf{uv} + V_3(\mathbf{f})\mathbf{u}^2\mathbf{v} + V_2(\mathbf{f})\mathbf{u}(\mathbf{v}_1\mathbf{u}) = 0.$$

Now the surface $\mathbf{s}$ has a subparabolic point at w if and only if at that point

$$\mathbf{s}_1\mathbf{u}\cdot(\mathbf{s}_2\mathbf{uv} + \mathbf{s}_1(\mathbf{v}_1\mathbf{u})) = 0.$$

From the last two equations this is the case if and only if

$$2(\mathbf{s}_1\mathbf{u}\cdot\mathbf{s}_2\mathbf{uv})V_2(\mathbf{f})\mathbf{u}^2 = (\mathbf{s}_1\mathbf{u}\cdot\mathbf{s}_1\mathbf{u})V_3(\mathbf{f})\mathbf{u}^2\mathbf{v},$$

as required, since $\mathbf{s}_1\mathbf{u}\cdot\mathbf{s}_1\mathbf{u} = 1$. Clearly in this form the equation holds without imposing the normalisation of $\mathbf{u}$ or $\mathbf{v}$. □

With this form of the condition for subparabolicity, free from the necessity of defining what is meant by differentiation in the direction $\mathbf{u}$, we are able to elucidate the geometry of subparabolic lines of the surface $\mathbf{s}$ in the neighbourhood of an umbilic.

Proposition 13.11 (Wilkinson (1991) and Morris (1990)) *Let* $\mathbf{s}$ *be a regular smooth surface in* $\mathbb{R}^3$ *with an ordinary umbilic at the origin. Then, in the case that there are three directions for lines of curvature at the umbilic, there are three subparabolic lines passing smoothly through the umbilic in the same three directions, while, in the case that there is just one direction for lines of curvature through the umbilic, there is just one subparabolic line passing smoothly through the umbilic in that same direction.*

Proof Our proof takes us into nine-dimensional space! Consider the map

$$\mathbb{R}^3 \times \mathbb{R}^2 \times \mathbb{R}^2 \times \mathbb{R}^2 \rightarrowtail \mathrm{L}(\mathbb{R}^2, \mathbb{R}) \times \mathrm{L}(\mathbb{R}^2, \mathbb{R}) \times \mathbb{R} \times \mathbb{R} \times \mathbb{R} \times \mathbb{R}$$

$(\mathbf{c}, w, \mathbf{u}, \mathbf{v}) \mapsto$

$$(V(\mathbf{c})_1, V(\mathbf{c})_2\mathbf{v}, \mathrm{I}_2\mathbf{u}^2, \mathrm{I}_2\mathbf{uv}, \mathrm{I}_2\mathbf{v}^2, 2(\mathbf{s}_1\mathbf{u} \cdot \mathbf{s}_2\mathbf{uv})V(\mathbf{c})_2\mathbf{u}^2 - V(\mathbf{c})_3\mathbf{u}^2\mathbf{v}).$$

The source vector space has dimension 9 and the target vector space has dimension 8. So the differential has everywhere kernel rank at least equal to 1. What we prove is that at a point of the fibre over $(0, 0, 1, 0, 1, 0)$ it has kernel rank equal to 1, the second component of the kernel being a multiple of the vector $\mathbf{u}$. Where such a point is an umbilic $V(\mathbf{c})_2 = 0$, implying that there also $V(\mathbf{c})_3\mathbf{u}^2\mathbf{v} = 0$.

Two of the equations satisfied by a kernel vector $(\mathbf{c}_1, w_1, \mathbf{u}_1, \mathbf{v}_1)$ of the differential are

$$\mathbf{c}_1 \cdot \mathbf{s}_1 + V(\mathbf{c})_1 w_1 = 0 \text{ and } \mathbf{c}_1 \cdot \mathbf{s}_2\mathbf{v} + V(\mathbf{c})_3\mathbf{v}w_1 + V(\mathbf{c})_2\mathbf{v}_1 = 0.$$

At an umbilic $V(\mathbf{c})_2 = 0$ and it then follows from these equations that $V_3(\mathbf{c})\mathbf{uv}w_1 = 0$. If the umbilic is ordinary this then determines w_1 up to a multiple, since $V_3(\mathbf{c})\mathbf{uv} \neq 0$. However, since $V_3(\mathbf{c})\mathbf{u}^2\mathbf{v} = 0$ it follows that w_1 is a multiple of $\mathbf{u}$. This last equation is, by Proposition 12.4, the equation for possible tangent vectors to smooth lines of curvature through the umbilic.

Accordingly on the model of the focal surface in $\mathbb{R}^9$ determined by the first five components of the map there is, for each ordinary umbilical centre (c, w) of the surface $\mathbf{s}$, and for each vector $\mathbf{u}$ satisfying the equation $V_3(\mathbf{c})\mathbf{u}^2\mathbf{v} = 0$, a smooth curve passing through the point $(\mathbf{c}, w, \mathbf{u}, \mathbf{v})$, the second component of a tangent vector to this line at the umbilic being a multiple of $\mathbf{u}$. Such curves project to subparabolic lines on the surface $\mathbf{s}$ having the required properties at ordinary umbilics of $\mathbf{s}$. □

Computer pictures made by Morris (Figures 13.2–13.4) show what happens to the local configuration of subparabolic lines before, at and after the birth of a pair of umbilics. After the birth there are of course three subparabolic lines through each umbilic and, not surprisingly, one of these is a sort of cord passing through both two umbilics. Accordingly (no puns intended!) at the moment of birth there are not three but *five* subparabolic lines, two in the directions of the lines of curvature, one in the direction of the unique ridge through the point (the black line in the figures) and two others mutually tangent to one another and transversal to the others. The simplest genesis of the configuration of subparabolic lines at the birth of umbilics, starting from a featureless surface patch, is not known.

Morris has remarked, as have Bruce and Wilkinson (1991; see also Bruce (1992)), that in general an ordinary cuspidal edge on an otherwise regular surface in $\mathbb{R}^3$ has points of opposite Gaussian curvature lying on either side of the edge. Where this is *not* the case, that is where the edge is planar, it should be regarded as being both cuspidal and part of the parabolic line. One reason for doing this is that in the case that such a surface is one sheet $\mathbf{e}$ (the blue one, say) of the focal surface of a regular smooth surface $\mathbf{s}$ in $\mathbb{R}^3$ then a (blue) ridge on the surface $\mathbf{s}$ that is a line of symmetry of the surface, and whose rib on $\mathbf{e}$ is

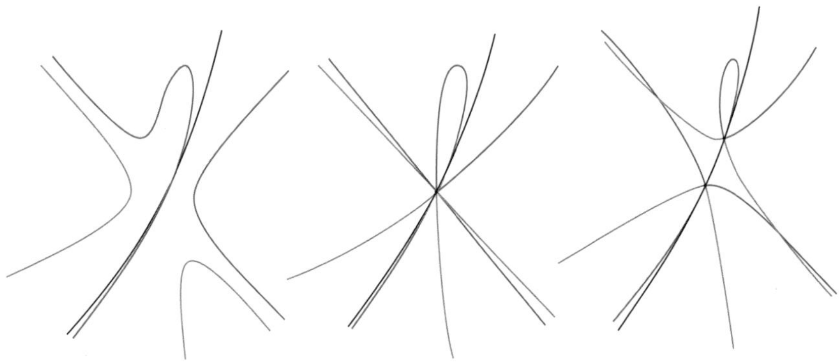

Figure 13.2 Figure 13.3 Figure 13.4

therefore everywhere planar, is also an everywhere geodesic (red) line of curvature and so should be regarded as a particular kind of (blue) subparabolic line.

Cuspidal edges of various types and interrelations between subparabolic lines and ridges are further discussed in Chapter 14. Surfaces with planes of symmetry are further discussed in Chapter 16.

13.5 Uses for inversion

Inversion of $\mathbb{R}^3$ with respect to a sphere in $\mathbb{R}^3$ is a smooth transformation of the space that maps spheres or planes to spheres or planes (cf. Section 2.6). The restriction of such an inversion to a regular surface in $\mathbb{R}^3$ preserves all the features of the surface that are defined in terms of contact with spheres or planes, in particular lines of curvature, ridges and their turning points, and umbilics. This can be used in either direction, either to allow us to avoid talking about the parabolic line at all or to justify the assumption that, without loss of generality, the point of the surface under study is a point of the parabolic line.

Suppose for example that we wish to describe the curve of intersection of a surface with a sphere whose centre lies on some normal to the surface, and which passes through the base of the normal. By inversion of $\mathbb{R}^3$ with respect to a point of the sphere away from the surface we can without loss of generality assume that the centre of the sphere is the point at infinity of the normal and that the sphere is the tangent plane to the surface at the base of the normal. Locally, taking the principal directions there as the x- and y-axes, the surface is the graph of a smooth function of the form $f : \mathbb{R}^2 \rightarrowtail \mathbb{R}; (x, y) \mapsto \frac{1}{2}\kappa x^2 + \frac{1}{2}\lambda y^2 + \frac{1}{6}(ax^3 + 3x^2y + 3xy^2 + y^3) + O(x, y)^4$, with the z-axis as normal to the surface at the origin, the curve of intersection of the surface with the plane being the zero set of the function. Where the Gaussian curvature $\kappa\lambda$ of the surface is positive, and both focal points on the normal lie on the same side of the surface, the point at infinity is an elliptic A_1 point of the normal and the curve of intersection of surface and plane has an acnode (an isolated node) at the origin. On the other hand where the Gaussian curvature is negative and the focal points on the normal lie on opposite sides of the surface, the point at infinity is a hyperbolic A_1 point of the normal and the curve of intersection of surface and plane has a crunode (a crossing node) at the origin, the tangents to the two branches there being the asymptotic directions to the surface at that point. The transitional case is where the Gaussian curvature is zero, one of the principal curvatures, say κ being equal to zero. Suppose then that the other principal curvature λ is non-zero, that is that the origin is not a flat umbilic of the surface. Then, where the origin is an A_2 parabolic point, that is not a cusp of Gauss, the curve of intersection of surface and plane has an ordinary cusp at the origin. By contrast at an A_3 parabolic point, a generic cusp of Gauss, where $a = 0$, but the coefficient, α say, of x^4 is non-zero, the significant terms of f near the origin are $\frac{1}{2}\lambda y^2 + \frac{1}{2}x^2 y + \alpha x^4$. At a hyperbolic A_3 parabolic point this is a hyperbolic quadratic form in y and x^2 and the curve of intersection of surface and plane resembles a pair of parabolas through the origin with common tangent there, while at an elliptic A_3 parabolic point this is an elliptic quadratic form in y and x^2 and the curve of intersection of surface and plane has an isolated singularity at the origin. In the transitional ordinary A_4 case, where the quadratic form is parabolic, the curve of intersection of surface and plane has an ordinary rhamphoid cusp at the origin.

Before the initial inversion the tangent plane was a sphere with centre on the normal. The various types of curve of intersection of this sphere with the surface are similar to those noted in the above discussion.

The above account is just an outline of what can occur. We return to these matters in Chapter 15.

Exercises

13.1 Let $\mathbf{s} : \mathbb{R}^2 \rightarrowtail \mathbb{R}^3$ be a regular surface, with Gaussian curvature K. Prove that $K = \det \mathrm{II}/\det \mathrm{I}$.

13.2 Let $\mathbf{r} : \mathbb{R} \rightarrowtail \mathbb{R}^3$ be a regular nowhere linear space curve and $\mathbf{m} : \mathbb{R} \rightarrowtail \mathbb{R}^3$ a regular curve on the unit sphere, that together define a *ruled surface* $\mathbf{s} : \mathbb{R}^2 \rightarrowtail \mathbb{R}^3$: $(t, u) \mapsto \mathbf{r}_1(t) + u\mathbf{m}(t)$. Verify that at regular points of $\mathbf{s}$, with unit normal $\mathbf{n}$, the Gaussian curvature K is given by the formula

$$K = \frac{(\mathbf{n} \cdot \mathbf{m}_1)^2}{\mathbf{r}_1 \cdot \mathbf{r}_1 + 2u\mathbf{r}_1 \cdot \mathbf{m}_1 + u^2 \mathbf{m}_1 \cdot \mathbf{m}_1 - (\mathbf{r}_1 \cdot \mathbf{m})^2}.$$

Prove that there is unique regular curve $t \mapsto \mathbf{q}(t) = \mathbf{r}(t) + \mathbf{u}(t)\mathbf{m}(t)$ on $\mathbf{s}$ such that everywhere $\mathbf{q}_1(t) \cdot \mathbf{m}_1(t) = 0$. This is the *line of striction* of $\mathbf{s}$. Verify that

(i) any point at which $\mathbf{s}$ fails to be regular lies on the line of striction;

(ii) at each point of the line of striction that is a regular point of $\mathbf{s}$ the restriction of the Gaussian curvature to the generator of the ruled surface is critical.

13.3 Show by an example that the line of striction of a ruled surface need not be a ridge of the surface.

13.4 Let $\mathbf{r} : \mathbb{R} \rightarrowtail \mathbb{R}^2$ be a regular smooth curve in the right half-plane $\{(r, z) \in \mathbb{R}^2 : r > 0\}$, and let $\mathbf{s}$ be the *surface of revolution*

$$\mathbf{s} : \mathbb{R}^2 \rightarrowtail \mathbb{R}^3;\ (t, \theta) \mapsto (r(t)\cos\theta,\ r(t)\sin\theta,\ z(t))$$

obtained by spinning the curve $\mathbf{r}$ about the z-axis. Prove that $\mathbf{s}$ is regular. Prove also that the parabolic lines of $\mathbf{s}$ are all *parallels of latitude* of the surface, each such parallel being swept out either by a point of inflection of $\mathbf{r}$ or by a point at which the tangent to $\mathbf{r}$ is perpendicular to the axis of rotation.

13.5 Show that the parabolic curve of the surface

$$(x, y) \mapsto (x, y, y^2 + x^3 + 3x^2y)$$

is represented in the (x, y)-plane by a parabola through the origin.

13.6 Find the parabolic curve of the tube $\mathbf{S}$ of radius δ and core $\mathbf{r}$ discussed in Exercise 10.7.

13.7 Let $\mathbf{a}_1$ be a principal vector at w to a regular smooth surface $\mathbf{s}$, and let $\mathbf{n}$ be the Gauss map of $\mathbf{s}$. Suppose also that, at w, $\mathbf{n}_1\mathbf{a}_1 = 0$. Prove that, at w, $\mathbf{n}_2{\mathbf{a}_1}^2 + \mathbf{n}_1\mathbf{a}_2 = 0$ for some $\mathbf{a}_2 \in \mathbb{R}^2$ if and only if $\mathbf{n} \cdot \mathbf{s}_3{\mathbf{a}_1}^2 + \mathbf{n} \cdot \mathbf{s}_2\mathbf{a}_2 = 0$ for some $\mathbf{a}_2 \in \mathbb{R}^2$ if and only if $\mathbf{n} \cdot \mathbf{s}_3{\mathbf{a}_1}^3 = 0$.

13.8 The Gauss map $\mathbf{n}$ will have a Whitney cusp at w if and only if not only $\mathbf{n}_1\mathbf{a}_1 = 0$ and $\mathbf{n}_2{\mathbf{a}_1}^2 + \mathbf{n}_1\mathbf{a}_2 = 0$ for some $\mathbf{a}_2 \in \mathbb{R}^2$ but also $\mathbf{n}_3{\mathbf{a}_1}^3 + 3\mathbf{n}_2\mathbf{a}_1\mathbf{a}_2 + \mathbf{n}_1\mathbf{a}_3$ is not equal to 0 for any $\mathbf{a}_3 \in \mathbb{R}^3$. Find the alternative expression for this last condition suggested by the previous exercise.

13.9 Let $\mathbf{r} : \mathbb{R} \rightarrowtail \mathbb{R}^2$ represent an asymptotic line on a regular smooth surface $\mathbf{s} : \mathbb{R}^2 \rightarrowtail \mathbb{R}^3$. Prove that where the asymptotic line meets the parabolic line at an ordinary (A_2) parabolic point it has an ordinary cusp. (Hint: Differentiate the equation $\mathbf{n} \cdot \mathbf{s}_2{\mathbf{r}_1}^2 = 0$ four times along $\mathbf{r}$ to deduce that where $\mathbf{n} \cdot \mathbf{s}_2\mathbf{r}_1 = 0$ then $\mathbf{r}_1 = 0$ and $\mathbf{n} \cdot \mathbf{s}_2{\mathbf{r}_2}^2 = 0$ and where also $\mathbf{r}_2 \neq 0$ then $\mathbf{r}_3$ is not a multiple of $\mathbf{r}_2$.)

13.10 Let $\mathbf{s}$ be a non-singular point of the surface given by the equation $F(\mathbf{s}) = 0$, where $F : \mathbb{R}^3 \rightarrowtail \mathbb{R}$ is smooth, suppose that $\mathbf{s}$

is not an umbilic, and let $\boldsymbol{\alpha}$ and $\boldsymbol{\beta}$ be principal tangent vectors to the surface at $\mathbf{s}$. Prove that if $\mathbf{s}$ lies on the parabolic line of the surface then either $F_2(\mathbf{s})\boldsymbol{\alpha}^2 = 0$ or $F_2(\mathbf{s})\boldsymbol{\beta}^2 = 0$, and conversely.

(Since $F_2(\mathbf{s})\boldsymbol{\alpha\beta} = 0$, $F_2(\mathbf{s})\boldsymbol{\alpha}^2 = 0 \Leftrightarrow F_2(\mathbf{s})\boldsymbol{\alpha\gamma} = 0$ for $\boldsymbol{\alpha}$ a principal tangent vector at $\mathbf{s}$, and $\boldsymbol{\gamma}$ any tangent vector at $\mathbf{s}$.)

13.11 Let $\mathbf{s}$ be a non-singular point of the surface given by the equation $F(\mathbf{s}) = 0$, where $F : \mathbb{R}^3 \rightarrowtail \mathbb{R}$ is smooth, suppose that $\mathbf{s}$ is not an umbilic, and let $\boldsymbol{\alpha}$ and $\boldsymbol{\beta}$ be mutually orthogonal principal tangent vectors and $\mathbf{n}$ a normal vector to the surface at $\mathbf{s}$. Prove that if $\mathbf{s}$ lies on a subparabolic line of the surface then either

$$(F_1(\mathbf{s})\mathbf{n})(F_3(\mathbf{s})\boldsymbol{\alpha}^2\boldsymbol{\beta}) - (F_2(\mathbf{s})\boldsymbol{\beta}\mathbf{n})(F_2(\mathbf{s})\boldsymbol{\alpha}^2) = 0$$

or

$$(F_1(\mathbf{s})\mathbf{n})(F_3(\mathbf{s})\boldsymbol{\beta}^2\boldsymbol{\alpha}) - (F_2(\mathbf{s})\boldsymbol{\alpha}\mathbf{n})(F_2(\mathbf{s})\boldsymbol{\beta}^2) = 0.$$

(This is to be compared with the result of Exercise 11.8, bearing in mind that $F_2(\mathbf{s})\boldsymbol{\alpha\beta} = 0$.)

14

Involutes of geodesic foliations

14.0 Introduction

In the case of a plane or space curve we saw in earlier chapters how to recover the original curve and its parallels from the evolute, in the former case by the Huygens process, and in the latter case by a natural analogue of this process. We discuss here the analogous process for surfaces.

Given a regular surface **s** in $\mathbb{R}^3$ we limit ourselves to considering the simple case where **s** is free of umbilics or parabolic points. Then the two sheets of the focal surface can be considered separately. Either sheet is foliated by focal curves of **s**, each of these being, by Propositions 10.7 and 10.5, a geodesic of the focal surface without any linear points. Moreover, by Proposition 10.8, the orthogonal trajectories of these focal curves are the level curves or fibres of the relevant curvature function on **s** lifted to the relevant sheet of the focal surface.

Let $\mathbf{e} : \mathbb{R}^2 \rightarrowtail \mathbb{R}^3$ be one of the sheets of the focal surface of **s** presented parametrically, and suppose that it is regular without cuspidal edges. It is, as we have just remarked, the union of a one-parameter family of disjoint geodesics. Now one can roll a straight line ruler in Huygens style on each of these geodesics separately so that at each moment it coincides with a tangent line to the geodesic on which it is rolling. To coordinate these rolling rulers we choose an orthogonal trajectory of the geodesics as a start-line. Then the points of the rulers that are at a fixed distance, δ say, from the start-line will trace out one family of lines of curvature of a surface in $\mathbb{R}^3$, any two such surfaces being parallel, one of them being the original surface **s**.

The geodesic foliation of **e** determines not only **s** and its parallels but also as a consequence the other sheet **f** of the focal surface of **s** complete with its geodesic foliation. The classic readily available account is that of Weatherburn (1927). The interest in what follows, much of which is of very recent date, lies in relationships between these two foliations when one of them violates the conditions imposed on them by the requirement that the original surface **s** is regular. Explicitly we study the involute surfaces of a geodesic foliation on a regular surface **e**, one geodesic of which has an ordinary linear point. Each neighbouring geodesic then necessarily also has an ordinary linear point. The tangent line to a geodesic at such a point necessarily must be a root line of the second fundamental form, that is either an asymptotic (hyperbolic) or parabolic tangent line, the Gauss curvature there necessarily being either respectively negative or zero. In either case each involute surface of the foliation has a rhamphoid cuspidal edge, the lines of curvature of either family in the hyperbolic case having ordinary cusps at the edge. In the parabolic case the line of curvature of the involute swept out by a ruler rolling on that geodesic has a rhamphoid cusp on the edge and through that point there not only passes a subparabolic line of the same colour as the line of curvature but also a ridge of the surface and a subparabolic line of the opposite colour. The conjugate sheet of the focal surface then has a cuspidal edge with pinch point. The parallel to the surface for which the other radius of curvature is zero has two cuspidal edges that cross, one rhamphoid having an ordinary cusp at the singularity and the other ordinary with a rhamphoid cusp at the singularity. The entire configuration is associated to the group H_4 of isometries of the hypericosahedron, a polyhedron in $\mathbb{R}^4$, and has been studied both by O.P. Shcherbak ((1988), but announced in (1984)) and by Alex Flegmann (1985).

14.1 Cuspidal edges

In what follows we shall encounter several cuspidal edges of surfaces and it is important to be quite clear in particular about the difference between an ordinary cuspidal edge and a rhamphoid cuspidal edge. There are one or two subtle points that have to be attended to. Proofs that require sophisticated results from singularity theory are omitted. A detailed study of cuspidal edges has been undertaken by David Mond ((1985a, b) and personal communication).

A smooth surface $\mathbf{s} : \mathbb{R}^2 \rightarrowtail \mathbb{R}^3$ has a *cuspidal edge* at $w = (u, v)$ if there is a non-singular smooth curve passing through w such that on the curve the differential of $\mathbf{e}$ has rank 1 but elsewhere has rank 2. The restriction of $\mathbf{s}$ to this curve is the *edge*.

Suppose first that the edge itself is a regular space curve. Then the edge is said to be *ordinary* or 3/2 at w if there is a smooth foliation of the domain of $\mathbf{s}$ such that the leaf $\mathbf{a} : \mathbb{R} \rightarrowtail \mathbb{R}^2$ of the foliation passing through $w = \mathbf{a}(t)$, and therefore also every nearby leaf, is transversal to the domain of the edge, and $\mathbf{sa}$ has an ordinary cusp at w, with the second and third derivatives of $\mathbf{sa}$ at t not coplanar with the one-dimensional image of the first derivative of $\mathbf{s}$ at w.

Example 14.1 The curve $v \mapsto (0, v) \mapsto (0, 0, v)$ is an ordinary cuspidal edge of the surface $(u, v) \mapsto (u^2, u^3, v)$. □

It may be proved that any ordinary cuspidal edge is $\mathcal{A}$-equivalent (locally diffeomorphic) to this example.

Proposition 14.2 Let $\mathbf{s}$ *be a smooth surface with an ordinary cuspidal edge. Then no smooth curve crossing the edge has a rhamphoid cusp at the edge. Nor does such a curve have a rhamphoid cusp* modulo the image of $\mathbf{s}_1$, *that is if the second derivative of the curve does not lie along the edge then the second and third derivatives of the curve and the image of* $\mathbf{s}_1$ *at the point in question are linearly independent.*

Proof We can work on the model surface. Suppose that $(u, v) = \phi(t)$ where $\phi : \mathbb{R} \rightarrowtail \mathbb{R}^2$ is smooth. Then the first three derivatives of the curve $t \mapsto (u^2, u^3, v)$ are $(2uu_1, 3u^2u_1, v_1)$, $(2u_1{}^2 + 2uu_2, 6uu_1{}^2 + 3u^2u_2, v_2)$ and $(6u_1u_2 + 2uu_3, 6u_1{}^3 + 18uu_1u_2 + 3u^2u_3, v_3)$, which, if both $u = 0$ and $v_1 = 0$, reduce to 0, $(2u_1{}^2, 0, v_2)$ and $(6u_1u_2, 6u_1{}^3, v_3)$. Clearly, if the second is non-zero modulo the vector $(0, 0, 1)$, then the second, the third and $(0, 0, 1)$ are linearly independent, this last vector spanning the image of the first derivative of the model. The assertions follow at once. □

In particular the intersection of such a surface with a plane in $\mathbb{R}^3$ is a curve in the plane with an ordinary cusp where the curve of intersection crosses the edge.

Proposition 14.3 A regular nowhere linear or planar space curve is an ordinary cuspidal edge of its tangent developable. □

Proposition 14.4 An A_3 rib of a regular surface **s** *is an ordinary cuspidal edge of the focal surface.* □

The standard model for a cuspidal edge of Example 14.1 and the example of Proposition 14.2 are both misleading to the extent that in either case away from the edge the surface has Gaussian curvature zero. In general, as one approaches an ordinary cuspidal edge one of the principal curvatures approaches infinity while the other remains finite and non-zero, so that the Gaussian curvature at the edge is infinite. The generic situation is that in which the surface has positive Gaussian curvature on one side of the edge and negative Gaussian curvature on the other side (Figure 14.1). In the case that the edge is a line of symmetry of the surface the generic situation is that the Gaussian curvature is non-zero on either side of the edge, being either positive on both sides (Figure 14.2) or negative on both sides (Figure 14.3).

A surface **s** may have a cuspidal edge that is ordinary except at

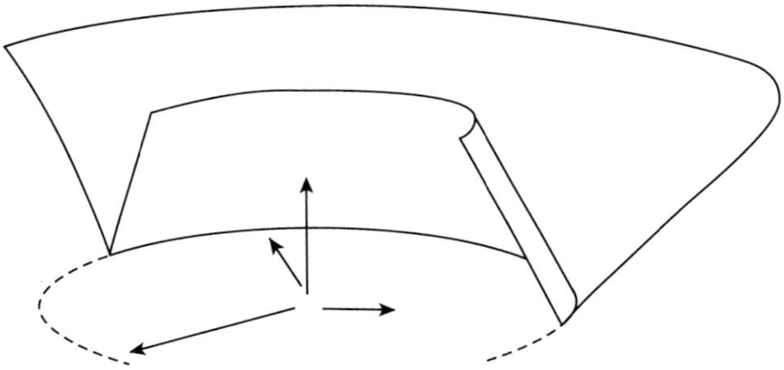

Figure 14.1

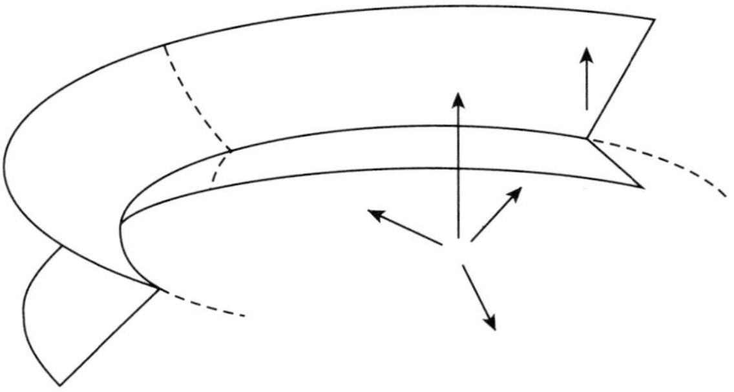

Figure 14.2

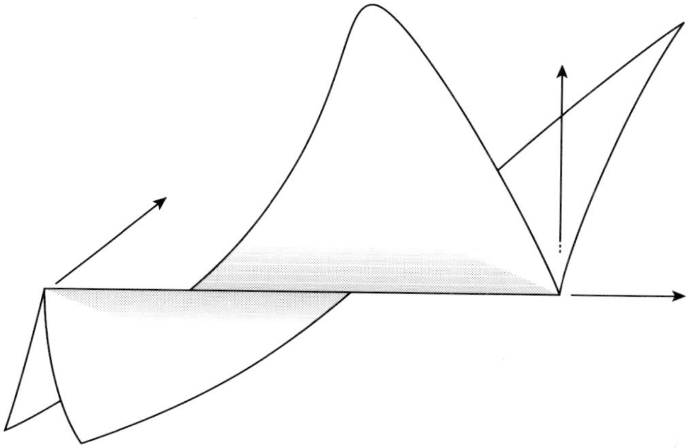

Figure 14.3

isolated points, which will happen if the edge fails to be regular at a point. This is the case with a *swallow-tail cusp* on a cuspidal edge, such as we have already encountered in various places (Figure 14.4). Such a cuspidal edge is characterised by having an ordinary cusp at a point w, there being a smooth foliation of the domain of $\mathbf{s}$ such that the leaf $\mathbf{a} : \mathbb{R} \rightarrowtail \mathbb{R}^2$ of the foliation passing through $w = \mathbf{a}(t)$ is simply tangent there to the edge and has there an ordinary kink where not only $(\mathbf{sa})_1 = 0$ and $(\mathbf{sa})_2 = 0$ but also $(\mathbf{sa})_3(t)$ and $(\mathbf{sa})_4(t)$ are not coplanar with the image of $\mathbf{s}_1(w)$.

It may be proved that any two swallow-tails are $\mathcal{A}$-equivalent.

Proposition 14.5 An ordinary A_4 point of the focal surface of a regular surface $\mathbf{s}$ *is a swallow-tail point of the focal surface.* □

A *regular* cuspidal edge may also fail to be ordinary in various ways. For example it may fail to be ordinary at isolated points, for example by having a single leaf of a smooth foliation transverse to the edge having a rhamphoid (5/2) cusp instead of an ordinary one. Such a point is called a *cuspidal pinch-point* of the surface.

Proposition 14.6 The tangent developable of a regular nowhere linear space curve has a cuspidal pinch-point at each ordinary planar point.

This is the result of Cleave (1980) that we have already encountered in Figure 6.2 and Exercise 6.18. □

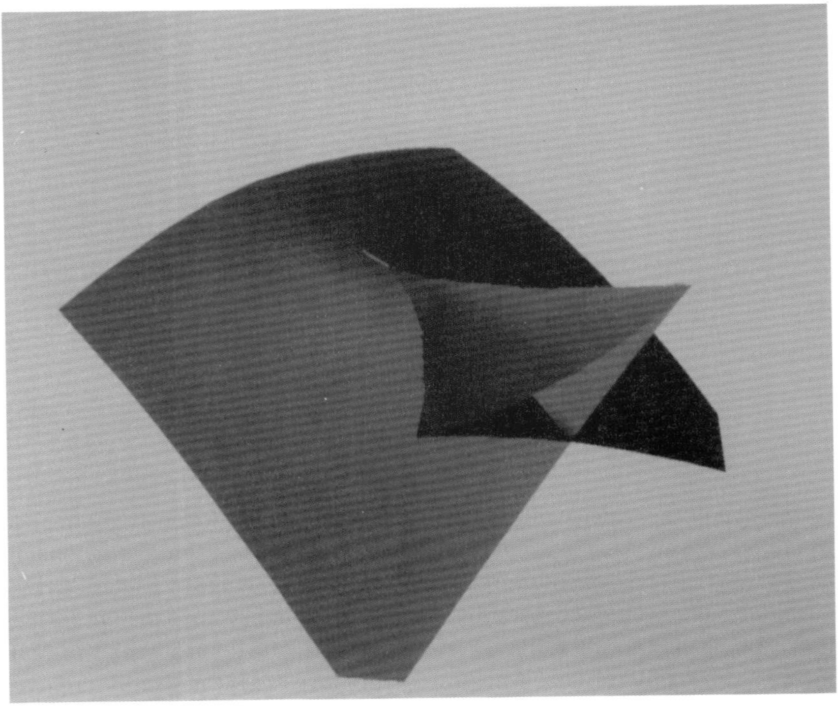

Figure 14.4

Example 14.7 (The cuspidal cross-cap.) The surface $(u, v) \rightarrowtail (u^2, u^3 v, v)$ has a cuspidal edge $v \rightarrowtail (0, 0, v)$ that is ordinary except at the origin. The restriction of $\mathbf{s}$ to a generic regular smooth curve through the origin in $\mathbb{R}^2$ is a curve with a 5/2 cusp at the origin. □

It may be proved that any smooth surface-germ with a cuspidal pinch-point is $\mathcal{A}$-equivalent to this example.

Both the example of Proposition 14.6 and the model of Example 14.7 are misleading in the same sense that the tangent developable of a nowhere planar space curve and the standard model of an ordinary cuspidal edge are misleading, the Gaussian curvature of either away from the edge being everywhere zero. We remarked earlier that generically an ordinary cuspidal edge separates points of positive Gaussian curvature from points of negative Gaussian curvature. Now the parabolic line of a surface, consisting of points of the surface at which the Gaussian curvature is zero, also has this property and in general there are points of an otherwise ordinary edge where it is crossed by the parabolic line, the edge and the parabolic line being mutually tangent at such a crossing, but with the representative curves in the parameter space being mutually transverse. The next proposition, due to Richard Morris, whose pictures illustrate this chapter, shows that this occurs in general at a cuspidal pinch-point (Figure 14.5).

Proposition 14.8 *Through a cuspidal pinch-point of a smooth surface with cuspidal edge there passes, in general, a parabolic line, the edge and the parabolic line being mutually tangent at such a crossing, but with the representative curves in the parameter space being mutually transverse.* □

In the example depicted in two separate views in Figures 14.5(a) and 14.5(b) most of the surface has negative Gaussian curvature, there being a narrow strip of positive Gaussian curvature between the cuspidal edge and the parabolic line.

A regular cuspidal edge of a smooth surface $\mathbf{s}$ is said to be *rhamphoid* or 5/2 at w if there is a smooth foliation of the domain of $\mathbf{s}$ such that the leaf $\mathbf{a} : \mathbb{R} \rightarrowtail \mathbb{R}^2$ of the foliation passing through $w = \mathbf{a}(t)$ is transversal to the domain of the edge, and $\mathbf{sa}$ has a cusp at w that may be ordinary and is at worst 5/2, but whose second and third derivatives at t are coplanar with the image of $\mathbf{s}_1(w)$, the same being true of every nearby leaf.

Example 14.9 The edge $v \mapsto (0, 0, v)$ of the surface $(u, v) \mapsto (u^2, u^5, v)$ is a 5/2 or rhamphoid cuspidal edge. □

Any 5/2 cuspidal edge is $\mathcal{A}$-equivalent to this example.

We must emphasise that a curve that crosses a rhamphoid edge on a surface $\mathbf{s}$ may have an *ordinary* cusp there. But the cusp must be rhamphoid modulo the image of $\mathbf{s}_1$, that is the second and third derivatives of the curve at the cusp must be coplanar with the image of $\mathbf{s}_1$ there.

A cuspidal edge may be 5/2 except at isolated points:

Example 14.10 (The 7/2 cross-cap.) The surface $(u, v) \rightarrowtail (u^2, u^5 v, v)$ has a cuspidal edge $v \rightarrowtail (0, 0, v)$ that is 5/2 except at the origin. The

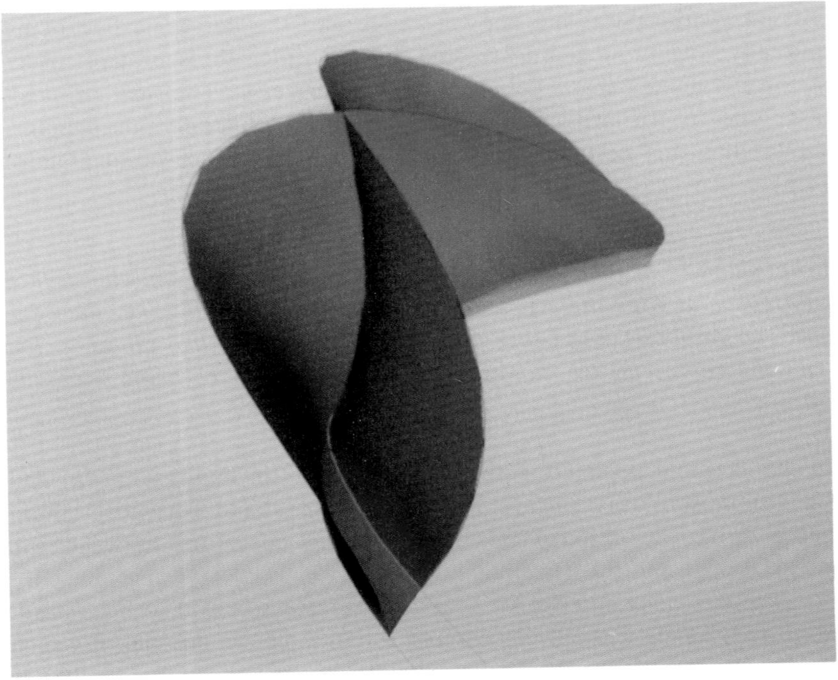

Figure 14.5(a)

restriction of **s** to a generic regular smooth curve through the origin is a curve with a 7/2 cusp there. □

Dare we say 'and so on'?

14.2 The involutes of a geodesic foliation

Suppose that **e** is a regular surface foliated by geodesics. That is the surface is the union of a one-parameter family of disjoint geodesics. We repeat what we outlined in the introduction to the chapter. One can roll a straight line ruler in Huygens style on each geodesic separately so that at each moment it coincides with a tangent line to the geodesic on which it is rolling. To coordinate these rolling rulers we choose an orthogonal trajectory of the geodesics as a start-line. Then the points of the rulers that are at a fixed distance, δ say, from the start-line will trace out one family of lines of curvature of a surface **s**, any two such surfaces being parallel. The following propositions are

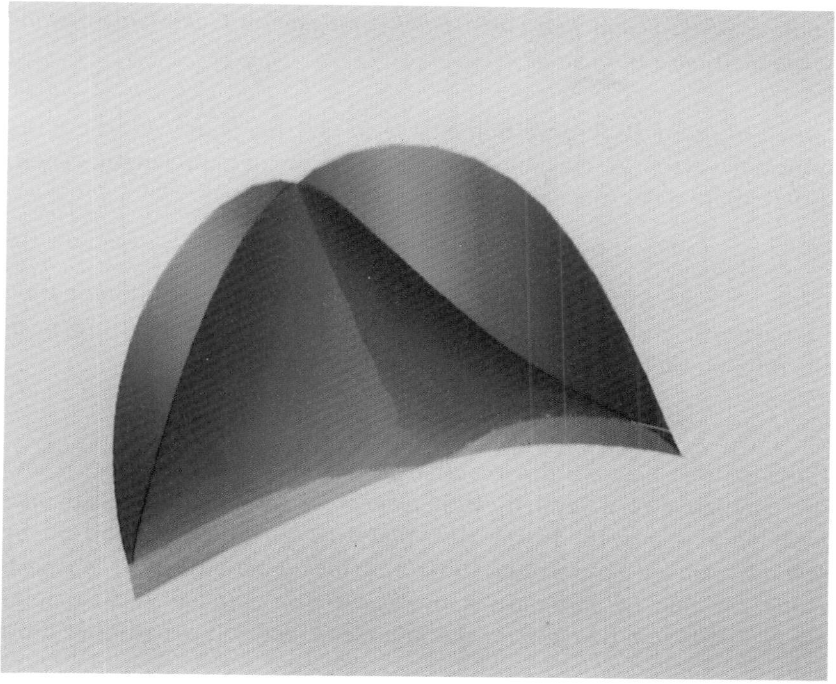

Figure 14.5(b)

concerned with the regularity or otherwise of such a surface **s** and of its two focal surfaces, one of which is the surface **e** from which we started.

The key to the geometry is the following proposition.

Proposition 14.11 *Let* **s** *be a regular smooth surface free of umbilic or parabolic points, with focal surfaces* **e** *and* **f**, *and principal vectors* $\mathbf{s}_1\mathbf{a}_1$ *and* $\mathbf{s}_1\mathbf{b}_1$ *at some point* w, $\mathbf{a} : \mathbb{R}^2 \rightarrowtail \mathbb{R}$ *and* $\mathbf{b} : \mathbb{R}^2 \rightarrowtail \mathbb{R}$ *representing the lines of curvature through* w. *Then the vectors* $\mathbf{e}_1\mathbf{a}_1$ *and* $\mathbf{e}_1\mathbf{b}_1$ *are* conjugate *on* **e** *at* w *with respect to the second fundamental form on* **e** *there; that is* $\mathbf{s}_1\mathbf{a}_1 \cdot \mathbf{e}_2\mathbf{a}_1\mathbf{b}_1 = 0$, $\mathbf{s}_1\mathbf{a}_1 = (\mathbf{sa})_1$ *being normal to* **e** *at* w. *Moreover, if* **s** *is not everywhere regular, but* **e** *is, then at a point of non-regularity of* **s** *where* $(\mathbf{sa})_1 = 0$ *but* $(\mathbf{sa})_2 \neq 0$ *the same is true of* $\mathbf{e}_1\mathbf{a}_1$ *and* $\mathbf{e}_1\mathbf{b}_1$; *that is* $(\mathbf{sa})_2 \cdot \mathbf{e}_2\mathbf{a}_1\mathbf{b}_1 = 0$, *since* $(\mathbf{sa})_2$ *now is normal to* **e**.

In the particular case that the focal line of curvature **ea** *is linear at* w *the tangent vector* $(\mathbf{ea})_1$ *is self-conjugate on* **e** *and therefore* $(\mathbf{eb})_1$ *is a real multiple of* $(\mathbf{ea})_1$. *If* **e** *is regular* $\mathbf{b}_1$ *is a real multiple of* $\mathbf{a}_1$, *both being mapped to 0 by* $\mathbf{s}_1$. *It is then also the case that* $(\mathbf{fb})_1$ *is a real* *multiple of* $(\mathbf{fa})_1$ *and that* $(\mathbf{fb})_1$ *is self-conjugate on* $\mathbf{f}$, *implying that the geodesic* $\mathbf{fb}$ *also is linear at* w.

Sketch of proof By Proposition 10.4, $\mathbf{e}_1\mathbf{a}_1 \cdot \mathbf{s}_1 = \mathbf{e}_1 \cdot \mathbf{s}_1\mathbf{a}_1 = 0$, where the principal vector $\mathbf{a}_1$ may be normalised to be of unit length. Then, differentiating the first form of Proposition 10.4,

$$\mathbf{e}_2\mathbf{a}_1 \cdot \mathbf{s}_1 + \mathbf{e}_1(\mathbf{a}_1)_1 \cdot \mathbf{s}_1 + \mathbf{e}_1\mathbf{a}_1 \cdot \mathbf{s}_2 = 0.$$

(Note that we write $(\mathbf{a}_1)_1$ here and not $\mathbf{a}_2$, since we are differentiating here over $\mathbb{R}^2$ and not along the line of curvature $\mathbf{a}$.) Now putting $\mathbf{a}_1$ in the original vacant slot and $\mathbf{b}_1$ in the new one we have

$$\mathbf{e}_2\mathbf{a}_1\mathbf{b}_1 \cdot \mathbf{s}_1\mathbf{a}_1 + \mathbf{e}_1\mathbf{a}_1 \cdot \mathbf{s}_2\mathbf{a}_1\mathbf{b}_1 = \mathbf{e}_2\mathbf{a}_1\mathbf{b}_1 \cdot \mathbf{s}_1\mathbf{a}_1 = 0.$$

Provided that $\mathbf{s}_1\mathbf{a}_1 \neq 0$ this vector is normal to the focal surface $\mathbf{e}$ and the conjugacy of $\mathbf{a}_1$ and $\mathbf{b}_1$ with respect to the second fundamental form on $\mathbf{e}$ is proved. In the case that $\mathbf{s}_1\mathbf{a}_1 = 0$ differentiation of the second form of Proposition 10.04 and the equation $\mathbf{e}_2\mathbf{a}_1\mathbf{b}_1 \cdot \mathbf{s}_1\mathbf{a}_1 = 0$ along the line of curvature $\mathbf{a}$ proves at once that at such a point $\mathbf{e}_2\mathbf{a}_1\mathbf{b}_1 \cdot (\mathbf{sa})_2 = 0$. The rest of the argument then follows without difficulty. □

If a focal curve on $\mathbf{e}$ is linear at w then any focal curve close to $\mathbf{e}$ has a linear point close to w, for the curvature of any geodesic on a surface can be assigned a sign, since the principal curvature vector always points along the surface normal. Then there is a curve on the surface in general across which the geodesics have zero curvature. As we shall presently verify, in such a case $\mathbf{s}$ in general has a rhamphoid edge at w. At isolated points along this edge we would expect the edge to be of type 7/2 rather than 5/2. Of greater interest are points of the edge at which the principal curvature σ along the line of curvature $\mathbf{b}$ has an ordinary critical point, implying that a ridge on $\mathbf{s}$ intersects the rhamphoid edge there and that the focal curve $\mathbf{fb}$ has an ordinary rhamphoid cusp there, $\mathbf{f}$ itself having a cuspidal pinch point at that point.

As a first step to showing all this we choose an appropriate parametrisation for the surface $\mathbf{e}$.

Proposition 14.12 Let $\mathbf{e} : (s, t) \mapsto \mathbf{e}(s, t)$ *be a regular surface in* $\mathbb{R}^3$, *such that for each* t *the curve* $s \mapsto \mathbf{e}(s, t)$ *is a unit-speed geodesic and* *such that the curve $t \mapsto \mathbf{e}(0, t)$ cuts each geodesic orthogonally. Then, for each s, the curve $t \mapsto \mathbf{e}(s, t)$ cuts each geodesic orthogonally. Moreover, the vectors $\mathbf{e}_{1,0}$, $\mathbf{e}_{0,1}$, $\mathbf{e}_{2,0}$ are mutually orthogonal.*

Proof From the unit-speed condition $\mathbf{e}_{1,0} \cdot \mathbf{e}_{2,0} = 0$ and $\mathbf{e}_{1,0} \cdot \mathbf{e}_{1,1} = 0$ and from the geodesic condition $\mathbf{e}_{2,0} \cdot \mathbf{e}_{0,1} = 0$, all for any (s, t). So for each $t \in \mathbb{R}$ the function $s \mapsto \mathbf{e}_{1,0}(s, t) \cdot \mathbf{e}_{0,1}(s, t)$ is constant. But it is zero for $s = 0$, so is zero for all s. □

Theorem 14.13 Let $\mathbf{e} : (s, t) \mapsto \mathbf{e}(s, t)$ be a regular surface in $\mathbb{R}^3$, such that for each t the curve $s \mapsto \mathbf{e}(s, t)$ is a unit-speed geodesic and such that the curve $t \mapsto \mathbf{e}(0, t)$ cuts each geodesic orthogonally. For some $\delta \in \mathbb{R}$ let $\mathbf{s}(s, t) = \mathbf{e}(s, t) + (\delta - s)\mathbf{e}_{1,0}(s, t)$. Then at a point (s, t) of regularity of $\mathbf{s}$ the principal radii of curvature of $\mathbf{s}$ are $\rho = -(\delta - s)$ and σ where

$$\sigma - \rho = -\frac{\mathbf{e}_{0,1} \cdot \mathbf{e}_{0,1}}{\mathbf{e}_{0,1} \cdot \mathbf{e}_{1.1}}$$

$$= \frac{\mathbf{e}_{0,1} \cdot (\mathbf{e}_{0,1} + (\delta - s)\mathbf{e}_{1,1})}{(\delta - s)(\mathbf{e}_{2,0} \cdot \mathbf{e}_{1,1})^2 - (\mathbf{e}_{0,1} + (\delta - \mathbf{s})e_{1,1}) \cdot \mathbf{e}_{1,1}\mathbf{e}_{2,0} \cdot \mathbf{e}_{2,0}}.$$

The surface $\mathbf{s}$ is regular unless:

(i) $\mathbf{e}_{2,0}(s, t) = 0$, *that is the geodesic through (s, t) is linear there, in which case the same holds for neighbouring geodesics and the surface has a rhamphoid cuspidal edge in general, or*
(ii) $s = \delta$ or $(\mathbf{e}_{0,1} + (\delta - s)\mathbf{e}_{1,1}) \cdot \mathbf{e}_{0,1} = 0$ *at (s, t), in the first of which cases the surface has an ordinary cuspidal edge, and in the second an ordinary cuspidal edge in general.*

Proof We have $\mathbf{s}(s, t) = \mathbf{e}(s, t) + (\delta - s)\mathbf{e}_{1,0}(s, t)$, implying that

$$\mathbf{s}_1 = ((\delta - s)\mathbf{e}_{2,0}, \mathbf{e}_{0,1} + (\delta - s)\mathbf{e}_{1,1}),$$

injective at (s, t) provided that $s \neq \delta$, $\mathbf{e}_{2,0} \neq 0$ and $\mathbf{e}_{0,1} + (\delta - s)\mathbf{e}_{1,1}$ not a multiple of $\mathbf{e}_{2,0}$. On the other hand injectivity fails if one or other of these conditions fails, the latter failing if and only if

$$(\mathbf{e}_{0,1} + (\delta - s)\mathbf{e}_{1,1}) \cdot \mathbf{e}_{0,1} = 0,$$

that is if and only if

$$\delta - s = -\frac{\mathbf{e}_{0,1} \cdot \mathbf{e}_{0,1}}{\mathbf{e}_{0,1} \cdot \mathbf{e}_{1,1}},$$

since everywhere $(\mathbf{e}_{0,1} + (\delta - s)\mathbf{e}_{1,1}) \cdot \mathbf{e}_{1,0} = 0$. From this the first form of $\sigma - \rho$ may at once be inferred.

The alternative form arises from direct computation of the first and second fundamental forms $\mathbf{s}_1 \cdot \mathbf{s}_1$ and $-\mathbf{n}_1 \cdot \mathbf{s}_1$ of $\mathbf{s}$, the unit normal of $\mathbf{s}$ being $\mathbf{e}_{0,1}$. The equality of the two forms may be directly checked.

Clearly $\mathbf{s}$ acquires a cuspidal edge where it intersects either sheet of the focal surface, one such sheet being the regular surface $\mathbf{e}$ from which we started. The other sheet of the focal surface may have cuspidal edges, that is ribs corresponding to ridges on $\mathbf{s}$, and at the intersection of $\mathbf{s}$ with such a rib the surface $\mathbf{s}$ will have a swallow-tail point, the cuspidal edge having a cusp.

The only other way in which $\mathbf{s}$ is non-regular is when the unit-speed geodesic $s \mapsto \mathbf{e}(s, t)$ is linear, that is where $\mathbf{e}_{2,0} = 0$. In this case the surface $\mathbf{s}$ has an ordinary rhamphoid cuspidal edge in general. What we prove is that, where $\mathbf{e}_{2,0} = 0$, and therefore where $\mathbf{s}_{1,0} = 0$, $\mathbf{s}_{2,0}$, $\mathbf{s}_{3,0}$ *and* $\mathbf{s}_{0,1}$ are all coplanar. Note the subtlety of this! We might have expected to have to show that $\mathbf{s}_{3,0}$ was a multiple of $\mathbf{s}_{2,0}$ but this need not be the case. Yet the edge is still rhamphoid. To see all this we remark that for each t the line of curvature $s \mapsto \mathbf{s}(s, t) = \mathbf{e} + (\delta - s)\mathbf{e}_{1,0}$, with parameter s arc-length along the corresponding *focal* curve from the start line, has derivatives

$$\mathbf{s}_{1,0} = (\delta - s)\mathbf{e}_{2,0},$$

$$\mathbf{s}_{2,0} = (\delta - s)\mathbf{e}_{3,0} - \mathbf{e}_{2,0},$$

$$\mathbf{s}_{3,0} = (\delta - s)\mathbf{e}_{4,0} - 2\mathbf{e}_{3,0},$$

$$\mathbf{s}_{4,0} = (\delta - s)\mathbf{e}_{5,0} - 3\mathbf{e}_{4,0},$$

$$\mathbf{s}_{5,0} = (\delta - s)\mathbf{e}_{6,0} - 4\mathbf{e}_{5,0},$$

where $\mathbf{e}_{1,0} \cdot \mathbf{e}_{1,0} = 1$, $\mathbf{e}_{2,0} \cdot \mathbf{e}_{1,0} = 0$, $\mathbf{e}_{3,0} \cdot \mathbf{e}_{1,0} + \mathbf{e}_{2,0} \cdot \mathbf{e}_{2,0} = 0$, $\mathbf{e}_{4,0} \cdot \mathbf{e}_{1,0} + 3\mathbf{e}_{3,0} \cdot \mathbf{e}_{2,0} = 0$, $\mathbf{e}_{5,0} \cdot \mathbf{e}_{1,0} + 4\mathbf{e}_{3,0} \cdot \mathbf{e}_{2,0} + 3\mathbf{e}_{2,0} \cdot \mathbf{e}_{3,0} = 0$ and $\mathbf{e}_{6,0} \cdot \mathbf{e}_{1,0} + 5\mathbf{e}_{5,0} \cdot \mathbf{e}_{2,0} + 10\mathbf{e}_{4,0} \cdot \mathbf{e}_{3,0} = 0$.

Accordingly at a point where the focal curve has an ordinary linear point, that is where $\mathbf{e}_{2,0} = 0$ but $\mathbf{e}_{3,0}$ is not a multiple of $\mathbf{e}_{1,0}$, the vectors $\mathbf{e}_{3,0}$, $\mathbf{e}_{4,0}$ and $\mathbf{s}_{0,1}$, being each orthogonal to the vector $\mathbf{e}_{1,0}$, are all coplanar, from which it follows at once that the vectors $\mathbf{s}_{2,0}$, $\mathbf{s}_{3,0}$ and $\mathbf{s}_{0,1}$ are all coplanar, and this is what we set out to show. Moreover, since $\mathbf{e}_{5,0} \cdot \mathbf{e}_{1,0} \neq 0$ the vector $\mathbf{s}_{4,0}$ does not lie in this plane. Accordingly the surface at such a point has a rhamphoid edge. Let λ be such that $\mathbf{s}_{3,0} = 3\lambda\mathbf{s}_{2,0} + \nu\mathbf{s}_{0,1}$, that is

$$(\delta - s)\mathbf{e}_{4,0} - 2\mathbf{e}_{3,0} = 3\lambda(\delta - s)\mathbf{e}_{3,0} + \nu(\mathbf{e}_{0,1} + (\delta - s)\mathbf{e}_{1,1}).$$

Then the edge will be ordinary unless

$$\mathbf{s}_{5,0} = 10\lambda\mathbf{s}_{4,0} + \mu\mathbf{s}_{2,0} + \nu'\mathbf{s}_{0,1},$$

that is

$$(\delta - s)\mathbf{e}_{6,0} - 4\mathbf{e}_{5,0} = 10\lambda((\delta - s)\mathbf{e}_{5,0} - 3\mathbf{e}_{4,0}) + \mu(\delta - s)\mathbf{e}_{3,0} + \nu'(\mathbf{e}_{0,1} + (\delta - s)\mathbf{e}_{1,1}).$$

But then, taking the dot product of this with $\mathbf{e}_{1,0}$, we find that

$$-(\delta - s)10\mathbf{e}_{4,0} \cdot \mathbf{e}_{3,0} + 12\mathbf{e}_{3,0} \cdot \mathbf{e}_{3,0} = 30\lambda(\delta - s)\mathbf{e}_{3,0} \cdot \mathbf{e}_{3,0},$$

that is

$$\nu(\mathbf{e}_{0,1} + (\delta - s)\mathbf{e}_{1,1}) \cdot \mathbf{e}_{3,0} = 8\mathbf{e}_{3,0} \cdot \mathbf{e}_{3,0}.$$

Moreover, since $\mathbf{e}_{2,0} \cdot \mathbf{e}_{0,1} = 0$ everywhere we have $\mathbf{e}_{3,0} \cdot \mathbf{e}_{0,1} = 0$ where $\mathbf{e}_{2,0} = 0$, so that finally we have that the cuspidal edge is an ordinary rhamphoid one provided that $(\delta - s)\mathbf{e}_{1,1} \cdot \mathbf{e}_{3,0} \neq 8\mathbf{e}_{3,0} \cdot \mathbf{e}_{3,0}$. Since $\mathbf{e}_{3,0} \neq 0$ it follows that the edge is ordinary rhamphoid except at one point on each normal, where it may not be. □

The climax to this story is the unravelling of the geometry in the special case that a geodesic of the foliation is linear at a point at which the tangent direction to the the geodesic is *parabolic*.

Theorem 14.14 Let $\mathbf{e}$ *be a blue regular surface foliated by geodesics one of which is linear at a point* w *at which the tangent direction to the geodesic is parabolic, neighbouring geodesics having a point at which the tangent direction to the geodesic is hyperbolic. Then the line of curvature on each involute surface* $\mathbf{s}$ *corresponding to this geodesic has a rhamphoid cusp at* w *and through this cusp there passes not only a blue subparabolic line but also a red ridge. Moreover, the red sheet* $\mathbf{f}$ *of the focal surface of* $\mathbf{s}$ *has a cuspidal pinch-point at* w.

Proof With the notations of Proposition 14.12 the parabolic condition on the geodesic is that $\mathbf{e}_{3,0} \cdot \mathbf{e}_2(1, 0) = 0$, that is not only $\mathbf{e}_{3,0} \cdot \mathbf{e}_{2,0} = 0$, clearly the case since $\mathbf{e}_{2,0} = 0$, but also $\mathbf{e}_{3,0} \cdot \mathbf{e}_{1,1} = 0$. However, differentiating twice the equation $\mathbf{e}_{2,0} \cdot \mathbf{e}_{0,1} = 0$ we have that everywhere $\mathbf{e}_{3,0} \cdot \mathbf{e}_{0,1} + \mathbf{e}_{2,0} \cdot \mathbf{e}_{1,1} = 0$ and $\mathbf{e}_{4,0} \cdot \mathbf{e}_{0,1} + 2\mathbf{e}_{3,0} \cdot \mathbf{e}_{1,1} + \mathbf{e}_{2,0} \cdot \mathbf{e}_{1,2} = 0$, so that where $\mathbf{e}_{2,0} = 0$ and $\mathbf{e}_{3,0} \cdot \mathbf{e}_{1,1} = 0$ we have not only $\mathbf{e}_{3,0} \cdot \mathbf{e}_{0,1} = 0$ but also $\mathbf{e}_{4,0} \cdot \mathbf{e}_{0,1} = 0$. Since also $\mathbf{e}_{3,0} \cdot \mathbf{e}_{1,0} = 0$ and $\mathbf{e}_{3,0} \cdot \mathbf{e}_{0,1} = 0$ it follows that $\mathbf{e}_{4,0}$ is a multiple of $\mathbf{e}_{3,0}$. But then not only is $\mathbf{s}_{1,0} = 0$ but also $\mathbf{s}_{3,0}$ is a multiple of $\mathbf{s}_{2,0}$, that is the blue line of curvature on $\mathbf{s}$ at w has a rhamphoid cusp there. Moreover since $\mathbf{e}$ is parabolic at w a blue subparabolic line passes through the rhamphoid edge of $\mathbf{s}$ at w.

Finally, and surprisingly, the red principal radius of curvature σ of $\mathbf{s}$ is critical at w, so that a red ridge passes through that point also.

To prove this we remark first that, where $\mathbf{e}_{2,0} = 0$ and $\mathbf{e}_{3,0} \cdot \mathbf{e}_{1,1} = 0$ but $\mathbf{e}_{3,0} \neq 0$, there $\mathbf{e}_{1,1}$ is a real multiple of $\mathbf{e}_{1,0}$. For each are orthogonal both to $\mathbf{e}_{0,1}$ and to $\mathbf{e}_{3,0}$. It is then enough to differentiate the formula for σ from Theorem 14.13 at w in the red direction, and to demonstrate that this derivative is zero at w. But, by the second part of Proposition 14.11, on the rhamphoid edge of $\mathbf{s}$ the red direction and the blue direction coincide, $\mathbf{b}_1$ being a multiple of $\mathbf{a}_1$ there, though of course both are mapped to zero by $\mathbf{s}_1$. Now by Theorem 14.13

$$\sigma = s - \delta - \frac{\mathbf{e}_{0,1} \cdot \mathbf{e}_{0,1}}{\mathbf{e}_{0,1} \cdot \mathbf{e}_{1,1}}.$$

Then $\sigma_{1,0} = 0$ if and only if

$$(\mathbf{e}_{0,1} \cdot \mathbf{e}_{1,1})^2 - 2(\mathbf{e}_{0,1} \cdot \mathbf{e}_{1,1})^2 + (\mathbf{e}_{0,1} \cdot \mathbf{e}_{0,1})(\mathbf{e}_{1,1} \cdot \mathbf{e}_{1,1} + \mathbf{e}_{0,1} \cdot \mathbf{e}_{2,1}) = 0.$$

But, since $\mathbf{e}_{0,1} \cdot \mathbf{e}_{2,0} = 0$ everywhere, it follows that $\mathbf{e}_{0,1} \cdot \mathbf{e}_{2,1} = 0$ where $\mathbf{e}_{2,0} = 0$. So $\sigma_{1,0} = 0$ if and only if

$$(\mathbf{e}_{0,1} \cdot \mathbf{e}_{1,1})^2 - (\mathbf{e}_{0,1} \cdot \mathbf{e}_{0,1})(\mathbf{e}_{1,1} \cdot \mathbf{e}_{1,1}) = 0.$$

But by the Cauchy–Schwarz *equality* (Exercise 2.1) this is so if and only if $\mathbf{e}_{0,1}$ and $\mathbf{e}_{1,1}$ are linearly dependent, which is the case, as we have just seen.

Finally $\mathbf{n} = \mathbf{e}_{1,0}$ differentiated along $\mathbf{b}$ at w is a multiple of $\mathbf{e}_{2,0}$, $\mathbf{b}_1$ being a multiple of $\mathbf{a}_1$ there. But $\mathbf{e}_{2,0} = 0$ at w. So on differentiating the equation $\mathbf{f}_1\mathbf{b}_1 = (\sigma\mathbf{b}_1)\mathbf{n}$ along $\mathbf{b}$ twice we find that while $(\mathbf{fb})_1 = 0$ at w both $(\mathbf{fb})_2$ and $(\mathbf{fb})_3$ are multiples of $\mathbf{n}$ there, implying that the red focal line on $\mathbf{f}$ has a rhamphoid cusp at w. □

There is a further surprise in store, although by hindsight and by Proposition 14.8 the additional subparabolic line was to be expected!

Theorem 14.15 (Morris) *With the set up and notations of the previous theorem there is also a red subparabolic line that passes through w (but not a blue ridge!).*

Proof Away from any cuspidal edge of the red sheet $\mathbf{f}$ the vector $\mathbf{e}_{0,1}$ is normal to $\mathbf{f}$ so the unit normal is equal to $\alpha\mathbf{e}_{0,1}$ where $1 = \alpha^2\mathbf{e}_{0,1} \cdot \mathbf{e}_{0,1}$.

Now, by Proposition 13.8, the kernel of the Gauss map of this sheet, if non-zero, is the blue principal tangent line of the surface **s**. So we shall be on a red subparabolic line of **s** exactly when $(\alpha \mathbf{e}_{0,1})_{1,0} = 0$. However,

$$(\alpha \mathbf{e}_{0,1})_{1,0} = \alpha_{1,0}\mathbf{e}_{0,1} + \alpha \mathbf{e}_{1,1},$$

while $0 = 2\alpha\alpha_{1,0}\mathbf{e}_{0,1} \cdot \mathbf{e}_{0,1} + 2\alpha^2\mathbf{e}_{1,1} \cdot \mathbf{e}_{0,1}$, from which it follows that

$$\mathbf{e}_{0,1} - \frac{\mathbf{e}_{0,1} \cdot \mathbf{e}_{0,1}}{\mathbf{e}_{0,1} \cdot \mathbf{e}_{1,1}}\mathbf{e}_{1,1} = 0.$$

This equation is satisfied in particular at the cuspidal pinch-point of the focal sheet **f**. So this point lies in the closure of the parabolic line on **f**, which is what had to be proved. □

Morris has shown (1991) that in general the parabolic line on the sheet **f** is regular at the pinch-point, its tangent at that point coinciding with the tangent to the edge. In the parameter space the generic situation is illustrated in Figure 14.6.

The curve representing the cuspidal edge on **f** (**f**∞) and the curve representing the parabolic line on **e** (**e**0) are not tangent, but the line (**rh**) representing the line of inflections on geodesics of either sheet, or equivalently the line representing the rhamphoid edge on any of the involutes, is tangent to that representing the parabolic line on **e** and

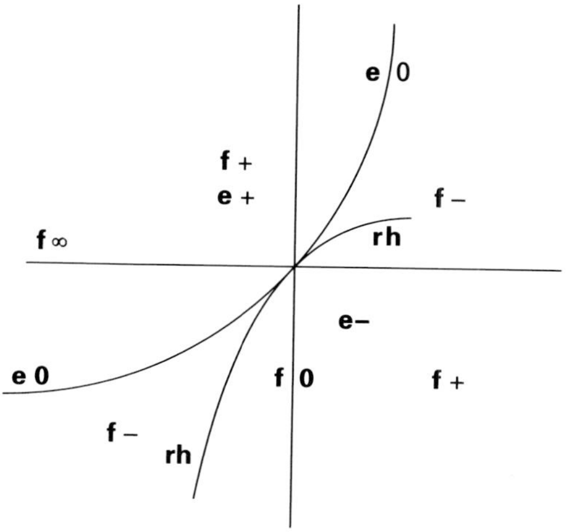

Figure 14.6

hence is transverse to that representing the cuspidal edge. The curve representing the parabolic line on **f** (**f**0) is transverse to all these curves. It and the curve representing the cuspidal edge on **f** divide the parameter space locally into four regions in two of which the Gaussian curvature of **f** is positive (**f**+) and in two of which the Gaussian curvature of **f** is negative (**f**−). The curve representing the line of inflections on geodesics of either sheet necessarily passes from one of these negative regions to the other. Indeed it was the need for this to be the case which led to the discovery of the red parabolic line. In fact Morris's remark holds for any cuspidal cross-cap. Through the pinch-point there passes in general a parabolic line tangent at the pinch-point to the cuspidal edge, but with their representatives in the parameter space transverse to each other.

Let **e** be as above foliated by geodesics each with a linear point. Then any point of intersection of any of the involutes **s** of **e** with either **e** or **f** is a point of intersection of an ordinary cuspidal edge and a rhamphoid cuspidal edge. It has been proved (see Arnol'd (1983) or (1990b), Bennequin (1984) or Shcherbak (1988)) that at such a point the image of **s** is $\mathcal{A}$-equivalent to the full involute of a plane curve with an ordinary inflection (Figure 14.7). For **e** as in Theorem 14.14 the involute surface that passes through the rhamphoid cusp of the focal line on the red surface **f** has two cuspidal edges, one rhamphoid with an ordinary cusp at w and the other ordinary with a rhamphoid cusp at w. Figure 14.8 shows what this surface looks like up to a local diffeomorphism. The proof of this depends on the deep connection that has recently been found between this configuration and a group generated by reflections of four-dimensional Euclidean space, the Coxeter group known as H_4. The actual surface has the subparabolic lines implied by the last few theorems.

14.3 Coxeter groups

A *Coxeter group* is a finite group that is generated by linear reflections of $\mathbb{R}^n$ for some n. Coxeter groups, named after H. S. M. Coxeter, have been completely classified and consist of several infinite series and several so-called exceptional groups. See, for example, Hiller (1982).

The easiest to describe are the groups A_k. The group A_k, isomorphic to the symmetric group $\mathcal{S}_{k+1}$ on $k+1$ letters, consists of the isometries

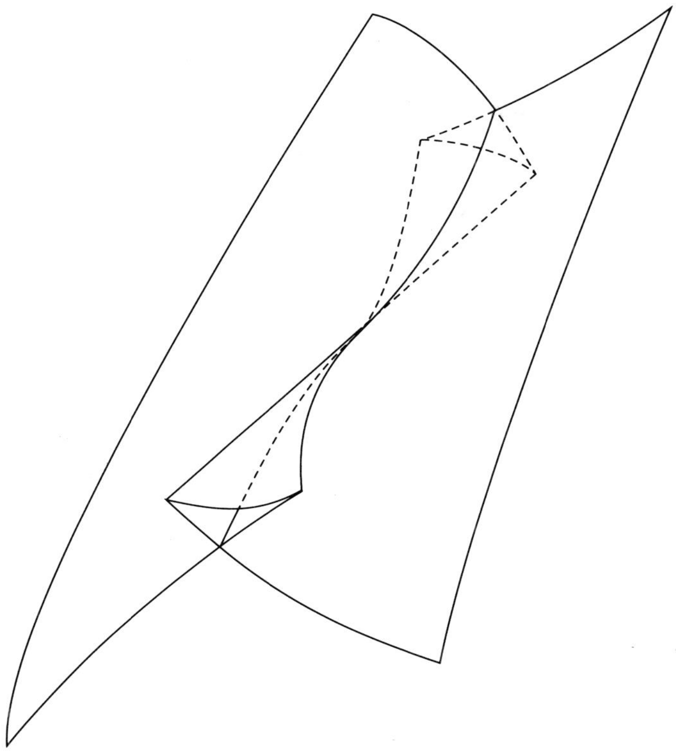

Figure 14.7

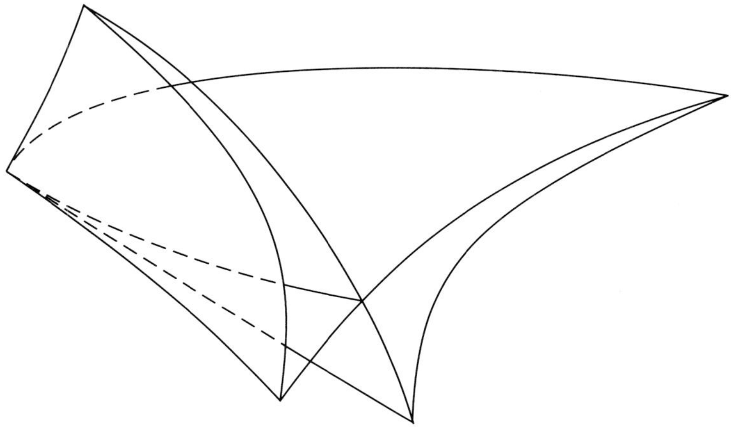

Figure 14.8

of the k-dimensional subspace of $\mathbb{R}^{k+1}$ with equation $x_0 + x_1 + \ldots + x_k = 0$ generated by the reflections that interchange adjacent coordinates.

Consider for example $A_2 = \mathcal{S}_3$. Then the polynomials $yz + zx + xy$ and xyz, where $x + y + z = 0$, are invariant under action of the group.

Proposition 14.16 The fibres of the map

$$\{(x, y, z) \in \mathbb{R}^3 : x + y + z = 0\} \to \mathbb{R}^2;$$

$$(x, y, z) \mapsto (yz + zx + xy, xyz)$$

are the orbits of the action of the group A_2. The map is regular except where $x = y$, $y = z$ or $z = x$, that is on the mirrors *of the group. The image of the non-regular points of this map is the image of the curve $\mathbb{R} \to \mathbb{R}^2$; $t \mapsto (-3t^2, -2t^3)$, which has an ordinary cusp at the origin.*

Proof That the fibres are the orbits of the action of the group follows from the fact that knowledge of the values of $x + y + z = \alpha$, $yz + zx + xy = \beta$ and $xyz = \gamma$ determines x, y and z up to a permutation, as roots of the cubic equation $t^3 - \alpha t^2 + \beta t - \gamma = 0$.

The composite of the given map with the bijective linear map

$$\mathbb{R}^2 \to \{(x, y, z) \in \mathbb{R}^3 : x + y + z = 0\}; (x, y) \mapsto (x, y, -x - y)$$

is the map

$$\mathbb{R}^2 \to \mathbb{R}^2; (x, y) \mapsto (-x^2 - xy - y^2, -x^2y - xy^2),$$

which has the Jacobian matrix at (x, y)

$$\begin{bmatrix} -2x - y & -x - 2y \\ -2xy - y^2 & -x^2 - 2xy \end{bmatrix},$$

of determinant $(2x + y)(x + 2y)(x - y) = (z - x)(y - z)(x - y) = 0$ when $z = x$, $y = z$ or $x = y$, on setting $z = -x - y$.

On setting $x = y = t$ the map reduces at once to the map

$$\mathbb{R} \to \mathbb{R}^2; t \mapsto (-3t^2, -2t^3),$$

the same being true if $x = z = t$, when $y = -2t$, or when $y = z = t$, when $x = -2t$. □

This cuspidal curve is the *variety of non-regular orbits* of the group A_2. Likewise the *variety of non-regular orbits* of the group A_3 may be shown to be a swallow-tail surface. Arnol'd (1972) showed that the full involute of a plane curve with an ordinary cusp is $\mathcal{A}$-equivalent to the *variety of non-regular orbits* of the group A_3, this being the reason for the notation A_3 to describe the centre of curvature of a plane curve at an ordinary inflection.

The notations A_4, D_4, D_5, E_6 and so on derive from analogous considerations involving the appropriate Coxeter groups, a full discussion requiring complexification of the polynomial maps that arise.

There is a complete list of Coxeter groups. Most, namely the groups A_k, $B_k \cong C_k$, D_k, E_6, E_7, E_8, F_4 and G_2, had already turned up in singularity theory, these being the groups with crystallographic extensions that also turn up in the classification of Lie groups and Lie algebras. Apart from the groups of symmetries of regular plane polygons there are only two other Coxeter groups, H_3, the full group of isometries of an icosahedron, and H_4, the full group of symmetries of an analogue of the icosahedron in $\mathbb{R}^4$, the hypericosahedron. Their role in the theory of involutes was unexpected. As we have already remarked, the variety of non-regular orbits of H_3 is $\mathcal{A}$-equivalent to the full involute of a plane curve with an ordinary inflection. What Shcherbak proved was that the full involute of the surface **e** of Theorem 14.14 is $\mathcal{A}$-equivalent to the variety of non-regular orbits of the group H_4. Shcherbak's paper (1988) is posthumous and somewhat vague as to the precise geometry at an H_4 point, stating only that it occurs on a *distinguished* asymptotic tangent line. Apart from an obscure announcement in (1984) Arnol'd gives the clear statement that this tangent line is *parabolic* only in (1990b). The H_4 configuration is discussed more from our point of view in the Liverpool thesis of Alex Flegmann (1985), though that account also is incomplete.

In conclusion we quote from Shcherbak's paper the parametrisation that he gives for the variety of nonregular orbits of the group H_4, namely

$$(a, b, c) \to (a, ac + \tfrac{1}{2}b^2, ab^3 + \tfrac{1}{2}c^2, ab^3c + \tfrac{1}{5}b^5 + \tfrac{1}{3}c^3),$$

a three-dimensional variety in $\mathbb{R}^4$, whose intersection with the hyperplane $a = 0$ is the two-dimensional variety in $\mathbb{R}^3$ with the parametrisation

$$(b, c) \to (\tfrac{1}{2}b^2, \tfrac{1}{2}c^2, \tfrac{1}{5}b^5 + \tfrac{1}{3}c^3).$$

Figures 14.9 and 14.8, taken from Flegmann's thesis, illustrate the sections of this variety in the case that $a = 1$ and the transitional case that $a = 0$, respectively. The latter is the surface previously referred to with two cuspidal edges intersecting at the origin, the one being an

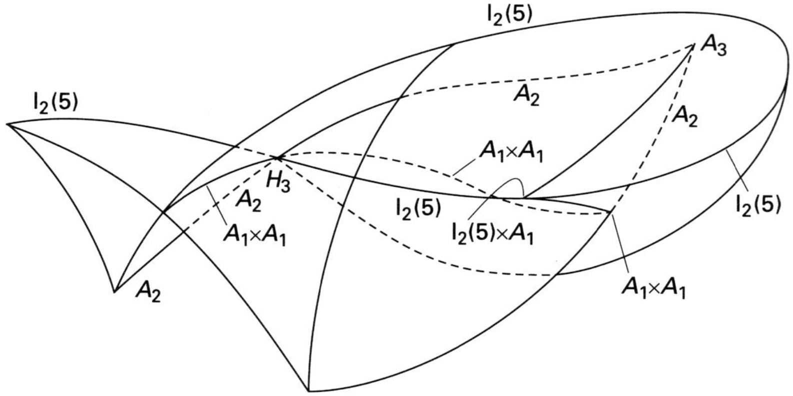

Figure 14.9

ordinary (3/2) edge with a 5/2 cusp there and the other a 5/2 edge with an ordinary (3/2) cusp there.

Exercises

14.1 Prove Proposition 14.3.

14.2 Complete the proof of Proposition 14.8.

14.3 Verify the equality of the two expressions for $\sigma - \rho$ in the statement of Theorem 14.10.

14.4 Verify that the variety of non-regular orbits of the Coxeter group A_3 is a swallow-tail surface, the orbits of $A_3 = \mathcal{S}_4$ being represented as image points in $\mathbb{R}^3$ of the map

$$\{(w, x, y, z) \in \mathbb{R}^4 : (w + x + y + z = 0\} \to \mathbb{R}^3$$

$(w, x, y, z) \mapsto$

$(wx + wy + wz + xy + xz + yz, xyz + wyz + wxz + wxy, wxyz).$

15

The circles of a surface

15.0 Introduction

Euler defined the principal curvatures of a smooth surface by means of the curvature of normal sections of the surface and Meusnier extended Euler's analysis to arbitrary sections. But apparently neither examined in detail whether or not there were sections with higher order circularity or linearity at the point of interest of the surface. Answering this question leads to an alternative approach to classical surface theory which emphasises the role of the *ridges* of a surface, the loci of points where a principal curvature is stationary in the principal direction, and their classification into *fertile* and *sterile* ridges. The behaviour of highly osculating circles in the neighbourhood of an umbilical point is especially entertaining. They curl up and die on approach to the umbilics, but whether by falling on their face or on their back depends on the umbilical index.

15.1 The theorems of Euler and Meusnier

We start with the theorems of Euler and Meusnier. These relate the principal curvatures κ and λ at a point w of a regular smooth surface $\mathbf{s} : \mathbb{R}^2 \rightarrowtail \mathbb{R}^3$ to the curvature of plane sections of the surface through w. Euler's Theorem we have, of course, already discussed in Section 10.1. It concerns normal sections to the surface. Meusnier's Theorem extends this to arbitrary sections.

Theorem 15.1 (Euler–Meusnier) *Let $\kappa_{\theta,\psi}$ be the curvature at w of the curve cut on a smooth surface* $\mathbf{s}$ *by a plane through w, whose normal makes an angle $\psi < \frac{1}{2}\pi$ with the normal to the surface and which cuts the tangent plane in the line that makes an angle θ with the principal* *tangent corresponding to the principal curvature κ, with λ denoting the second principal curvature. Then*

$$\kappa_{\theta,\psi} = \kappa_\theta \sec \psi \quad \textit{(Meusnier, 1785)}$$

where

$$\kappa_\theta = \kappa \cos^2 \theta + \lambda \sin^2 \theta \quad \textit{(Euler, 1760)}.$$

An alternative statement of Meusnier's part of the theorem is that, whatever the angle ψ, the circle of curvature of the section lies on the sphere through w with centre $\rho_\theta \mathbf{n}$, where $\rho_\theta = \kappa_\theta{}^{-1}$, or, if $\kappa_\theta = 0$, on the tangent plane at w, when it reduces to the tangent line.

Proof Since $\psi \leqslant \frac{1}{2}\pi$ the curve of section is non-singular at w. So we may choose a regular representation $\mathbf{q} : \mathbb{R} \rightarrowtail \mathbb{R}^2$ for it, with $\mathbf{q}(0) = w$, and such that $\mathbf{s}_1\mathbf{q}_1 \cdot \mathbf{s}_1\mathbf{q}_1 = 1$ at w. Let $\mathbf{a}_1$ and $\mathbf{b}_1$ represent principal vectors at w such that $\mathbf{s}_1\mathbf{a}_1 \cdot \mathbf{s}_1\mathbf{a}_1 = \mathbf{s}_1\mathbf{b}_1 \cdot \mathbf{s}_1\mathbf{b}_1 = 1$ and let $\mathbf{n}$ be the unit normal vector. Then

$$\mathbf{n} \cdot \mathbf{s}_2(w)\mathbf{a}_1 = \kappa \mathbf{s}_1(w)\mathbf{a}_1 \cdot \mathbf{s}_1(w)$$

$$\mathbf{n} \cdot \mathbf{s}_2(w)\mathbf{b}_1 = \lambda \mathbf{s}_1(w)\mathbf{b}_1 \cdot \mathbf{s}_1(w)$$

and

$$\mathbf{q}_1 = \mathbf{a}_1 \cos \theta + \mathbf{b}_1 \sin \theta.$$

We regard the curve of section $\mathbf{sq}$ as a curve in space. Then the points $\mathbf{c}$ of its focal line satisfy the equations

$$(\mathbf{c} - s) \cdot \mathbf{s}_1\mathbf{q}_1 = 0,$$

$$(\mathbf{c} - \mathbf{s}) \cdot (\mathbf{s}_2\mathbf{q}_1{}^2 + \mathbf{s}_1\mathbf{q}_2) = \mathbf{s}_1\mathbf{q}_1 \cdot \mathbf{s}_1\mathbf{q}_1 = 1.$$

The normal line to the surface lies in the normal plane to the curve and intersects the focal line (possibly at ∞) where $(\mathbf{c} - \mathbf{s}) \cdot \mathbf{s}_1 = 0$.

So for the point of intersection $(\mathbf{c} - \mathbf{s}) \cdot \mathbf{s}_2\mathbf{q}_1{}^2 = 1$. Indeed this determines $\mathbf{c}$ provided that $\mathbf{q}_2$ is independent of $\mathbf{q}_1$.

Let κ_θ be defined by $\kappa_\theta(\mathbf{c} - \mathbf{s}) = \mathbf{n}$, with $\kappa_\theta = 0$ if $\mathbf{c}$ is at ∞. Then

$$\begin{aligned} \kappa_\theta &= \mathbf{n} \cdot \mathbf{s}_2\mathbf{q}_1{}^2 \\ &= \mathbf{n} \cdot \mathbf{s}_2(\mathbf{a}_1 \cos \theta + \mathbf{b}_1 \sin \theta)^2 \\ &= \mathbf{n} \cdot \mathbf{s}_2\mathbf{a}_1{}^2 \cos^2 \theta + \mathbf{n} \cdot \mathbf{s}_2\mathbf{b}_1{}^2 \sin^2 \theta \\ &= \kappa \cos^2 \theta + \lambda \sin^2 \theta. \end{aligned}$$

This is the Euler formula.

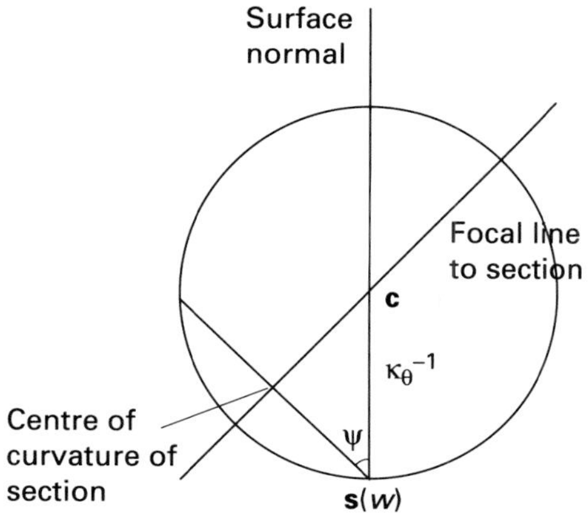

Figure 15.1

Now the focal line of a plane curve passes through its centre of curvature. It follows that

$$\kappa_{\theta,\psi}{}^{-1} = \kappa_\theta{}^{-1} \cos \psi$$

or, equivalently, that

$$\kappa_{\theta,\psi} = \kappa_\theta \sec \psi.$$

See Figure 15.1. This is the Meusnier formula. □

We shall refer to the sphere with centre $\rho_\theta \mathbf{n}$ as the *Meusnier sphere* at w for the direction θ. It reduces to the tangent plane for $\kappa_\theta = 0$.

15.2 Osculating circles

The Euler–Meusnier Theorem can be extended further to the description of points of the surface $\mathbf{s}$ through which there pass circles or lines with higher order contact than merely three point (A_2) contact.

Theorem 15.2 Let a tangent line be chosen through the point $\mathbf{s}(w)$ *of a regular smooth surface* $\mathbf{s}$ *at the parameter point* w, *not a point of the parabolic line. Then if*:

(i) *the tangent line is not one of the principal tangent lines of the surface at* w *there is exactly one plane in* $\mathbb{R}^3$ *through this line that cuts out a plane curve on the surface with a point of stationary curvature at* w;

(ii) *the tangent line is one of the principal lines at w then either there is no plane section through the line, transverse to the tangent plane, with a point of stationary curvature at w, or every such plane through the line has a point of stationary curvature at w, the latter being the case if and only if the corresponding principal centre of curvature is at least an A_3 or D_4 focal point of the surface, that is if and only if w lies on a relevant ridge, possibly at an umbilic*;

(iii) *w lies on an ordinary ridge then there is a quadratic equation for plane sections through the relevant principal tangent line, having at least an A_4 point of stationary curvature at w, there being two such planes if w is a fertile ridge point, but none if it is a sterile point, and one (double one) if it is an ordinary turning point, or A_4, turning point of the ridge*;

(iv) *w is an ordinary umbilic then the relevant principal directions are root directions of the principal cubic associated to the umbilic, there being exactly one section for each of the three principal directions in the elliptic case and the one principal direction in the hyperbolic case, having at least an A_4 point of stationary curvature at w. Moreover, as one approaches the umbilic along a ridge then one of the two circles 'curls up and dies' either by falling on its face or on its back according as the index of the umbilic is equal to $\frac{1}{2}$ or $-\frac{1}{2}$.*

Proof Since we are only considering planes transverse to the tangent plane, the curve of intersection may be represented parametrically near w by a regular map $\mathbf{q} : \mathbb{R} \rightarrowtail \mathbb{R}^2$ with $\mathbf{q}(0) = w$, where as before we may suppose that $\mathbf{s}_1\mathbf{q}_1 \cdot \mathbf{s}_1\mathbf{q}_1 = 1$ at w. Without loss of generality we may also suppose that at that point $\mathbf{s}_1\mathbf{q}_1 \cdot \mathbf{s}_1\mathbf{q}_2 = 0$.

For 0 to be a vertex (possibly an undulation or a circular or linear point of higher order) of the section we require to have at 0

$$2 = \operatorname{rk}\begin{bmatrix} (\mathbf{s}_1\mathbf{q}_1)\cdot & 0 \\ (\mathbf{s}_2\mathbf{q}_1{}^2 + \mathbf{s}_1\mathbf{q}_2)\cdot & \mathbf{s}_1\mathbf{q}_1 \cdot \mathbf{s}_1\mathbf{q}_1 \\ (\mathbf{s}_3\mathbf{q}_1{}^3 + 3\mathbf{s}_2\mathbf{q}_1\mathbf{q}_2 + \mathbf{s}_1\mathbf{q}_3)\cdot & 3(\mathbf{s}_2\mathbf{q}_1{}^2 + \mathbf{s}_1\mathbf{q}_2)\cdot\mathbf{s}_1\mathbf{q}_1 \end{bmatrix},$$

the matrix having three rows and four columns, and this will be so if we can find λ, μ, ν not all zero such that

$$\lambda\mathbf{s}_1\mathbf{q}_1 + \mu(\mathbf{s}_2\mathbf{q}_1{}^2 + \mathbf{s}_1\mathbf{q}_2) + \nu(\mathbf{s}_3\mathbf{q}_1{}^3 + 3\mathbf{s}_2\mathbf{q}_1\mathbf{q}_2 + \mathbf{s}_1\mathbf{q}_3) = 0$$

and

$$\mu(\mathbf{s}_1\mathbf{q}_1 \cdot \mathbf{s}_1\mathbf{q}_1) + 3\nu(\mathbf{s}_2\mathbf{q}_1{}^2 + \mathbf{s}_1\mathbf{q}_2)\cdot\mathbf{s}_1\mathbf{q}_1 = 0.$$

Since $\mathbf{s}_1\mathbf{q}_1 \neq 0$ it follows that $\nu \neq 0$. So choose $\nu = 1$. Then, applying $\mathbf{n}\cdot$ to the first of these equations, we get

$$\mu\mathbf{n}\cdot\mathbf{s}_2\mathbf{q}_1{}^2 + \mathbf{n}\cdot\mathbf{s}_3\mathbf{q}_1{}^3 + 3\mathbf{n}\cdot\mathbf{s}_2\mathbf{q}_1\mathbf{q}_2 = 0.$$

With $\mathbf{s}_1\mathbf{q}_1\cdot\mathbf{s}_1\mathbf{q}_1 = 1$ and $\mathbf{s}_1\mathbf{q}_2\cdot\mathbf{s}_1\mathbf{q}_1 = 0$, $\mu = -3\mathbf{s}_2\mathbf{q}_1{}^2\cdot\mathbf{s}_1\mathbf{q}_2$ and

$$-3\mathbf{s}_2\mathbf{q}_1{}^2\cdot\mathbf{s}_1\mathbf{q}_1\,\mathbf{n}\cdot\mathbf{s}_2\mathbf{q}_1{}^2 + \mathbf{n}\cdot\mathbf{s}_3\mathbf{q}_1{}^3 + 3\mathbf{n}\cdot\mathbf{s}_2\mathbf{q}_1\mathbf{q}_2 = 0,$$

which determines $\mathbf{q}_2$ uniquely, provided that $\mathbf{q}_1$ is *not* a principal vector at w. Then, whatever $\mathbf{q}_3$ may be, λ is given by the first equation, the three vectors $\mathbf{s}_1\mathbf{q}_1$, $\mathbf{s}_2\mathbf{q}_1{}^2 + \mathbf{s}_1\mathbf{q}_2$ and $\mathbf{s}_3\mathbf{q}_1{}^3 + 3\mathbf{s}_2\mathbf{q}_1\mathbf{q}_2 + \mathbf{s}_1\mathbf{q}_3$ being linearly dependent since they are coplanar.

If $\mathbf{q}_1$ *is* a principal vector with curvature κ, say, then $\mathbf{n}\cdot\mathbf{s}_2\mathbf{q}_1{}^2 = \kappa$ and the section does not have a vertex at w unless

$$\mathbf{n}\cdot\mathbf{s}_3\mathbf{q}_1{}^3 = 3\kappa\mathbf{s}_2\mathbf{q}_1{}^2\cdot\mathbf{s}_1\mathbf{q}_1$$

when it does have, whatever $\mathbf{q}_2$ may be.

This proves (i) and (ii) for w not an umbilic of $\mathbf{s}$.

At an umbilic, $\mathbf{n}\cdot\mathbf{s}_2\mathbf{q}_1\mathbf{q}_2 = 0$ for any mutually orthogonal $\mathbf{q}_1$ and $\mathbf{q}_2$ and in this case the equation

$$\mathbf{n}\cdot\mathbf{s}_3\mathbf{q}_1{}^3 = 3\kappa\mathbf{s}_2\mathbf{q}_1{}^2\cdot\mathbf{s}_1\mathbf{q}_1$$

determines the principal directions at the umbilic.

For the plane section of the surface $\mathbf{s}$ to have at least an A_4 vertex at w a necessary condition is that at 0 the 4×4 square matrix

$$\begin{bmatrix} (\mathbf{sq})_1 & 0 \\ (\mathbf{sq})_2 & (\mathbf{sq})_1\cdot(\mathbf{sq})_1 \\ (\mathbf{sq})_3 & 3(\mathbf{sq})_1\cdot(\mathbf{sq})_2 \\ (\mathbf{sq})_4 & 4(\mathbf{sq})_1\cdot(\mathbf{sq})_3 + 3(\mathbf{sq})_2\cdot(\mathbf{sq})_2 \end{bmatrix}$$

has $(\mathbf{n}, -\kappa)$ as a kernel vector, for this vector is already killed by each of the first three rows. The condition that it is killed by the last one reduces to

$$V_4\mathbf{q}_1{}^4 + 6V_3\mathbf{q}_1{}^2\mathbf{q}_2 + 3V_2\mathbf{q}_2{}^2 + 4V_2\mathbf{q}_1\mathbf{q}_3 + V_1\mathbf{q}_4 = 0,$$

or more simply, since $V_1 = 0$ and $V_2\mathbf{q}_1 = 0$, to

$$V_4\mathbf{q}_1{}^4 + 6V_3\mathbf{q}_1{}^2\mathbf{q}_2 + 3V_2\mathbf{q}_2{}^2 = 0,$$

V_i being either the ith derivative at w of $-\frac{1}{2}(\mathbf{c}-\mathbf{s})\cdot(\mathbf{c}-\mathbf{s})$ if $\kappa \neq 0$, with $\mathbf{c}$ afterwards put equal to $\kappa^{-1}\mathbf{n}$, or $\mathbf{n}\cdot\mathbf{s}_i$ if $\kappa = 0$.

This is the quadratic equation that characterises the sterile and fertile A_3 ridge points of $\mathbf{s}$, with equal roots only where the A_4 condition holds.

At an umbilic $V_2 = 0$, and in place of the quadratic equation we have the linear equation

$$V_4\mathbf{q}_1{}^4 + 6V_3\mathbf{q}_1{}^2\mathbf{q}_2 = 0,$$

which has a unique solution for each solution of the equation $V_3\mathbf{q}_1{}^3 = 0$, except at an ordinary parabolic umbilic where, for one of the solutions $\mathbf{q}_1$, $V_3\mathbf{q}_1{}^2 = 0$, but $V_4\mathbf{q}_1{}^4 \neq 0$, in which case no solution $\mathbf{q}_2$ exists.

Here are the answers to the questions raised earlier in Section 11.5 as to the identity of the probes entering into these formulas. They are the circles having highest order of contact with the surface.

Perhaps the easiest approach to part (iv) is to observe that, as one approaches an umbilic along a ridge, the length of one of the solutions $\mathbf{q}_2$ of the above quadratic equation tends to infinity and, as it does so, it approaches more and more closely the vector $\mathbf{q}_2$ given by the equation

$$2V_3\mathbf{q}_1{}^2 + V_2\mathbf{q}_2 = 0,$$

this vector being twice the second derivative $\mathbf{a}_2$ of the relevant line of curvature given, as we are here on a ridge, by the equation

$$V_3\mathbf{a}_1{}^2 + V_2\mathbf{a}_2 = 0,$$

where $\mathbf{a}_1 = \mathbf{q}_1$. The result then follows from an examination of the typical patterns of lines of curvature round umbilics of different index. □

Corollary 15.3 Where a ridge of a surface passes through an ordinary umbilic all points of the ridge near the umbilic are fertile. □

Things are different in the case that the Meusnier 'sphere' for the chosen tangent line to the surface reduces to the tangent plane at the point of contact. In such a case the only plane of interest passing through this line is the tangent plane itself, and all the circles with high contact with the surface at the point and having this line as tangent lie in this plane. Now, as we have already remarked towards the end of Chapter 13, any tangent plane to a smooth regular surface intersects the surface in a curve with a singularity at the point of contact, generically either a *crunode*, that is a cross node, or an *acnode*, that is an isolated node, or in the case that the point of the surface lies on the parabolic line and the tangent line is the relevant principal tangent line, an ordinary cusp or singularity of higher order. We state the generic possibilities as a series of propositions.

Proposition 15.4 Let a tangent line be chosen through the point $\mathbf{s}(w)$ *of a regular smooth surface* $\mathbf{s}$ *with parameter point* w*, not a principal line, and such that the 'Meusnier sphere' for that line reduces to the tangent plane at* w*. Then the Gaussian curvature of* $\mathbf{s}$ *at* w *is negative, the curve of intersection of the surface and tangent plane has a crunode at* w*, one of whose tangents is the tangent line originally chosen, and there is a unique circle in the plane through* w *and having that line as its tangent line, having at least* A_3 *contact with the surface there, namely the circle of curvature to the relevant branch of the curve of intersection.*

Proof A straightforward exercise, with the surface taken in Monge form. □

Proposition 15.5 Let $\mathbf{s}$ *be a regular smooth surface with* w *the parameter point representing an* A_2 *point of the parabolic line, that is a point of that line that is not a cusp of Gauss. Then the tangent plane to* $\mathbf{s}$ *at* w *intersects the surface in a curve with an ordinary cusp at* w *and there is no circle in the plane with at least* A_3 *contact with the surface at* w*.*

Proof Exercise. □

Proposition 15.6 Let $\mathbf{s}$ *be a regular smooth surface with* w *the parameter point representing an* A_3 *point* $\mathbf{s}(w)$ *of the parabolic line, that is a cusp of Gauss. There are two cases. In the hyperbolic case the tangent plane to* $\mathbf{s}$ *at* w *intersects the surface in a curve with a* tacnode *at* $\mathbf{s}(w)$*, that is in a curve with two tangential branches at* $\mathbf{s}(w)$*, every circle in the plane through* $\mathbf{s}(w)$ *with tangent line the principal line having* A_3 *contact with the surface at* $\mathbf{s}(w)$*, with the exception of the circle of curvature of either branch at* w *which each have at least* A_4 *contact with the surface there. In the elliptic case the tangent plane intersects the surface in a curve with an isolated singularity at* w*. Now every circle in the plane through* $\mathbf{s}(w)$ *and with tangent line the principal line has* A_3 *contact with the surface at* $\mathbf{s}(w)$*, there being no such circle with* A_4 *contact.*

Proof Exercise. □

Proposition 15.7 Let **s** *be a regular smooth surface with* w *the parameter point representing an ordinary* A_4 *point* $\mathbf{s}(w)$ *of the parabolic line. Then the tangent plane to* **s** *at* w *intersects the surface in a curve with a rhamphoid cusp at* $\mathbf{s}(w)$. *Every circle in the plane through* $\mathbf{s}(w)$ *with tangent line the principal line has* A_3 *contact with the surface at* $\mathbf{s}(w)$, *with the exception of the limiting circle of curvature of the curve of intersection at the cusp, which has at least* A_4 *contact with the surface there.*

Proof Exercise. □

Proposition 15.8 Let **s** *be a regular smooth surface with* w *the parameter point representing a flat* D_4 *umbilic* $\mathbf{s}(w)$ *of the surface. Then the tangent plane at the umbilic intersects the surface in a curve with either three branches or one at the umbilic, according as it is elliptic or hyperbolic, the circles of curvature at* $\mathbf{s}(w)$ *of each branch having at least* A_4 *contact with* **s** *there.*

Proof Exercise. □

In all the above propositions it is, of course, possible that exceptionally a circle reduces to a line. The reader can supply the necessary modification to the statement in each such case.

Theorem 15.2 can be seen in a fresh light if one recalls the remarks made at the end of Chapter 13. Under inversion of the ambient space $\mathbb{R}^3$ with respect to some point of the Meusnier sphere corresponding to some direction at a point w of a regular smooth surface **s**, the sphere reduces to the tangent plane at w, with all the various possibilities that we have just been enumerating. So conversely the various circles with high order contact with the surface may be interpreted as the principal circles of curvature of the branches of the curve of intersection with the Meusnier sphere.

15.3 Contours and umbilical hill-tops

Consider the form of the contours of a surface near a hill-top or the bottom of a depression. With the surface in Monge form

$$z = \tfrac{1}{2}\kappa x^2 + \tfrac{1}{2}\lambda y^2 + \tfrac{1}{6}C_3(x, y)^3 + O(x, y)^4$$

its contours are the curves in $\mathbb{R}^2$ obtained by fixing the value of z. For z small the curve will approximate near the origin to the ellipse

$$z = \tfrac{1}{2}\kappa x^2 + \tfrac{1}{2}\lambda y^2$$

and so will have vertices approximately on the lines of curvature through the origin. The locus of all vertices on the contours will be a curve in $\mathbb{R}^2$ with a crunode at the origin, the two branches crossing orthogonally at the hill-top or bottom of the depression, the tangent lines to the two branches coinciding with the principal tangent lines at the origin. (Near a saddle-point one has a similar result, the contours near the saddle being approximately hyperbolas.)

At an umbilical hill-top we have $\kappa = \lambda$ so that the contours are then approximately circular. Indeed for an ellipsoid they actually are circles. See the detailed discussion of the ellipsoid in Chapter 16. However, in general this is not the case. The question arises how many vertices the contours then have and how these are distributed around the bumpy circle. One might expect the answer to depend on the nature of the principal cubic form associated to the umbilic. This is indeed so, and the first guess might be that there are either two or six vertices, depending on the nature of the cubic form, though recollection of the four-vertex theorem, Theorem 1.30, immediately puts the lie to this.

The answer is perhaps a little surprising and depends on the *harmonic part* of the principal cubic at the umbilic. See the end of Section 7.4 for the definition. In the family of cubic forms obtained by adding to the principal form the product of the first fundamental form by an arbitrary linear form there is just one which satisfies the Laplace equation. It has all its root directions real and distributed evenly around the circle, the acute angle between any two being $\pi/3$. This form is the required harmonic part. Exceptionally it may reduce to zero, in the case that the umbilic is a pure lemon.

Theorem 15.9 With the above notations, and provided the harmonic part of the principal cubic associated to the umbilic is not zero, the locus of all vertices of contours near an umbilic has three branches which cross at the umbilic, the acute angle between any two of the lines being $\pi/3$, the limiting tangent lines at the vertices being the root lines of the harmonic part of $V(\mathbf{c})_3$ and the tangent lines to the branches of the locus of vertices being orthogonal to these.

Sketch of proof We take the equation of a contour to be

$$\tfrac{1}{2}\kappa((x, y)\cdot(x, y)) + \tfrac{1}{6}C_3(x, y)^3 + O(x, y)^4 = \tfrac{1}{2}\kappa^3\varepsilon^2,$$

where ε is small, where C_3 is a thrice linear form, this being as we have seen a form that is equivalent to the form $V(\mathbf{c})_3$. Under the radial transformation $(x, y) = \kappa\varepsilon(x', y')$ this transforms (immediately dropping the primes) to

$$(x, y)\cdot(x, y) + (\varepsilon/3)C_3(x, y)^3 + \varepsilon^2 O(x, y)^4 = 1,$$

or, neglecting the ε^2 term, for ε small, to

$$2F(\mathbf{r}) = \mathbf{r}\cdot\mathbf{r} + (\varepsilon/3)C_3\mathbf{r}^3 = 1, \text{ where } \mathbf{r} = (x, y).$$

Our task is to locate any vertices of this *bumpy circle* (ignoring any faraway non-compact components of the curve). For any parametrisation $t \mapsto \mathbf{r}(t)$ of this bumpy circle the vertex centres $\mathbf{c}$ are given by the equations

$$\begin{aligned}
(\mathbf{c} - \mathbf{r})\cdot\mathbf{r}_1 &= 0 & & & F_1\mathbf{r}_1 &= 0 \\
(\mathbf{c} - \mathbf{r})\cdot\mathbf{r}_2 &= \mathbf{r}_1\cdot\mathbf{r}_1, & &\text{where} & F_2\mathbf{r}_1{}^2 + F_1\mathbf{r}_2 &= 0 \\
(\mathbf{c} - \mathbf{r})\cdot\mathbf{r}_3 &= 3\mathbf{r}_1\cdot\mathbf{r}_2 & & & F_3\mathbf{r}_1{}^3 + 3F_2\mathbf{r}_1\mathbf{r}_2 + F_1\mathbf{r}_3 &= 0.
\end{aligned}$$

Writing $F_1 = \lambda(\mathbf{c} - \mathbf{r})\cdot$ and then eliminating λ, and noting that $F_2 = \cdot + \varepsilon C_3\mathbf{r}$ and $F_3 = \varepsilon C_3$, we find that

$$(\mathbf{r}_1\cdot\mathbf{r}_1)(\varepsilon C_3\mathbf{r}_1{}^3 + 3\mathbf{r}_1\cdot\mathbf{r}_2 + 3\varepsilon C_3\mathbf{r}\mathbf{r}_1\mathbf{r}_2) = 3(\mathbf{r}_1\cdot\mathbf{r}_2)(\mathbf{r}_1\cdot\mathbf{r}_1 + \varepsilon C_3\mathbf{r}\mathbf{r}_1{}^2),$$

that is

$$(\mathbf{r}_1\cdot\mathbf{r}_1)(C_3\mathbf{r}_1{}^3 + 3C_3\mathbf{r}\mathbf{r}_1\mathbf{r}_2) = 3(\mathbf{r}_1\cdot\mathbf{r}_2)(C_3\mathbf{r}\mathbf{r}_1{}^2).$$

Now we may take the parametrisation of the bumpy circle to be

$$t \mapsto \mathbf{r}(t) = \mathrm{e}^{\mathrm{i}t} + \text{a small correction term},$$

where we identify $\mathbb{R}^2$ with the complex numbers $\mathbb{C}$. Then

$$\mathbf{r}_1(t) = \mathrm{i}\mathrm{e}^{\mathrm{i}t} + \ldots \text{ and } \mathbf{r}_2 = -\,\mathrm{e}^{\mathrm{i}t} + \ldots = -\mathbf{r} + \ldots.$$

So to the first order $\mathbf{r}_1\cdot\mathbf{r}_2 = 0$, with $\mathbf{r}_2\cdot\mathbf{r}_2 = \mathbf{r}_1\cdot\mathbf{r}_1$. Let $\mathbf{u}$, $\mathbf{v}$ be the limiting values of $\mathbf{r}_1$, $\mathbf{r}_2$ as ε reaches 0. Then

$$C_3\mathbf{u}^3 - 3C_3\mathbf{u}\mathbf{v}^2 = 0,$$

where $\mathbf{u}\cdot\mathbf{v} = 0$ and $\mathbf{u}\cdot\mathbf{u} = \mathbf{v}\cdot\mathbf{v}$. By Proposition 7.12 the left-hand side of this equation is just four times the harmonic part of $C_3\mathbf{u}^3$.

The various assertions we have made follow at once from this. □

Thus the smallest contours at a generic umbilical hill-top are bumpy circles with *six* vertices – a place perhaps to recall the comment of Osserman quoted at the end of Chapter 1.

For an indication of an alternative approach to hill-top contours see Exercise 8.2.

15.4 Higher order osculating circles

A second application of harmonic cubic forms arises in the search for circles with higher order contact with the surface than five-point (A_4), as briefly announced in Porteous (1983a). James Montaldi has shown (1983, 1986b) that at any point of a surface there are homogeneous polynomials of degrees 3 and 6 in three variables, ten of whose eighteen common zeros determine possible tangent directions of circles with five-point (A_4) contact at least with the surface. It is, so far as we are aware, still an open question as to whether all ten can be real. Six certainly can be, as follows from an example of R. Blum (1980) (a cyclide, a surface each of whose lines of curvature is a circle), and it seems likely that this is the maximum number.

As for circles with six-point (A_5) contact with a generic surface these will exist at points along certain curves on the surface, with seven-point (A_6) contact circles existing at isolated points. What Montaldi has shown is that through any umbilic (other than one of pure lemon type, where the harmonic form reduces to zero) there are always exactly three such A_5 curves and that as one approaches an umbilic along such a curve the six-point contact circle curls up and dies, just as for circles with A_3 contact in principal directions along ridges, the limit tangent line of the circle (*not* the limit tangent of the curve) being a root line of the harmonic cubic. That is the osculating circle shrinks to zero at the umbilic, either by falling onto its face or onto its back, according as the index of the umbilic is $\frac{1}{2}$ or $-\frac{1}{2}$ (not $-\frac{1}{2}$ or $\frac{1}{2}$ as Montaldi mistakenly says, due to a sign error in his argument). For the actual formulas determining the limit tangent lines to the three curves see equations (10) on p. 124 of Montaldi (1986b).

Exercises

15.1 Verify the assertion in the proof of Theorem 15.2, part (iii), that the condition that the vector $(\mathbf{n}, -\kappa)$ is killed by the last row of the 4×4 matrix of that proof reduces to

$$V_4\mathbf{q}_1{}^4 + 6V_3\mathbf{q}_1{}^2\mathbf{q}_2 + 3V_2\mathbf{q}_2{}^2 + 4V_2\mathbf{q}_1\mathbf{q}_3 + V_1\mathbf{q}_4 = 0,$$

where V_i is the ith derivative at w of $-\frac{1}{2}(\mathbf{c}-\mathbf{s})\cdot(\mathbf{c}-\mathbf{s})$ if $\kappa \neq 0$ and $\mathbf{c} = \kappa^{-1}\mathbf{n}$, or of $\mathbf{n}\cdot\mathbf{s}$ if $\kappa = 0$.

15.2 Prove Proposition 15.4.

15.3 Prove Proposition 15.5.

15.4 Prove Proposition 15.6.

15.5 Prove Proposition 15.7.

15.6 Prove Proposition 15.8.

15.7 In each of the above exercises consider the possibility that one or more of the circles involved reduces to a line.

15.8 Explore the alternative approach to hill-top contours suggested by Exercise 8.2.

16
Examples of surfaces

16.0 Introduction

Most of the surfaces considered in this chapter have already appeared in exercises. Though many possess symmetries and are to that extent certainly non-generic they exemplify well in a readily computable form much of the theory of the preceding chapters. Tubes are a half-way house between space curves and surfaces, ellipsoids possess lots of beautiful geometry, while the bumpy spheres of Stelios Markatis and the minimal monkey saddle are fun to play with. These are pregnant examples, all well worth detailed study. The touch here is light, leaving much of the detail to the reader to fill in.

16.1 Tubes

Consider the *tube* **s** with *core* a regular smooth space curve **r** and radius δ. If we parametrise **r** by arc-length then a parametrisation for the tube is $(s, \theta) \mapsto \mathbf{s}(s, \theta) = \mathbf{r}(s) + \delta(\mathbf{n}(s)\cos\theta + \mathbf{b}(s)\sin\theta)$, the unit normal to the tube at (s, θ) being $\mathbf{N}(s, \theta) = \mathbf{n}(s)\cos\theta + \mathbf{b}(s)\sin\theta$. By a straightforward computation involving the Serret–Frenet formulas

$$\mathbf{s}_1 \cdot \mathbf{s}_1 = [(1 - \kappa\delta\cos\theta)^2 + \tau^2\delta^2,\ \tau\delta^2,\ \delta^2]_2$$

and

$$\mathbf{N} \cdot \mathbf{s}_2 = -\mathbf{N}_1 \cdot \mathbf{s}_1 = [(1 - \kappa\delta\cos\theta)\kappa\delta\cos\theta - \tau^2\delta,\ -\tau\delta,\ -\delta]_2.$$

(In this case there is a simple expression for **N** so that the form $-\mathbf{N}_1 \cdot \mathbf{s}_1$ of the second fundamental form is the easier of the two to deal with.) The principal directions are then represented by

$$\begin{bmatrix} 0 \\ 1 \end{bmatrix} \text{ and } \begin{bmatrix} 1 \\ -\tau \end{bmatrix},$$

with images the tangent vectors $\delta\partial\mathbf{N}/\partial\theta = -\mathbf{n}\delta\sin\theta + \mathbf{b}\delta\cos\theta$ and $(1-\kappa\delta\cos\theta)\mathbf{t}$, the latter being zero only where $\rho = \kappa^{-1} = \delta\cos\theta$. The corresponding principal curvatures are $-\rho$ and $\cos\theta(\rho - \delta\cos\theta)^{-1}$, the latter being zero where $\cos\theta = 0$, that is where $\theta \equiv \pm\frac{1}{2}\pi$, mod 2π. It is easy to verify that an equivalent condition for this is that $\partial\mathbf{N}/\partial s = 0$. What follows from this is just what one would expect, namely that provided that everywhere $\delta < \rho$ the tube is regular, with two parabolic lines, its lines of curvature being the circles in which it is cut by the normal planes to $\mathbf{r}$ and parallel curves to $\mathbf{r}$ at distance δ. The evolute surface is degenerate in that one of its sheets collapses to the original space curve $\mathbf{r}$, the core of the tube. The other coincides with the focal surface of $\mathbf{r}$, covered twice, since the focal points of 'mutually antipodal points' of the tube coincide. The focal curves of the tube on the non-degenerate sheet of the focal surface coincide with the focal curves of the core and provide the proof of their existence that was lacking in Chapter 6. The tube has two 'mutually antipodal' ridges, each a trace on the tube of the space evolute of the curve $\mathbf{r}$. These ridges are generically of type A_3 but are of ordinary type A_4 where the space evolute of the core of the tube has an ordinary cusp.

In the special case that the core has a special linear point at which the entire focal line satisfies the A_3 condition, the point of tangency with the space evolute being a special A_4 point, the same is true for each line of curvature of the tube, in particular for that line of curvature that passes through either ridge point of the tube there. Then the meridian circle at that point also is a ridge of the tube of the same colour as the others, providing an example of a regular smooth surface whose ridges exhibit crunodes.

Are the ridges on a tube sterile or fertile? Certainly umbilics cannot occur since the focal points on any normal can never coincide. However, ordinary A_4 points may occur, corresponding to ordinary cusps on the space evolute, the doubly covered rib of the tube, and it follows that both types of ridge can occur. In fact it may be shown that antipodal ridge points are of opposite type.

16.2 Ellipsoids

The general ellipsoid is the quadratic surface in $\mathbb{R}^3$ with equation

$$\frac{x^2}{a^2} + \frac{y^2}{b^2} + \frac{z^2}{c^2} = 1, \text{ where } a > b > c > 0.$$

This can be explored locally by various choices of parametric representation, or directly from the equation. A useful model of such an ellipsoid may be constructed on the egg-box pattern from a number of interlocking cardboard circles, as in Figure 16.1 (see Cundy and Rollett (1961), Section 4.3.6, for details). Every section of an ellipsoid is an ellipse and every non-circular ellipse has exactly four vertices, at each of which the circle of curvature has exactly four point (A_3) contact.

The ellipsoid has three ridges (Figure 16.2), the *major* and *minor* sections of symmetry, which are both seen from the model in Figure 16.1 to be sterile since none of the sections normal to either ridge is circular, and the *intermediate* or *major–minor* section of symmetry, which is fertile and which contains the four umbilical points, whose positions are very clear in the model! These are at the points ($a \cos \varphi$, 0, $c \sin \varphi$), where $a^2 \sin^2 \varphi + c^2 \cos^2 \varphi = b^2$. The planes parallel to the tangent planes at these points which intersect the ellipsoid do so in circles. It is also clear from Figure 16.1 how as one approaches one of these umbilics along the fertile ridge one of the two circular sections falls on to its face and dies, as the theory presented in Chapter 15 predicts.

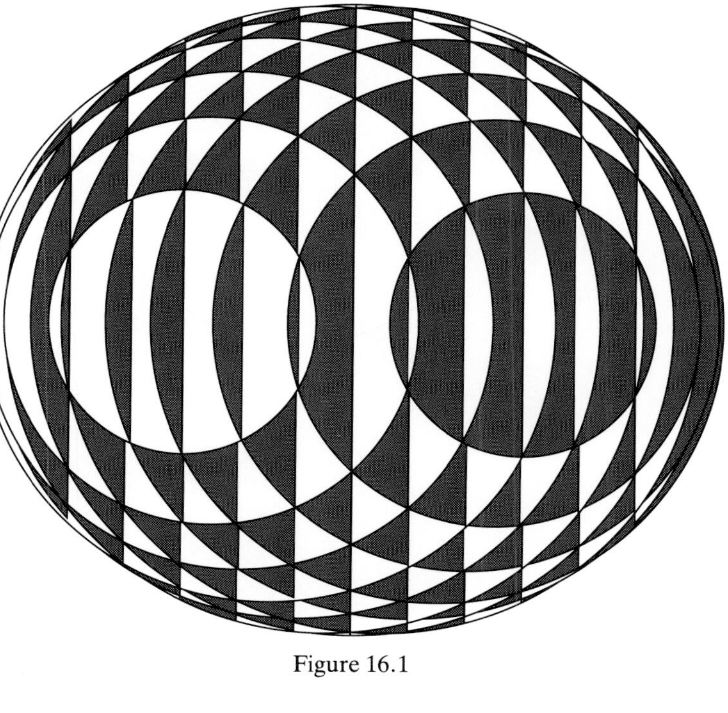

Figure 16.1

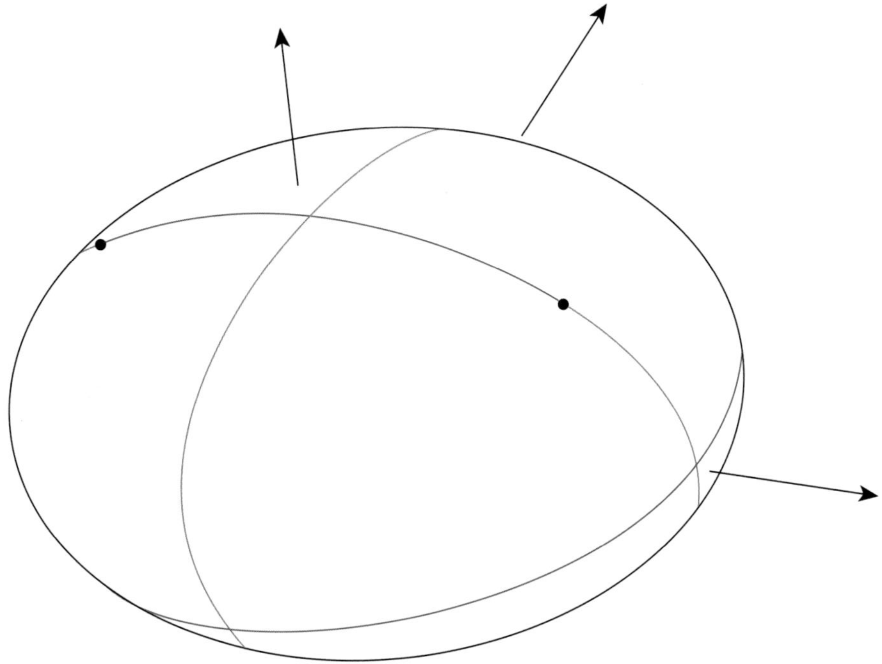

Figure 16.2

The ridge through the umbilics changes colour at each one. Of the other two one is wholly red and the other wholly blue. Each of these ridges crosses each of the other two at a vertex of that ridge when regarded as a plane curve, as the theory predicts. At such a point the ridges that cross are of opposite colour, so that the crossing point itself may be regarded as a *purple* point. It can be verified that there are no other ridges other than these three. The four umbilics are all of pure lemon type, each of index $\frac{1}{2}$. This may be verified directly but follows indirectly as a corollary of the study of the 'contours' round an umbilic that we made in Chapter 15, these contours being for an ellipsoid exact circles.

In the special case that two of the semiaxes are equal, say $b = c$, the ellipsoid becomes an ellipsoid of revolution and then has just two umbilics, at the 'poles' $(\pm a, 0, 0)$, each being of index 1. In this case one of the sheets of the focal surface collapses to the interval of the axis of revolution with end-point the poles, while the other has a rib lying over the equator, cut out on the ellipsoid by the plane $x = 0$.

Gaspard Monge, a friend of Napoleon, who was the first Professor of Mathematics at the École Polytechnique, the military university founded by Napoleon in Paris in 1794, described the lines of curvature of an ellipsoid in his lectures. There was at that time an architectural competition for a new High Court building and Monge imagined that the ideal floor plan would be elliptical, with the dock at one focus and the judge's throne at the other. The natural form for the roof would then naturally be half an ellipsoid, and Monge gives reasons why the semiaxes should be such that two of the umbilics would be visible in the roof. The edges of the roof tiles would ideally follow the lines of curvature of the roof (Figure 16.3). His conclusion, from his lecture notes at the École Polytechnique, is worth quoting verbatim:

> *Enfin, deux lustres suspendus aux ombiliques de la voûte, et à la suspension desquel la voûte entière semblerait concourir, serviraient à éclairer la salle pendant la nuit.*
>
> *Nous n'entrerons pas de plus grand détails à cet égard; il nous suffit d'avoir indiqué aux artistes un objet simple, et dont la décoration, quoique très riche, pourrait n'avoir rien d'arbitraire, puisqu'elle consisterait principalement à dévoiler à tous les yeux une ordonnance très-gracieuse, qui est dans la nature même de cet objet.*
>
> G. Monge: *Feuilles d'Analyse* – 1795

Monge died shortly after his expulsion from the Académie Française following Napoleon's banishment to St Helena.

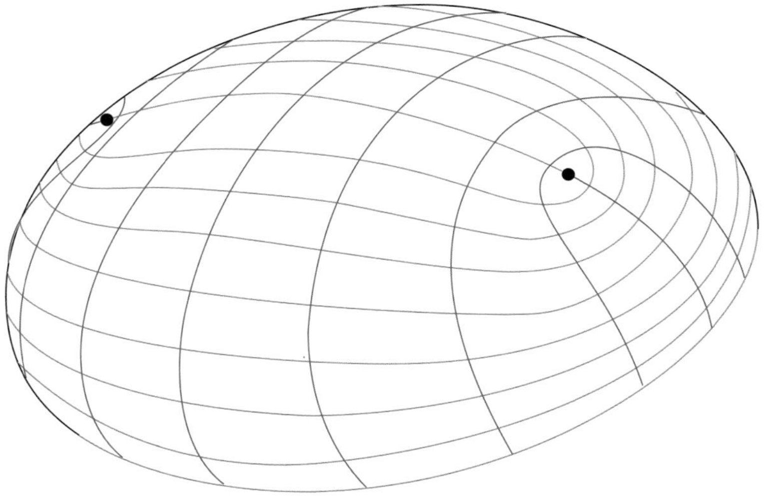

Figure 16.3

16.3 Symmetrical singularities

As we have seen for an ellipsoid in $\mathbb{R}^3$ there are reflections of $\mathbb{R}^3$ that map the ellipsoid to itself and in such a case the fixed point set of such a reflection is a ridge of the ellipsoid. Each such reflectional symmetry can be classified as an action of the cyclic group $\mathscr{C}(2)$ of order 2 on $\mathbb{R}^3$ on the ellipsoid. Several singularities of the distance-squared function for a surface in $\mathbb{R}^3$ can occur in such a way and it is convenient to recode these, in accord with Slodowy (1980). Explicitly we recode A_3 with reflectional symmetry as B_2, A_5 as B_3, D_4 as C_3, D_5 as C_4 and E_6 as F_4. We shall also encounter examples of surfaces invariant under actions of the group $\mathscr{S}(3)$, of order 6, on $\mathbb{R}^3$. A fixed point of the surface under such an action will in general be an elliptic umbilic whose classifying cubic form is harmonic, represented in the β-plane by $\beta = 0$. Such an umbilic will be said to be of type G_2.

As far as the ellipsoid is concerned its ridges are each of type B_2 and its umbilics are of type C_3.

As was pointed out in Chapter 13 a B_2 ridge is also a subparabolic line of the same colour and an everywhere geodesic line of curvature of the opposite colour. As a plane curve it has in general isolated points that are ordinary vertices of the curve. It is an easy exercise (Exercise 16.4) to show that at such a point such a ridge is crossed at right angles by a ridge of the opposite colour.

On a generic regular surface one does not expect to have points where ridges of the *same* colour cross each other, though of course such crossing points can occur on particular surfaces in generic families of surfaces. However, for a surface with reflectional symmetry there are in general points on the line of symmetry where another ridge of the same colour crosses, necessarily at right angles. Such points, being symmetric A_5 points, may be labelled as B_3 points. There are no such points on ellipsoids but they do occur in other examples that we deal with in the next section.

In his work on the accommodation of the eye lens, for which he was awarded the Nobel Prize in 1911, Gullstrand studied umbilical points of surfaces in detail (1904). After describing in detail the patterns of lines of curvature and ridges around ordinary umbilics of a surface he proceeds to detail what happens at umbilics with rectangular symmetry. In the final section of his paper he also gives examples of surfaces having lines of umbilical points, with both variable and constant radius of curvature.

16.4 Bumpy spheres

In our study of the contours of a regular surface around an umbilical hill-top in Chapter 15 we made acquaintance with bumpy circles. Moving up a dimension Stelios Markatis (1980) considered the level surfaces around a local maximum or minimum of a function $h : \mathbb{R}^3 \rightarrowtail \mathbb{R}$. The most interesting case is that least likely to occur, when these level surfaces are bumpy spheres. By the sort of radial transformation that we made in Chapter 15 he was led to study surfaces of the form

$$x^2 + y^2 + z^2 + \tfrac{1}{3}\varepsilon C_3(x, y, z)^3 = 1,$$

where C_3 is a thrice linear form on $\mathbb{R}^3$, ε is small and faraway non-compact components are once again disregarded. Markatis has shown that in the family of cubic forms

$$C_3(x, y, z)^3 + A(x, y, z)(x^2 + y^2 + z^2),$$

where A is linear, there is a unique harmonic cubic and also a unique cubic of the form LMN, where L, M and N are real linear forms. The geometry of the bumpy sphere, namely its configuration of ridges and umbilics, is uniquely determined by either of these. Technically the *LMN* form is the easiest to handle.

Obvious examples to look at initially are:

$x^2 + y^2 + z^2 + \frac{1}{3}\varepsilon xyz = 1$ (the *bumpy cube*, or more properly the *bumpy tetrahedron*, since the symmetries of the surface are those of the tetrahedron with vertices $(1, 1, 1)$, $(-1, 1, 1)$, $(1, -1, 1)$, $(1, 1, -1)$, rather than the full group of symmetries of the cube inside which the tetrahedron lies),

$x^2 + y^2 + z^2 + \frac{1}{3}\varepsilon(x^3 - 3xy^2) = 1$ (the *bumpy orange*),

$x^2 + y^2 + z^2 + \frac{1}{3}\varepsilon xy^2 = 1$ (the *bumpy tennis ball*), and

$x^2 + y^2 + z^2 + \frac{1}{3}\varepsilon x^3 = 1$ (the *bumpy sphere of revolution*),

these four types being represented by the vertices of a solid three-dimensional tetrahedron, representing all possible types.

It is not difficult to figure out that in the case of the bumpy cube the sections of symmetry illustrated in Figure 16.4 are all ridges, there being elliptic (in fact G_2) umbilics, each of index $-\frac{1}{2}$, at the eight

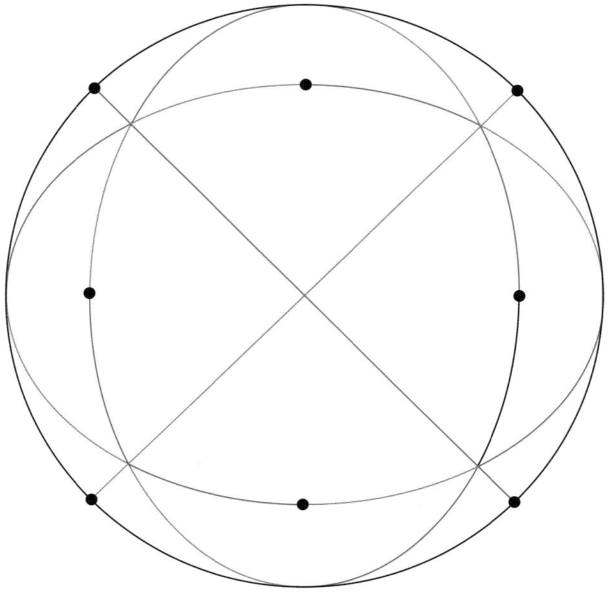

Figure 16.4

vertices of the cube and hyperbolic (in fact lemon B_3) umbilics, each of index $\frac{1}{2}$, at, or rather near, the mid-points of the twelve edges of the cube. The total index is equal to the Euler characteristic of the sphere, namely 2, as the index theorem predicts. At the mid-point of each face of the cube there is a purple flyover point where a red ridge crosses a blue one. Moreover, each of these ridges is a bumpy circle, with six vertices equally spaced round the circle in the limit, as follows from the argument used in the proof of Theorem 15.4. Now by Exercise 16.4, already referred to, at each of these vertices the bumpy circle is crossed by a ridge of the opposite colour, and so we can mark four additional flyovers on each of these circles, lying close to but not at the vertices of the cube. Thus we can predict the existence of eight small almost circular ridges each surrounding one of the G_2 points (Figure 16.5). These were in fact first detected in a computer print-out! Four of these ridges are blue and four are red. Each crosses in turn the various ridges emanating from the vertex, alternately by a flyover and by a crossing on the level, this last being a symmetric A_5 or C_3 point. All other points of the ridges of symmetry are C_2 points.

For the bumpy orange it is clear that the equator and three equally

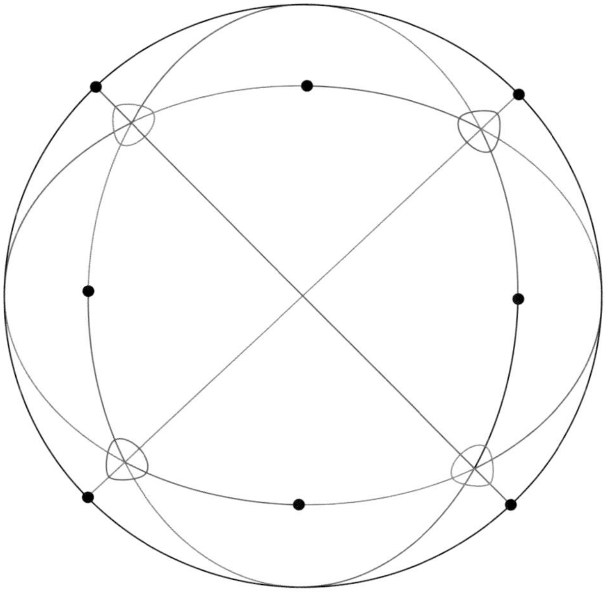

Figure 16.5

spaced meridian circles through the poles are ridges, these all being lines of symmetry. All are bumpy. There are also two nearly circular ridges, one blue and one red, entwining around the north pole, and likewise similar ridges in the southern hemisphere. These also were first detected in a computer print-out. The positions of the two G_2 umbilics, each of index $-\frac{1}{2}$, one at each pole, and of six lemon B_3 umbilics, each of index $\frac{1}{2}$, are indicated on Figure 16.6, which is an orthogonal projection of the northern hemisphere. There are in this case eight umbilics in all, as opposed to the twenty in the case of the bumpy cube, the sum of the indices being once more equal to 2.

The bumpy tennis ball (Figure 16.7) also has eight umbilics, two rather special parabolic umbilics of index $-\frac{1}{2}$ and six hyperbolic umbilics of index $\frac{1}{2}$, the unique ridge through the parabolic umbilics being triple, while every point of the equator of the bumpy sphere of revolution is a non-generic umbilic, of index 0, the poles being non-generic umbilics of index 1.

The problem that Markatis resolved was how any one of these patterns transformed to another as the cubic form was transformed

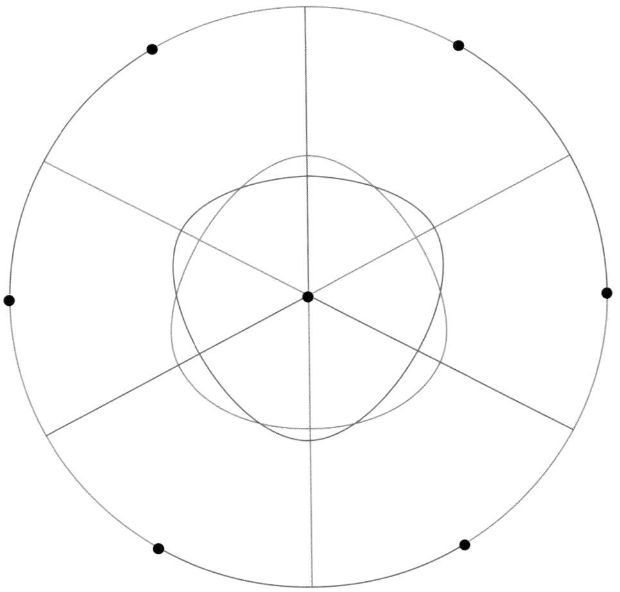

Figure 16.6

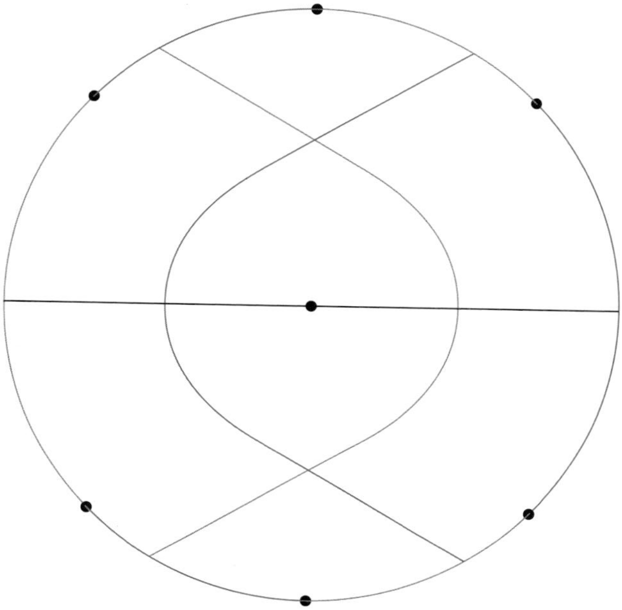

Figure 16.7

gradually from the one to the other. He studied in particular how the twenty umbilics of the bumpy cube reduce to the eight of the bumpy orange or tennis ball. What he has shown is that no matter how the deformation is performed, all deaths or amalgamations occur simultaneously, when either six pairs annihilate each other or six triples become six singletons. The latter possibility occurs at an E_6 (in fact F_4) point in the right hand of the two sequences of Figure 16.8 which illustrate the unknotting of a 'G_2 circle', showing two ways in which the pattern of ridges round a G_2 umbilic may break up. It is of interest that in this context at least the symmetry breaks in two stages, from the full triangular $\mathcal{S}(3)$ symmetry to the cyclic symmetry $\mathcal{C}(2)$ first of all and not to the 'Manx' symmetry $\mathcal{C}(3)$ of Figure 16.9, before all symmetry disappears, as in the sequences of Figure 16.10.

Computer generated pictures by Markatis (1980) showing all types of bumpy sphere, with their ridges, are shown in Figures 16.11–16.20. The points where three ridges cross are elliptic umbilics. The hyperbolic umbilics are not shown.

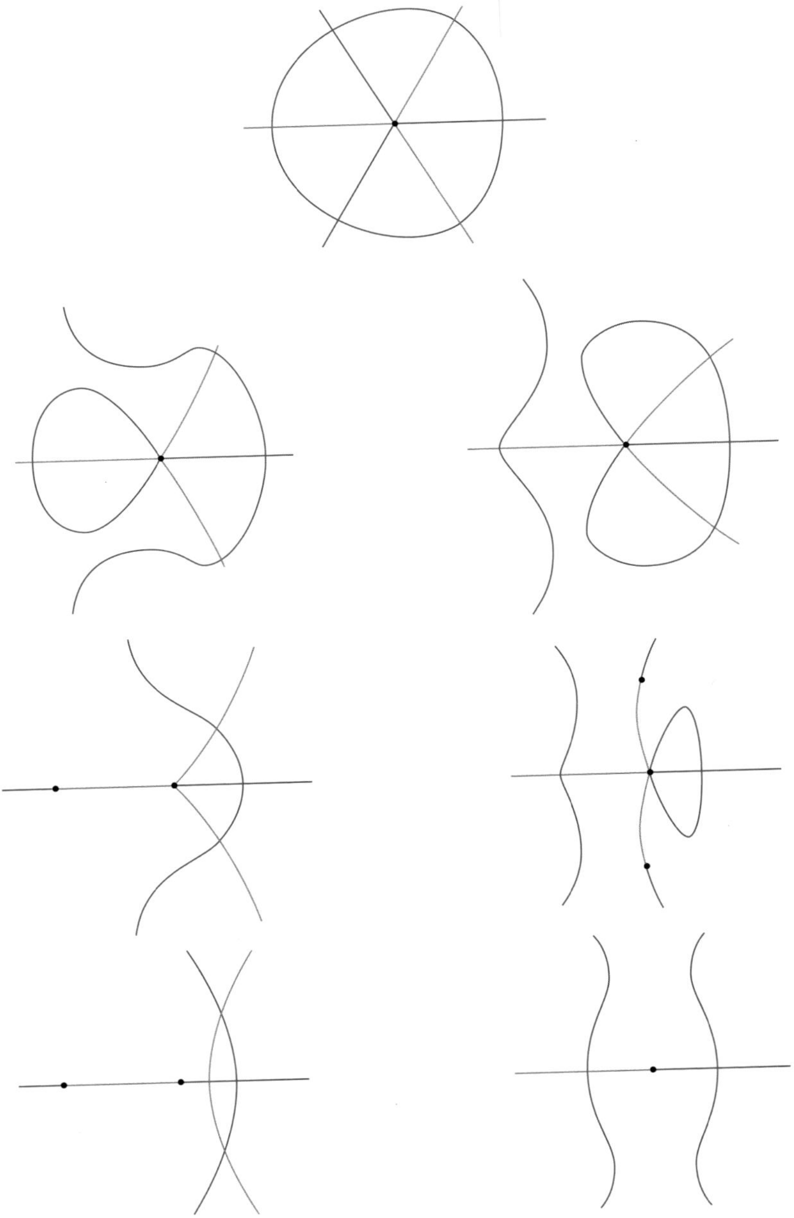

Figure 16.8

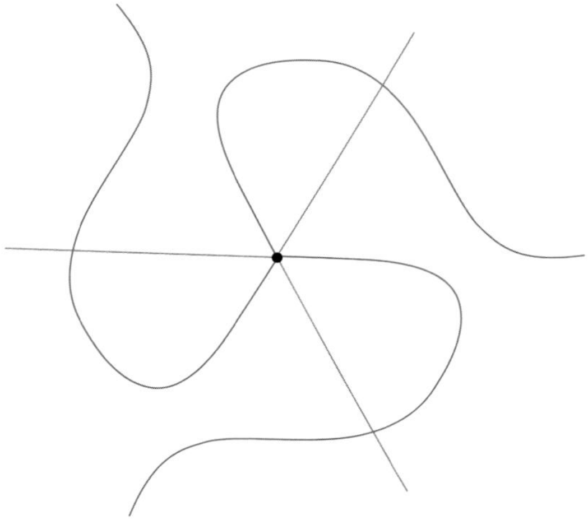

Figure 16.9

Figure 16.10

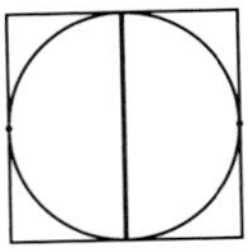

Figure 16.11

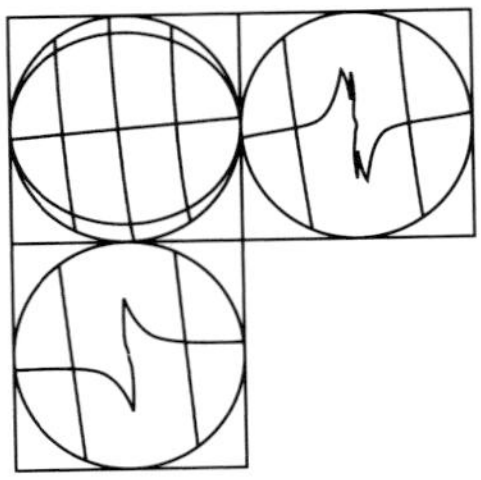

Figure 16.12

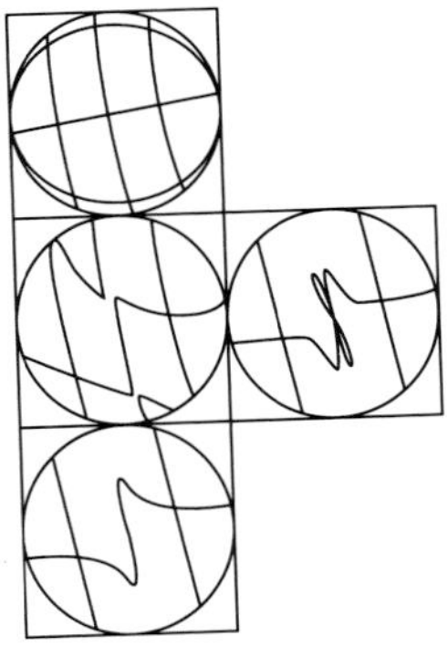

Figure 16.13

Figure 16.14

Figure 16.15

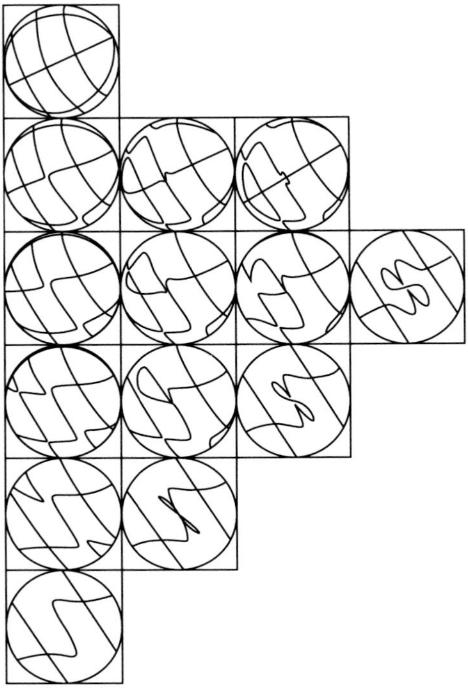

Figure 16.16

16.5 The minimal monkey-saddle

The question arises whether one can construct a surface with a G_2 umbilic for which the red and blue sterile ridges circling it coincide. This is answered in the affirmative by the *minimal monkey saddle*. A *minimal* surface is one for which the two principal curvatures are everywhere equal but of opposite sign. Weierstrass last century gave a construction for such surfaces and the minimal monkey saddle is such a surface, namely the surface $(x, y) \mapsto (u, v, w)$, where

$$u = x - \tfrac{1}{5}(x^5 - 10x^3y^2 + 5xy^4),$$
$$v = y + \tfrac{1}{5}(5x^4y - 10x^2y^3 + y^5),$$
$$w = \tfrac{2}{3}(x^3 - 3xy^2).$$

(Notice here the appearance of the real and imaginary parts of z^5 and the real part of z^3, where z is the complex number $x + \mathrm{i}y$.)

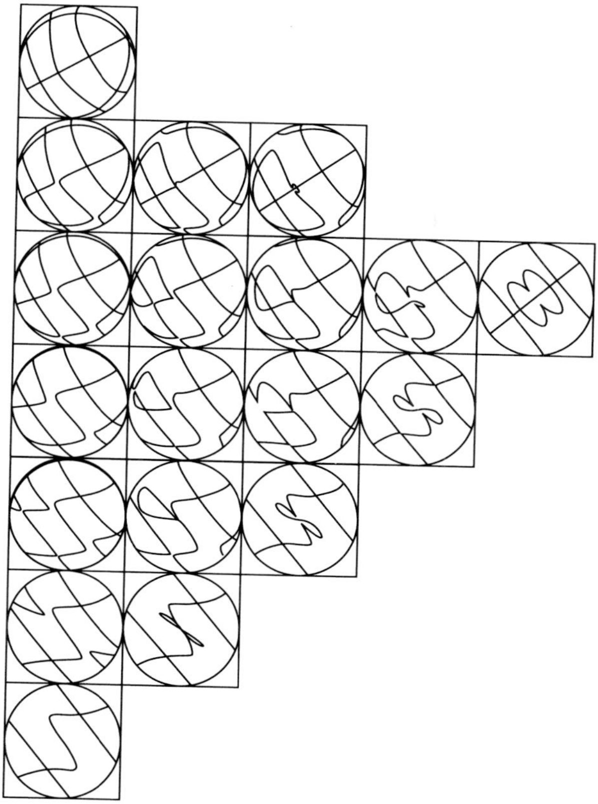

Figure 16.17

The argument is simple. Computation shows that the unit normal to the surface at (x, y) is $(1 + r^4)^{-1}(-2x^2 + 2y^2,\ 4xy,\ 1 - r^4)$ where $r^2 = x^2 + y^2$. Moreover, the fundamental forms at (x, y) are

$$\mathrm{I}_2(x, y) = (1 + r^4)^2[1\ \ 0\ \ 1]_2 \text{ and } \mathrm{II}_2(x, y) = 4(1 + r^4)[x\ \ -y\ \ -x]_2,$$

from which it follows that the principal curvatures there are $\pm 4r/(1 + r^4)$. That is, the surface is a minimal surface, while the Gaussian curvature is $-16r^2/(1 + r^4)^2$, which depends only on the distance r from the origin in the parameter space, this being critical (in fact a maximum) all round the circle in the parameter space with centre the origin and radius $r = 3^{-1/4}$. The image of this circle on the monkey saddle surface is the desired purple ridge.

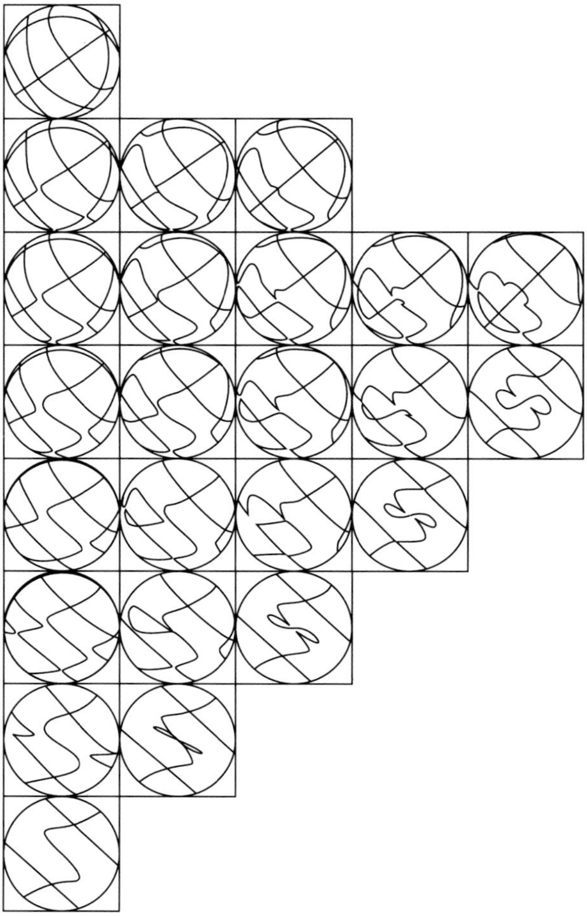

Figure 16.18

In the example the curvatures are zero at the umbilic, implying that the umbilic is a flat umbilic. One can construct an example with any desired curvature at the umbilic simply by inverting the above example in a suitable sphere, with centre on the z-axis, the normal at the umbilic. Of course the inverted surface will not be a minimal surface.

For recent work on minimal surfaces see, for example, Osserman (1989) and Hoffman and Meeks (1990).

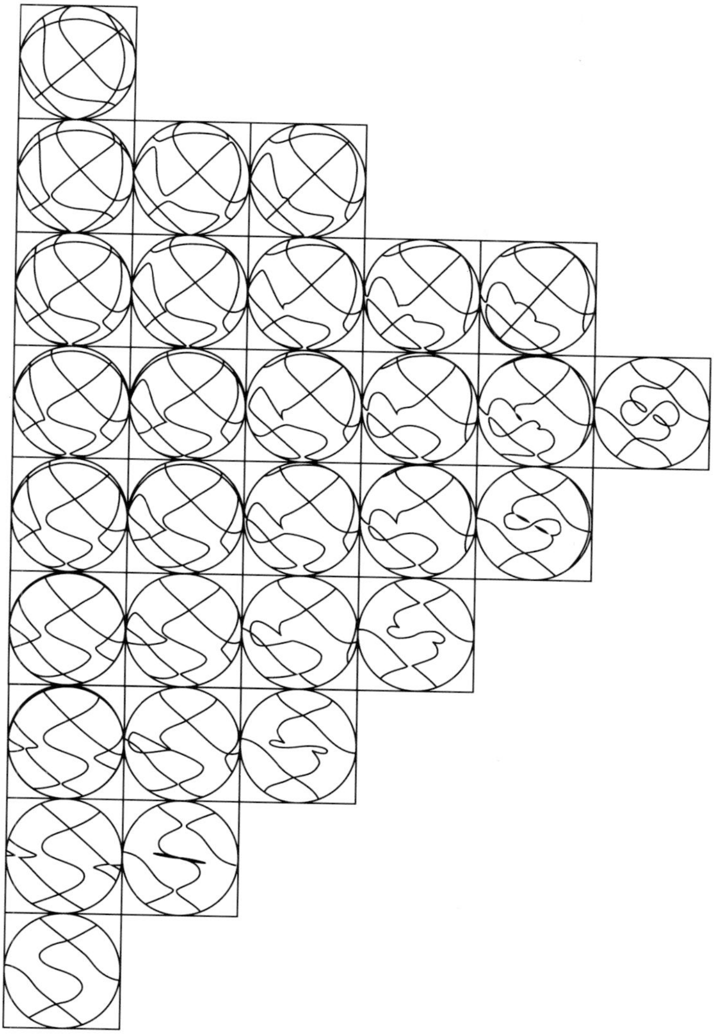

Figure 16.19

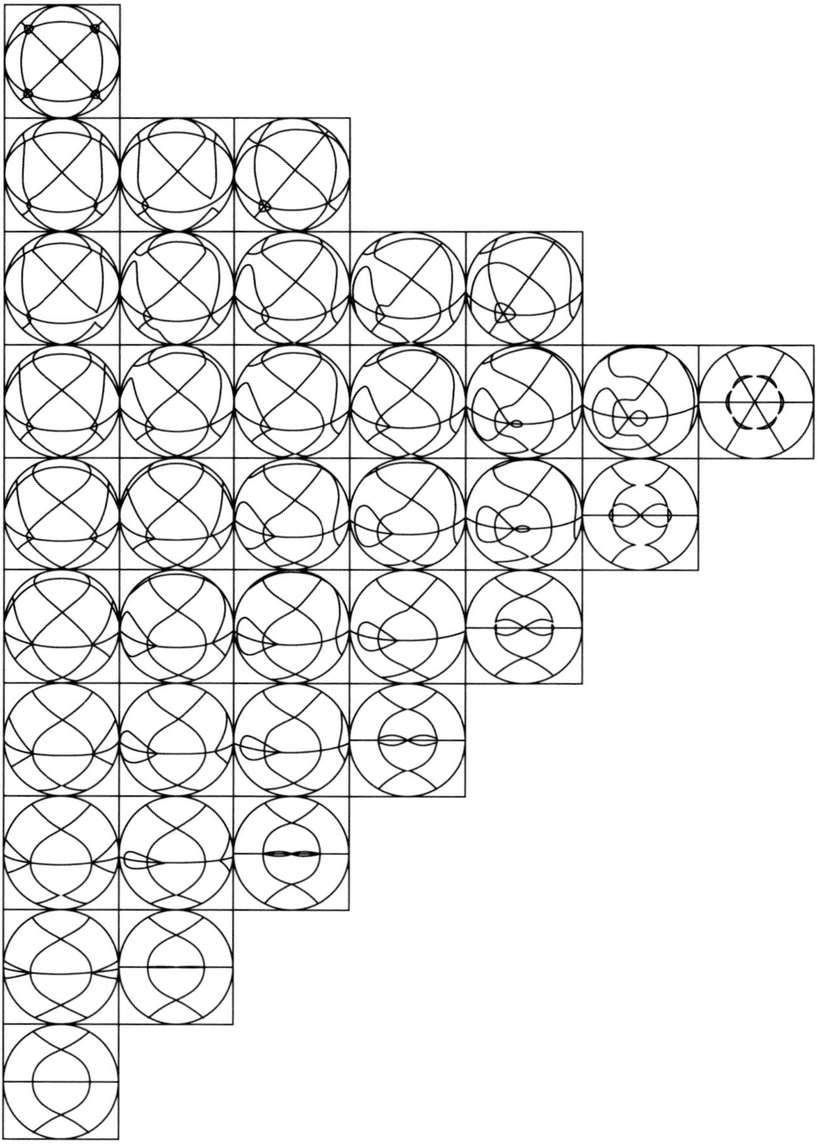

Figure 16.20

Exercises

16.1 Prove that antipodal ridge points of a tubular surface are of opposite type, one fertile and the other sterile.

16.2 Let $F : \mathbb{R}^3 \to \mathbb{R}$ by a homogeneous polynomial map of degree n. Clearly every point of the cone with equation $F = 0$ and vertex at the origin is parabolic, each point of a generator being flat if and only if the corresponding point of the projective curve is a linear point. Hence derive a fresh proof of the Hessian condition for the linearity of a real projective curve. (Exercise 2.5 is relevant here.)

16.3 Make an 'egg-box' ellipsoid.

16.4 Suppose that a regular smooth surface in $\mathbb{R}^3$ has reflectional symmetry. Prove that at any vertex of the curve of intersection of the surface with the plane of symmetry, regarded as a ridge of the surface, the curve is crossed at right angles by a ridge of the opposite 'colour'.

16.5 Prove that the line of intersection of the surface with equation

$$z = \frac{1}{2}\left[x^2 + \frac{y^2}{x^2 + 1}\right]$$

with the plane $y = 0$ consists of umbilic points at each of which the classifying cubic form is right-angled (Gullstrand, 1904).

16.6 Let α, β and γ each be angles in the interval $[0, \pi]$. Prove that in one and only one of the triples (α, β, γ), $(\alpha, \pi - \beta, \pi - \gamma)$, $(\pi - \alpha, \beta, \pi - \gamma)$, $(\pi - \alpha, \pi - \beta, \gamma)$ is the sum of the two largest angles less than or equal to π.

16.7 Prove that the line of intersection of the bumpy sphere with equation

$$x^2 + y^2 + z^2 + \varepsilon z^3 = 1 \ (\varepsilon \text{ small})$$

with the equatorial plane $z = 0$ consists entirely of umbilics.

16.8 Verify the stated positions of the umbilics of the bumpy cube, and bumpy orange.

16.9 Verify that the umbilics of the bumpy tennis ball with equation

$$x^2 + y^2 + z^2 + \tfrac{1}{3}\varepsilon xy^2 = 1$$

are at $(\pm 1, 0, 0)$, and near $(\pm\sqrt{(\frac{1}{3})}, \pm\sqrt{(\frac{2}{3})}, 0)$ (hyperbolic of index $\frac{1}{2}$), and at the poles $(0, 0, \pm 1)$ (hyperbolic of index $-\frac{1}{2}$).

16.10 Determine and plot the subparabolic lines of the bumpy spheres.

Further reading

First of all, it is time to come clean on the subject of genericity! Strictly speaking there is no such thing as a *generic curve* or *generic surface*, but individual *properties* of curves or surfaces may be generic. For example to say that a particular property of a curve of a particular class of curves is *generic* is to say that in the *space of all curves of that class* the curves exhibiting this property form an *open dense* set, implying that *arbitrarily close* to any curve of the class there is a curve of the class possessing the property.

For example, to say that for a generic regular smooth space curve the curvature is nowhere zero is to say that arbitrarily close to any regular smooth space curve in the space of all regular smooth space curves there is one for which the curvature is everywhere non-zero.

For the above statements to be meaningful the various phrases in italics have to be well defined. The first phrase that demands definition is the *space of all curves of a class*. Clearly this is too vague, but even if one narrows this down to the space of regular smooth maps of the interval $]0, 1[$ to the plane $\mathbb{R}^2$ or to the space $\mathbb{R}^3$ there are choices to be made. The problem is what *topology* to put on the set of all such curves. When are two curves to be considered as *close*? Must their derivatives to arbitrarily high order be close as well as corresponding points, or may we, or should we, settle for something less? The early answers were given by Whitney and Thom, and a huge volume of work has appeared since.

It was the original intention to conclude this book with some details and also to review related fundamental concepts of singularity theory such as transversality, stability, determinacy, unfoldings, etc., but I am in fact not going to do this but rather take the easy way out and refer the reader to other places. The only problem is what to recommend.

One very accessible book is Poston and Stewart (1978). Particularly helpful is Chapter 8 on determinacy and unfoldings. Slightly earlier is the review by Wall (1977). The presentation of part of the thesis of Ed Looijenga in the early part of that review is particularly relevant. Also recommended is Bröcker and Lander (1975). A much more recent book that complements very well the material of this book is Bruce and Giblin (1984, second edition 1992).

There is a large Russian school of singularity theorists under the leadership of V.I. Arnol'd. See Arnol'd, Gusein-Zade and Varchenko (1985) for a very wide account of singularity theory in general and Arnol'd (1990b) for Lagrangian and Legendrian maps and a survey of the *Russian* work on singularities of apparent contours in particular. The place to start for a comprehensive account of the latter subject is Bruce and Giblin (1985).

Within the last few years Bruce has developed several alternative approaches to surface theory. One relates the geometry of a surface to that of its *dual*. For this approach see Bruce (1981a). Another involves the concept of *folding maps*. For this see Bruce and Wilkinson (1991). A recent review article contains many other references (Bruce, 1992).

An additional reference for the classical work on mechanisms is Hartenberg and Denavit (1964). The historical notes in this book are especially useful in indicating some of the practical things that the mathematicians of the late 19th century were involved with. In recent years it has become clear that singularity theory has much to offer in this area. See, for example, Gibson and Newstead (1986).

Recent advances in depth measurement, magnetic resonance imagery and computer vision have prompted fresh interest in such robust features of a smooth surface as its parabolic line and its umbilics and ridges, these, as well as its subparabolic lines being features which, in contrast to such fragile features as lines of curvature or geodesics, have individual identities under deformation of the surface. Of these the parabolic lines have been studied the most. See, for example, Koenderink (1990). On the other hand until now the ridges of a surface have received remarkably little attention, doubtless because they essentially involve third derivatives, but very recent rediscoveries of them include work by Gaile Gordan (1991) in collaboration with David Mumford, on 'face recognition from depth and curvature' and a group at the Institut National de Recherche en Informatique et en Automatique, at Le Chesnay, France, are concerned not only with face recognition but also magnetic resonance scans of the surface of the brain, in which latter work identifications have been made between certain ridges (or *crest* lines, as they are called at INRIA) and certain circumvolutions (*gyri* and *sulci*, that is crests and troughs) of the brain. For details see, for example, Guéziec (1992) and Thirion and Gourdon (1993).

Of all the textbooks on elementary differential geometry published in the last fifty years the most readable is one of the earliest, namely that by D.J. Struik (1950). He is the only one to mention Gullstrand (in a footnote). None of them give the Darboux classification of umbilics and none mention ridges!

For the mathematical work of Christiaan Huygens see a paper by H.J.M. Bos (1980). Finally, the historical romp of Arnol'd (1990a) is an especial joy!

References

Arnol'd, V.I. (1972) Normal forms of functions near degenerate critical points, the Weyl groups A_k, D_k, E_k and Lagrangean singularities, *Funktsional'. Anal. i Prilozhen.*, **6 : 4**, 3–25 = *Functional Anal. Appl.*, **6**, 254–72.

Arnol'd, V.I. (1975) Critical points of smooth functions and their normal forms, *Uspekhi Mat. Nauk* **30 : 5**, 3–65 = *Russian Math. Surveys* **30 : 5**, 1–75.

Arnol'd, V.I. (1978) Critical points of functions on a manifold with boundary, the simple Lie groups B_k, C_k, F_4, and singularities of evolutes, *Uspekhi Mat. Nauk* **33 : 5**, 91–105 = *Russian Math. Surveys* **33 : 5**, 99–106.

Arnol'd, V.I. (1983) Singularities of systems of rays, *Uspekhi Mat. Nauk* **38 : 2**, 77–147 = *Russian Math. Surveys* **38 : 2**, 87–176.

Arnol'd, V.I. (1984) Report of the meeting of 18 April of the Petrovskiĭ seminar on differential equations and mathematical problems of physics, *Uspekhi Mat. Nauk* **39:5**, 255–6 (*not* in *Russian Math Surveys*).

Arnol'd, V.I. (1990a) *Huygens and Barrow, Newton and Hooke*, Birkhäuser, Basel.

Arnol'd, V.I. (1990b) *Singularities of caustics and wave fronts*, Kluwer Academic Publishers, Dordrecht.

Arnol'd, V.I., Gusein-Zade S.M. and Varchenko, A.N. (1985) *Singularities of differentiable maps*, Vol I, Birkhaüser, Boston. (Original Russian edition Nauka, Moscow, 1982.)

Ball, R.S. (1871) Notes on applied mechanics, I, Parallel motion, *Proc. Irish Acad. (Series 2)*, **1**, 243–5.

Banchoff, T., Gaffney, T. and McCrory, C. (1982) *Cusps of Gauss mappings*, Research Notes in Mathematics, 55, Pitman, London.

Barnes, D.W. (1967) On the existence of umbilics on compact surfaces, *Arch. Math. (Basel)*, **18**, 320–4.

Bennequin, D. (1984) *Caustique mystique*, Séminaire Bourbaki, Exposé 634, Paris.

Berry, M.V. and Hannay, J.H. (1977) Umbilic points on Gaussian random surfaces, *J. Phys. A* **10**, 1809–21.

Berry, M.V. and Upstill, C. (1980) *Catastrophe optics: morphologies of caustics and their diffraction patterns*, Progress in Optics XVIII, North Holland, Amsterdam.

Blum, R. (1980) *Circles on surfaces in Euclidean 3-space*, Lecture Notes in Math. 792, Springer, Berlin, pp. 213–21.

Boardman, J.M. (1967) Singularities of differentiable maps, *Inst. Hautes Études Sci. Publ. Math.* **33**, 21–57.

Bos, H.J.M. (1980) *Huygens and mathematics*, Studies on Christiaan Huygens, Swets and Zeitlinger, Lisse, p. 126–46.

Bottema, O. (1961) On instantaneous invariants, *Proc. International Conference on Mechanisms, Yale University*. The Shoe String Press, New Haven, Connecticut, pp. 157–64.

Bottema, O. and Roth, B. (1979) *Theoretical kinematics*, North-Holland series in Applied Mathematics and Mechanics 24, North-Holland Publishing Company, Amsterdam.

Brieskorn, E. (1979) Die Hierarchie der 1-modularen Singularitäten, *Manuscripta Math.* **27**, 183–219.

Brieskorn, E. and Knörrer, H. (1986) *Plane algebraic curves*, Birkhäuser, Basel.

Bröcker, Th. and Lander, L.C. (1975) *Differentiable germs and catastrophes*, London Math. Soc., Lecture Note Series 17, Cambridge University Press, Cambridge.

Bruce, J.W. (1981a) The duals of generic hypersurfaces, *Math. Scand.* **54**, 262–78.

Bruce, J.W. (1981b) On singularities, envelopes and elementary differential geometry, *Math. Proc. Camb. Phil. Soc.* **89**, 43–8.

Bruce, J.W. (1983) Wavefronts and parallels in Euclidean space, *Math. Proc. Camb. Phil. Soc.* **93**, 323–33.

Bruce, J.W. (1992) Generic geometry and duality, – to appear.

Bruce, J.W. and Fidal, D.L. (1989) On binary differential equations and umbilics, *Proc. Roy. Soc. Edinburgh*, **111A**, 147–68.

Bruce, J.W. and Giblin, P.J. (1981) Generic curves and surfaces, *J. London Math. Soc.* (2) **24**, 555–61.

Bruce, J.W. and Giblin, P.J. (1983) Generic geometry, *Amer. Math. Monthly*, **90**, 529–45.

Bruce, J.W. and Giblin, P.J. (1984, second edition 1992) *Curves and singularities*, Cambridge University Press, Cambridge.

Bruce, J.W. and Giblin, P.J. (1985) Outlines and their duals, *Proc. London Math. Soc.* (3) **50**, 552–70.

Bruce, J.W., Giblin, P.J. and Gibson, C.G. (1981) On caustics of plane curves, *Amer Math. Monthly*, **88**, 651–67.

Bruce, J.W. and Wilkinson, T.C. (1991) Folding maps and focal sets, *Proc. of Warwick Symposium on Singularities*, Lecture Notes in Math. 1462, Springer, Berlin, pp. 63–72.

Burmester, L. (1888) *Lehrbuch der Kinematik*, Felix, Leipzig.

Camacho, C. and Neto, L (1985) *Geometric theory of foliations*, Birkhäuser, Boston.

Cayley, A. (1870) Note on Mr Frost's paper on the direction of lines of curvature in the neighbourhood of an umbilicus, *Quart. J. Pure Appl. Math.*, **10**, 111–13.

Chasles, M. (1830) Note sur les propriétés générales du système de deux corps. . . , *Bull. Sci. Math. Ferrusac*, **14**, 321–6.

Clairaut, A.C. (1831) *Recherche sur les courbes à double courbure*, Paris.

Cleave, J.P. (1980) The form of the tangent developable at points of zero torsion on space curves, *Math. Proc. Camb. Phil. Soc.* **88**, 403–7.

Coolidge, J.L. (1940) *A history of geometrical methods*, Oxford University Press, Oxford, second edition, Dover, New York (1963).

Coxeter, H.S.M. (1948) *Regular polytopes*, Methuen, London, third edition, Dover, New York (1973).
Cundy, H.M. and Rollett, A.P. (1961) *Mathematical models*, second edition, Oxford University Press, Oxford, third edition, Tarquin Press, Diss (1981).
Darboux, G. (1896) *Leçons sur la théorie générale des surfaces*, 4me partie, Gauthiers-Villars et Fils, Paris.
David, J.M.S. (1983) Projection generic curves, *J. Lond. Math. Soc.* (2) **27**, 552–62.
Euler, L. (1760) Recherches sur la courbure des surfaces, *Mém, Ac. Berlin*, 119–43.
Faà de Bruno (1857), Note sur une nouvelle formule de calcul differentiel, *Quart. J. Math.* **1**, 359–60.
Flegmann, A.M. (1985) Evolutes, involutes and the Coxeter group H_4, Thesis, University of Liverpool.
Frenet, J.F. (1852) Sur quelques propriétés des courbes à double courbure, *J. Math. Pure Appl.* **17**, 437–47.
Frost, P. (1870) On the direction of lines of curvature in the neighbourhood of an umbilicus. *Quart J. Pure and Appl. Math.*, **10**, 78–86.
Gaffney, T.J. (1983) *The structure of TA(f), classification and an application to differential geometry*, *Singularities*, Part 1, (Arcata, Calif., 1981), Proc. Symp. Pure Math. 40 (American Mathematical Society, Providence, R.I.) pp. 409–27.
Gauss, C.F. (1828) Disquisitiones generales circa superficies curvas, *Commentationes Societatis Regiae Scientiarum Gottingensis Recentiores*, VI, Gottingae. For a translation of this article into English and an assessment of it 150 years later by P. Dombrowski see *Astérisque, Société mathématique de France*, **62** (1979).
Giblin, P.J. and Porteous, I.R. (editors) (1990) *Challenging mathematics*, Oxford University Press, Oxford.
Gibson, C.G. (1979) *Singular points of smooth mappings*, Research Notes in Mathematics, No 25, Pitman, London.
Gibson, C.G. and Newstead, P.E. (1986) On the geometry of the planar 4-bar mechanism, *Acta applicandae mathematicicae*, **7**, 113–35.
Gordan, G.G. (1991) Face recognition from depth maps and surface curvature, in *Proceedings of SPIE Conference on Geometric Methods in Computer Vision*, San Diego, CA, July 1991.
Guéziec, A. (1992) Large deformable splines, crest lines and matching, *Rapports de Recherche*, No 1782, INRIA, Le Chesnay, France.
Gullstrand, A. (1904) Zur Kenntnis der Kreispunkte, *Acta Mathematica*, **29**, 59–100.
Gullstrand, A. (1911) How I found the mechanism of intracapsular accommodation, *Nobel Lecture, Physiology or Medicine 1901–1921*, published for the Nobel Foundation in 1967 by Elsevier Publishing Company, Amsterdam, pp. 414–31.
Gutierrez, C. and Sotomayor Teilo, J. (1991) *Lines of curvature and umbilical points of surfaces*, IMPA, Rio de Janeiro.
Hartenberg, R.S. and Denavit J. (1964) *Kinematic synthesis of linkages*, McGraw-Hill, New York.
Hiller, H. (1982) *Geometry of Coxeter groups*, Research Notes in Mathematics, No 54, Pitman, London.
Hoffman, W. and Meeks, W.H. III (1990) Embedded minimal surfaces of

finite topology, *Ann. Math.* (2) **131**, 1–34.
Hôpital, G.F.A. Marquis de l' (1696) *L'analyse des infiniments petits, pour l'intelligence des courbes planes*, Imprimerie Royale, Paris.
Hunt, K.H. (1978) *Kinematic geometry of mechanisms*, Clarendon Press, Oxford.
Hutchison, H.J., Nye, J.F. and Salmon, P.S. (1983) The classification of isotropic points in stress fields. *J. Struct. Mech.* **11(3)**, 371–81.
Koenderink, J.J. (1984) What does the occluding contour tell us about solid shape? *Perception* **13**, 321–330.
Koenderink, J.J. (1990) *Solid shape*, The MIT Press, Cambridge, Mass.
Koenderink, J.J. and van Doorn, A.J. (1976) The singularities of the visual mapping, *Biol. Cybernetics*, **24**, 51–9.
Krause, M. (1920) *Analysis der ebenen Bewegung*, Vereinigung Wissenschaftlicher Verleger, Berlin.
Lancret, M.-A. (1806) Mémoire sur les courbes à double courbure, *Mémoires présenté à l'Institut par divers Savans*, **1**, 416–54.
Landis, E.E. (1981) Tangential singularities, *Funktsional'. Anal. i Prilozhen.*, **15 : 2**, 36–49 = *Functional Anal. Appl.*, **15**, 103–14.
Lawrence, J.D. (1972) *A catalog of special plane curves*, Dover, New York.
Markatis, S. (1980) Some generic phenomena in families of surfaces in $\mathbb{R}^3$, Thesis, University of Liverpool.
Mather, J. (1970) Stability of C^∞ mappings, IV: Classification of stable map-germs by R-algebras, *Publ. Math. IHES* **3**, 223–48.
Meusnier, J.-B.-M.-C. (1785) Mémoire sur la courbure des surfaces, *Mémoire Div. Sav.*, **10**, 477–510.
Mond, D.M.Q. (1985a) On the classification of germs of maps from $\mathbb{R}^2$ to $\mathbb{R}^3$, *Proc. Lond. Math. Soc.* **50 : 2**, 333–69.
Mond, D.M.Q. (1985b) Normal forms for the singularities of the tangent developable of a space curve, Preprint, University of Warwick.
Monge, G. (1784) Mémoire sur la théorie des déblais et des remblais, *Hist. Ac. Roy. Sci.*, 2ᵉ partie, 555–704.
Monge, G. (1795) *Feuilles d'analyse appliqué à la Géométrie à l'usage de l'École Polytechnique*.
Monge, G. (1850) *Application de l'analyse à l'géométrie*, 5 éd. par Liouville, Paris.
Montaldi, J. (1983) Contact, with applications to submanifolds of $\mathbb{R}^n$, Thesis, University of Liverpool.
Montaldi, J. (1986a) On contact between submanifolds, *Michigan Math. J.*, **33**, 195–9.
Montaldi, J.A. (1986b) Surfaces in 3-space and their contact with circles, *J. Diff. Geom.*, **23**, 109–26.
Morris, R. (1990) Symmetry of curves and the geometry of surfaces: two explorations with the aid of computer graphics, Thesis, University of Liverpool.
Morris, R. (1991) Preprint, Liverpool.
Nye, J.F. (1983) Monstars on glaciers, *J. Glaciology*, **29**, 70–7.
Nye, J.F. (1986) Isotropic points on glaciers, *J. Glaciology*, **32**, 363–5.
Osserman, R. (1985) The four-or-more vertex theorem, *Amer. Math. Monthly*, **92**, 332–7.
Osserman, R. (1989) Minimal surfaces in $\mathbb{R}^3$, global differential geometry, *MAA Studies in Mathematics*, **27**, 73–98.
Porteous, I.R. (1962) *Simple singularities of maps*, Columbia University Notes,

reprinted in Proc. Liverpool Singularities Sympos. I, Lecture Notes in Math., 192 (C.T.C. Wall, ed.), Springer-Verlag, Berlin and New York, 1971, pp. 286–307.

Porteous, I.R. (1971) The normal singularities of a submanifold, *J. Diff. Geom.*, **5**, 543–64.

Porteous, I.R. (1981) *Topological geometry*, second edition, Cambridge University Press, Cambridge.

Porteous, I.R. (1983a) The normal singularities of surfaces in $\mathbb{R}^3$, *Singularities*, Part 2, (Arcata, Calif., 1981), Proc. Symp. Pure Math. 40, American Mathmatical Society, Providence, R.I., pp. 379–93.

Porteous, I.R. (1983b) Probing singularities, *Singularities*, Part 2, (Arcata, Calif., 1981), Proc. Symp. Pure Math. 40, American Mathematical Society, Providence, R.I., pp. 395–406.

Porteous, I.R. (1987a) The intelligence of curves, *The mathematics of surfaces*, II, (Cardiff, 1986), Clarendon Press, Oxford, pp. 1–16.

Porteous, I.R. (1987b) Ridges and umbilics of surfaces, *The mathematics of surfaces*, II (Cardiff, 1986), Clarendon Press, Oxford, pp. 447–58.

Porteous, I.R. (1989) The circles of a surface, *The mathematics of surfaces*, III, (Oxford, 1988), Clarendon Press, Oxford, pp. 135–43.

Poston, T. and Stewart, I.N. (1978), *Catastrophe theory and its applications*, Pitman, London.

Procter, R.A. (1878) *A treatise on the cycloid and all forms of cycloidal curves, and on the use of such curves in dealing with the motions of planets, comets, &c. and of matter projected from the sun*, Longmans, Green, and Co., London.

Ramsay, J.G. (1967) *Folding and fracturing of rocks*, McGraw Hill, New York.

Robinson, J. (1992) *Symbolic Sculpture*, Édition Limitée, Carouge-Geneva.

Ronga, F. (1983) A new look at Faà de Bruno's formula for higher derivatives of composite functions and the expression of some intrinsic derivatives, *Singularities*, Part 2 (Arcata, Calif., 1981), Proc. Symp. Pure Math. 40, American Mathematical Society, Providence, R.I. pp. 421–31.

Schönflies, A. (1886) *Geometrie der Bewegung in synthetischer Darstellung*, Teubner, Leipzig.

Senff, C.E. (1831) *Theoremata principaliae theoriae curvarum et superficierum*, Dorpat Univ.

Serret, J. (1851) Mémoire sur quelques formules relatives à la théorie des courbes à double courbure, *J. Math. Pure Appl.* **16**, 193–207.

Shcherbak, O.P. (1983) Singularities of a family of evolvents in the neighbourhood of a point of inflection of a curve, and the group H_3 generated by reflections, *Funktsional'. Anal. i Prilozhen.*, **17 : 4**, 70–2 = *Functional Anal. Appl.*, **17**, 301–3.

Shcherbak, O.P. (1984) H_4 in the problem of avoiding an obstacle, *Uspekhi Mat. Nauk*, **39 : 5**, 256 (*not* in *Russian Math. Surveys*).

Shcherbak, O.P. (1988) Wavefronts and reflection groups, *Uspekhi Mat. Nauk*, **43 : 3**, 125–60 = *Russian Math. Surveys*, **43 : 3**, 149–94.

Slodowy, P. (1980) *Simple singularities and simple algebraic groups*. Lecture Notes in Mathematics, 815, Springer-Verlag, Berlin-New York.

Sotomayor Teilo., J. and Gutierrez, C. (1982) Structurally stable configurations of lines of principal curvature, *Astérisque*, **98–99**, 195–215.

Struik, D.J. (1950) *Lectures on classical differential geometry*, second edition, Addison-Wesley, Mass.

Taton, R. (1951) *L'oeuvre scientifique de Monge*, Presses Univ. de France, Paris.

Thirion, J.-P., Gourdon, A. (1993) The Marching Lines Algorithm: new results and proofs, *Rapports de Recherche*, No 1881, INRIA, Le Chesnay, France.

Thom, R. (1975) *Structural stability and morphogenesis*, W.A. Benjamin, Reading, Mass.

Thompson, J.M.T. and Hunt, G.W. (1975) Towards a unified bifurcation theory, *J. Appl. Math. Phys.*, **26**, 581–603.

Thomson, Sir William (Lord Kelvin) and Tait, P.G. (1879) *Natural Philosophy*, Vol I, Part I, Cambridge. (For the 'muslin clamps' see Section 149, page 114.)

Thorndike, A.S., Cooley, C.R. and Nye, J.F. (1978) The structure and evolution of flow fields and other fields. *J. Phys. A: Math. Gen.* **11**, 1455–90.

Truesdell, C. (1958) The new Bernoulli edition, *Isis*, **49**, 54–62.

Veldkamp, G.R. (1964) Applications of the Bottema invariants in plane kinematics, *Proc. Koninkl. Nederl. Akademie van Wetenschappen*, Series A, **67**, 430–40.

Wall, C.T.C. (1977) *Geometric properties of generic differentiable manifolds*, Lecture Notes, 597, Springer, Berlin, pp. 707–74.

Weatherburn, C.E. (1927) *Differential geometry of three dimensions*, I, Cambridge University Press, Cambridge (fourth impression, with corrections, 1947.)

Whitney, H. (1955) On singularities of mappings of Euclidean space, I, Mappings of the plane into the plane, *Ann. Math.* (2) **62**, 374–410.

Wilkinson, T. (1991), The geometry of folding maps, Thesis, University of Newcastle.

Zeeman, E.C. (1976) *The umbilic bracelet and the double-cusp catastrophe*, Lecture Notes, 525, Springer, Berlin, pp. 328–366.

Index